# The Professional Writer

## A GUIDE FOR ADVANCED TECHNICAL WRITING

# The Professional Writer

## A GUIDE FOR ADVANCED TECHNICAL WRITING

Gerald J. Alred
University of Wisconsin—Milwaukee

Walter E. Oliu
U.S. Nuclear Regulatory Commission

Charles T. Brusaw
NCR Corporation (retired)

St. Martin's Press
New York

**Senior editor:** Mark Gallaher
**Development editor:** Joyce Hinnefeld
**Managing editor:** Patricia Mansfield
**Project editor:** Cheryl Friedman
**Production supervisor:** Alan Fischer
**Text design:** Paula Goldstein
**Graphics:** G&H Soho, Ltd.
**Cover design:** Darby Downey

Library of Congress Catalog Card Number: 90–71628

Manufactured in the United States of America.
65432
fedcba

For information, write:
St. Martin's Press, Inc.
175 Fifth Avenue
New York, NY 10010

ISBN: 0–312–00248–3

 The text of this book has been printed on recycled paper.

*For our wives and families*

# Preface

*The Professional Writer* represents nearly a decade of research to provide a guide for undergraduate and graduate students who are preparing for writing careers in business, industry, and government. Over the years, many of our colleagues have complained about the lack of a book of appropriate focus and sufficient scholarship, suited to the complexities of advanced technical and professional writing courses. Although such courses represent the fastest growing area in the writing curriculum, most of those who teach them must assign as readings articles and book excerpts. All the existing texts and anthologies we reviewed, regardless of their merits, aim at technical professionals who write as only a *part* of their jobs: this book's purpose is to help prepare the professional whose *primary* job is writing. In addition, *The Professional Writer* can serve engineers, scientists, and many others who may not think of themselves as professional writers but whose jobs regularly require them to produce writing of comparable quality. Whatever their career plans, we hope this book will continue as a resource long after students have finished the course.

The growth of advanced technical and professional writing is the result of a significant cultural and economic shift away from a labor-intensive manufacturing economy toward an economy driven by information-intensive high technology. The ubiquitous use of computers in the workplace and at home, for example, requires well-written user manuals and technical support documentation. As society becomes more dependent on technology—from computer systems to communication networks to advanced methods for the delivery of medical care—professional writers who can translate information for those who build, use, and maintain products and systems will be in increasingly greater demand. The focus of *The Professional Writer* is, throughout, on the kinds of writing required in such an environment.

## WHAT MAKES THIS TEXT ADVANCED

Five things, we believe, set this text apart from the less advanced books currently available on the topic:

- the interplay of theory and practice;
- the assumptions we make about our readers;
- the special topics covered;
- the focus on the writing process; and
- the case history documenting the preparation of an actual publication.

First, any book designed to prepare professional writers and advanced students must be firmly grounded in *both* theory and practice. *Theory* organizes and validates common sense, intuition, and individual experience; however, *practice* tests, fulfills, and reforms theory. Because it transcends particular techniques and forms, theory enables students to respond to changing communications demands and to the new technologies they will inevitably face during their careers. On the other hand, if it is not reinforced by the experience of seasoned practitioners, theory can seem abstract and irrelevant to the needs of the workplace. So, throughout this book, we present the concepts and research findings that underpin the practices of the professional writer. In other words, the chapters stress both *why* writers make decisions as well as *how* they handle particular problems.

We hope that our combined experience in academe, business, and government has enabled us to provide students with an appropriate and useful balance of theory and practice. Instructors will find the Works Cited and Further Readings sections at the end of each chapter particularly useful in demonstrating to students that theory and practice are inseparable. These sections should also enable instructors to enrich their courses by furthering students' knowledge of topics that will be important in meeting their specific career needs.

Second, we assume that students using this book see themselves as future professional writers or editors—or believe that they will need to produce writing of professional quality. We assume, therefore, that our readers possess basic fluency, having successfully completed one or more technical writing or other advanced writing courses, such as business writing, research writing, or intermediate composition. We also assume that these students will be familiar with the writing strategies and typical formats that appear in standard textbooks for these courses. For that reason, and because the documents that writers produce vary so greatly from organization to organization, we have focused on principles rather than products. We illustrate, for example, how writers use audience, purpose, and institutional needs to determine both the overall and page-level design of any document. Students can go on to adapt such decision-making strategies to particular formats on the job, and to the challenges of ever-changing technology.

We have not organized the text around a specific technical or vocational area, such as computer documentation or proposal writing, though we do include examples from these and many other areas. Further, we understand that those using this book will not have uniform academic preparation: some will be English or journalism majors, while others will come from engineering and the sciences. Some may even be working already as technical writers. We have provided abundant references in the Works Cited and Further Readings sections for those readers who wish to study an area in depth. For students who need to review the basics of writing or formats, we recommend our text *Handbook of Technical Writing* (also published by St. Martin's Press).

Third, as a quick review of the table of contents will reveal, *The Professional Writer* covers in depth topics that introductory texts cover only superficially (or not at all). Among these are collaborative work, reading theory and special audience needs, computer graphics, layout and typography, electronic publishing, review and evaluation, document standards, and online documentation. We also cover such topics as project scheduling and time management techniques for writers, and explore the process for establishing document usability testing procedures. Further, throughout the chapters—and especially in the case history—we cover many of the ethical, legal, corporate culture, and interpersonal issues a professional writer faces.

Fourth, one of the most important features of this book is its focus on the processes that professional writers use to produce documents. Even the sequence of the chapters roughly follows the order in which a document is generated—from the point a writer receives the assignment and assesses audience needs to the final phases of production and printing. Chapter 2 discusses this process, presenting it graphically in a Flowchart of the Document Process. Chapter 2 also serves as a window to subsequent chapters by linking them to the seven steps of the process: Document Need Assessment, Document Analysis, Research, Document Drafts, Review and Evaluation, Editing, and Production.

Our fifth distinguishing feature, which illustrates most vividly an individual writer's process and function within an organization, is our case history following a working writer at a large regional bank through the development and production of an actual technical publication. Based on in-depth interviews with the writer and illustrated with her notes, memos, outlines, and drafts, the case history sections in each chapter portray one writer's decision-making and writing processes. The culmination of the writer's efforts—an international banking software user's manual—is shown in complete, final form in the Appendix.

The case history should be useful to students as they work on their own projects, and it will help them visualize their future careers as professional writers. Of course, since the case history is based on an actual project, some of the writer's experiences are unique to her particular work setting and may not directly transfer to either the classroom or another professional environment. But, much like an ethnographic study, the case history provides a rich resource for students and instructors who wish to examine the writer's professional and organizational environment.

## INSTRUCTOR'S MANUAL

*The Professional Writer* is accompanied by a helpful Instructor's Manual, which presents pertinent suggestions and exercises for each chapter in the book. Written by William Van Pelt of the University of Wisconsin–Milwaukee, the manual can help provide both a theoretical and pedagogical framework for an advanced course. Both experienced instructors and those teaching an advanced technical writing course for the first time should find it an unusually important tool in adapting the book to the classroom.

The Instructor's Manual assumes that students will be required to produce lengthy and complex documents, often in collaboration with others in the class. Thus, many of the exercises focus on major projects and require workshop groups and peer reviews that mirror the experiences of working professionals. Because it presents project ideas that are designed specifically for the advanced student, the Instructor's Manual should provide rich resource material for advanced classes.

## ACKNOWLEDGMENTS

The authors owe a great debt to many individuals in the preparation of this book. Special thanks must begin with Bill Van Pelt, University of Wisconsin–Milwaukee, for his multiple contributions, foremost as a valued colleague, then as a reviewer for St. Martin's Press, and, finally, as the author of a truly advanced instructor's manual. We also owe a large debt to Mary Mullins, formerly of First Wisconsin National Bank of Milwaukee, for her invaluable collaboration in making the case history a worthwhile story to read. We wish to acknowledge Judith Early Sheldon for writing Chapter 14, "Creating Document Standards," and reviewing several other chapters, and Lynn Collins for coauthoring Chapter 15, "Online Documentation."

We are particularly grateful to L. W. Denton, IBM Corporation, for reviews both informally and for St. Martin's Press, and to Gail Atkins, First Wisconsin National Bank of Milwaukee, for professional reviews for St. Martin's Press and special help with the case history. To the other professional reviewers whose advice and support of the project enabled its completion we owe an important debt: William O. Coggin, Bowling Green State University; Lois Rew, San Jose State University; Fred Reynolds, Old Dominion University; Jack Selzer, Pennsylvania State University; and Thomas L. Warren, Oklahoma State University.

For their important contributions, we wish to recognize Bob MacKenzie, Mike Bartlett, Bill Stuckey, and Bob Horn, all of NCR Corporation; and Marshel Baggett, Janet Thot-Thompson, and Karen Vanduser, all of the Nuclear Regulatory Commission; Susan Ladwig, Milwaukee County Museum; Frank Hubbard, Marquette University; Richard Zauft, Graphic Design Department, and William Holahan, Economics Department, University of Wisconsin–Milwaukee; and Charlotte Ruenzel, Deluxe Data Systems.

We thank First Wisconsin National Bank of Milwaukee for permission to show their International Banking Manual and for allowing us to interview Mary Mullins. Thanks also to NCR Corporation for the use of many examples, and Ken Cook Company of Milwaukee, particularly Ken Cook, Jr., and John Fier. We wish to acknowledge in several ways the University of Wisconsin–Milwaukee: the encouragement and reviews of faculty, teaching assistants, and staff of the Rhetoric and Composition Program; the contributions of English 435 students over the past 15 years; and a sabbatical leave to help finish this book.

We would like to thank Mark Gallaher, English Editor, Kristin Bowen, Associate Editor, and Cheryl Friedman, Project Editor, at St. Martin's Press. We owe a special debt to Joyce Hinnefeld, Development Editor at St. Martin's Press, who coached this book through the final stages and skillfully edited the case history. Thanks also to Janice Alred for her valued advice and many hours of secretarial assistance. Finally, we must acknowledge the support and infinite patience of our families for more years than we might wish to admit.

G. J. A.

W. E. O.

C. T. B.

# Contents

## 5   Gathering Information        77

## 6   Organizing Information for a Document        103

## 7    Integrating Visuals    124

## 8    Layout and Design    160

## 9 Drafting a Document 203

## 10 Revising Your Writing 224

## 11      Review and Evaluation      246

## 15    Online Documentation    366

# The Professional Writer

## A GUIDE FOR ADVANCED TECHNICAL WRITING

# 1

# The Professional Writer

Definitions of Technical Writing

The Origins and Growth of Technical Writing

The Products of Technical Writing

Professional Roles and Settings

The Technical Writer's Responsibilities

The Technical Writer's Competencies

Academic Preparation

Finding a Job

Professional Organizations

No book can substitute for the intensive writing practice and expert coaching you need to develop professional writing skills. This book provides a focal point, however, around which you can develop the skills and competencies of the professional technical writer.

This chapter provides an overview of the functions, work settings, responsibilities, competencies, and preparation required of the professional writer in business, industry, and government. Topics that are more fully developed in other chapters are cross-referenced throughout this chapter in parentheses. The "Works Cited" and "Further Reading" sections at the end of each chapter provide references to additional materials that can also help increase your knowledge of important topics.

## DEFINITIONS OF TECHNICAL WRITING

The 1988–1989 edition of the *Occupational Outlook Handbook* states:

Technical writers put scientific and technical information into readily under-standable language. They prepare manuals, catalogs, parts lists, and instructional materials used by sales representatives to sell a wide variety of machinery and equipment and by technicians to install, maintain, and service it (U.S. Dept. of Labor 186).

However, this definition is only a starting point; it does not suggest the evolving role of technical writers in an "information age." Our society has witnessed an economic transition—a transition from a labor-intensive, manufacturing economy to an economy that is driven by information-intensive high technology. This transition has created a demand for writers with the skills, according to one definition, to "accommodate" the increasing volumes of information and technology to their users (Dobrin 242). In other words, technical writing serves as a "bridge" between technology and the people who must use it. Technical writers are the bridge builders because they draw on complex technical information to construct printed and on-screen (called *online*) materials that help readers use, learn, repair, or build equipment or systems.[1]

Obviously, technical writing offers challenges—not only must writers make complex concepts clear, but they must also make sure that technical information is accurate without oversimplifying or distorting it. At the same time, technical writers find the work rewarding because they enjoy using words to communicate effectively; they take pride in making complex information accessible to readers. Further, technical writing is fascinating because writers work as partners with people on the "cutting edge" of technical and scientific advances; writers must thoroughly understand these advances in order to explain them to readers.

## THE ORIGINS AND GROWTH OF TECHNICAL WRITING

Many historians find the origin of technical writing in ancient documents like *Aqueducts of Rome* (ca. A.D. 97). This "technical manual," written by Frontinus, the water commissioner of Rome, served as a guide for building and maintaining those marvels of ancient engineering (Moran 26). Most say that modern technical writing began as a special occupation during World War II. During that time, writers were employed to prepare documents describing the maintenance, construction, and operation of increasingly complex military and technical equipment. The *Occupational Outlook Handbook* first listed technical writing as a career in 1951, and the U.S. Bureau of Labor Statistics first compiled individual statistics for technical writers in 1983.

As of 1990, the U.S. Bureau of Labor Statistics reported that over 70,000 individuals were employed as technical writers. And according to the *Occupational Outlook Handbook*, that figure will grow significantly:

Demand for technical writers is expected to increase because of the continuing expansion of scientific and technical information and the continued need to

[1]Technical writing is sometimes referred to as *documentation*. Originally to *document* meant to teach or to inform. Although this use of the word *document* is obsolete, it does perhaps describe the role of a technical writer more precisely than the term *technical writing*.

communicate it to researchers, corporate managers, sales representatives, and technicians. With the increasing complexity of industrial and scientific equipment, more users will depend on the technical writer's ability to prepare precise but simple explanations and instructions (187).

The Society for Technical Communication (STC), the largest professional association of technical writers, listed its 1977 membership as near 3,000. By 1990, its membership had grown to over 13,000. Statistics like these suggest that technical writing is currently one of the fastest-growing professions.

One reason for this growth is that technical writing is crucial to the livelihood of many businesses. For example, many firms that manufacture materials (from aircraft to medical supplies) depend on proposals, usually prepared by technical writers, to win government procurement contracts. In the computer industry, technical documentation (in both printed and online form) has become key to the successful marketing of hardware and software. For the many other commercial businesses that depend heavily on computers, such as banks, materials written for employees and customers are essential to a smooth-running operation.

## THE PRODUCTS OF TECHNICAL WRITING

As a technical writer, you will write for a wide range of audiences (see Chapter 4) and may produce any of the following types of writing:

| | |
|---|---|
| Annual reports | Parts documentation (lists) |
| Audiovisual scripts | Patient care literature |
| Brochures (sales and promotional) | Policies and procedures |
| Commercial proposals | Position statements and papers |
| Educational material | Product literature |
| Employee publications | Question and answer documents |
| Ghost-written books, articles, letters | Software tutorials |
| | Speeches for executives |
| Government and grant proposals | System descriptions |
| Instruction manuals | Technical advertising |
| Industrial catalogs | Technical manuals for products |
| Letters to consumers | Trade and technical magazine articles |
| News and press releases | |
| Newsletter articles | Training manuals |
| Online documentation | |

At the start of your career as a technical writer, you might create parts lists or revise outdated documents. As you advance, you may work independently on large projects, such as developing training materials for company personnel or customers. As you gain even more experience, you may go on to supervise other writers or teach in-house writing courses.

## PROFESSIONAL ROLES AND SETTINGS

One survey found the following distribution of professional roles among the membership of the Society for Technical Communication (STC 8):

> Writing (64 percent)
> Editing (49 percent)
> Managing and supervising (32 percent)
> Consulting and freelance (13 percent)
> Graphic arts (13 percent)
> Education (12 percent)

Notice, as the total of 183 percent indicates, that many technical writers responding to the survey marked more than one category because they often assume more than one role. Another survey revealed the time spent on specific activities by graduates of a technical writing program (Kalmbach, Jobst, and Meese 23):

> Writing (30 percent)
> Research (23 percent)
> Editing (14 percent)
> Meetings (13 percent)
> Document design (11 percent)
> Field testing (7 percent)
> Other (2 percent)

These writers spend less than one-third of their time writing and revising a draft, devoting much of their time as well to research, design, field testing, and meeting with others in the planning of documents.

While performing such duties, writers work in a wide variety of settings. The survey of STC members found the following general types of employers among the respondents (STC 8):

> Private industry (63 percent)
> Government and military (12 percent)
> Self-employed (9 percent)
> Education (8 percent)
> Other (8 percent)

Although this survey shows only 12 percent of these writers working for the government directly, other writers work indirectly for the government by doing government contract work in private industries, producing such documents as proposals, technical manuals, and reports. Furthermore, government-related research and development organizations like Argonne and Lawrence Livermore National Laboratories employ many writers and editors.

In private industry, writers in large companies work in departments identified by various names, such as Technical Writing, Editing, Corporate Communication, Editorial Services, Technical Support Services, and Public Relations.

Respondents to the STC survey reported that they work in the following technologies and professional settings (STC 5):

Computer software (28 percent)
Research and development (21 percent)
Computer manufacturing (16 percent)
Electronics (16 percent)
Training (12 percent)
Scientific (10 percent)
Telecommunications (10 percent)
Aerospace (10 percent)
Construction (10 percent)
Other (17 percent)

Notice that the computer industry (software and manufacturing) is the largest corporate employer of technical writers.

Writers also work for independent technical publishing agencies that prepare manuals and other documentation for companies that do not have their own technical writing departments. Other writers work for professional associations, such as the Society for Quality Control, and at universities, where they often work on research and grant proposals. Writers proficient in more than one language may work as technical translators.

Some technical writers function as freelancers, working under contract for many different organizations. Freelancing requires extra marketing skills and perseverance (see Meyers).

As shown in the STC survey, the technical writing profession includes many editors, who review and revise the work of specialists in such areas as science, engineering, medicine, and economics. Editors may work in any of the settings described earlier. However, they often work for large scientific organizations where they collaborate with technical specialists to produce major reports and other large projects (see Chapter 12).

Finally, many individuals who may not think of themselves as professional technical writers have jobs that require professional writing skills. They may be engineers, scientists, graduate students, or others who must regularly produce writing of professional quality on technical topics.

## THE TECHNICAL WRITER'S RESPONSIBILITIES

The professional technical writer's first responsibility can be summarized in three words: **HELP THE READER**. These three words are a charge for the technical writer to concentrate on writing for the needs of specific readers (rather than merely about certain subjects or merely to produce certain documents). In fact, every topic in this book is concerned in one way or another with helping the reader, from such global writing principles as organization to such particular choices as the use of certain typefaces to highlight key information. Technical writers must use every device available to meet their readers' needs.

For many technical writers, an important part of this responsibility is ensuring that readers who work with potentially hazardous devices (such as X-ray machines) are adequately warned about and understand all possible dangers. And the greater the potential harm to readers, the greater the writer's obligation to be vigilant about warning them (Wicclair and Farkas). Like engineers who design bridges and other structures on which people's lives depend, technical writers must maintain high ethical and professional standards, accepting their responsibility to protect readers from harm.

Writers also have a responsibility to be careful and thorough in order to prevent financial harm to their employers through product liability (see Chapter 14 for a discussion of the use of warnings). Product liability means that a manufacturer can be sued if, for example, a technical manual does not adequately warn users of the potential hazards of equipment or materials.

The STC "Code for Communicators" (Figure 1.1) nicely sums up these responsibilities. As this code suggests, technical writers should put their readers' needs ahead of their own. In a larger sense, technical writers serve society by helping the public understand important but complex technical issues, such as acquired immunodeficiency syndrome (AIDS) and the Three Mile Island nuclear accident (Schmelzer 221).

## THE TECHNICAL WRITER'S COMPETENCIES

Some of the competencies required of technical writers are illustrated in the typical "Help Wanted" ad shown in Figure 1.2.

### Writing Skills

Obviously, writing ability (or aptitude) is the basic competency of a technical writer. So it is not surprising that students considering this career often wonder if their writing skill is strong enough. One way to answer this question for yourself is to consider your *interest* in writing. Aptitude is often a product of interest, and interest often reflects an underlying aptitude (Weber 65). You should also assess your performance in both technical writing and other writing courses.

Besides having a strong interest in and aptitude for writing, you must be firmly grounded in the basics of language and writing and must master the advanced rhetorical skills of technical writing before you seek employment (Little and McLaren 23). One necessary advanced concept you must learn is rhetorical organization—the ability to plan your document and organize your writing around the readers' needs and the document's purpose (discussed in Chapter 6). Professional writers often cite the ability to organize and outline as essential to handling major projects.

In addition to organization, you must know the fundamentals of page layout (presented in Chapter 8) and integrating visuals (Chapter 7). In recent years, computers have made possible (through electronic publishing) the blending of language and visual elements, requiring writers to assume duties once in the

## Code for Communicators

As a technical communicator, I am the bridge between those who create ideas and those who use them. Because I recognize that the quality of my services directly affects how well ideas are understood, I am committed to excellence in performance and the highest standards of ethical behavior.

I value the worth of the ideas I am transmitting and the cost of developing and communicating those ideas. I also value the time and effort spent by those who read or see or hear my communication.

I therefore recognize my responsibility to communicate technical information truthfully, clearly, and economically.

My commitment to professional excellence and ethical behavior means that I will

- Use language and visuals with precision.
- Prefer simple, direct expression of ideas.
- Satisfy the audience's need for information, not my own need for self-expression.
- Hold myself responsible for how well my audience understands my message.
- Respect the work of colleagues, knowing that a communication problem may have more than one solution.
- Strive continually to improve my professional competence.
- Promote a climate that encourages the exercise of professional judgment and that attracts talented individuals to careers in technical communication.

**Figure 1.1** Society for Technical Communication Code for Communicators

**TECHNICAL WRITERS**

Metro Data Systems is one of the leading data processing companies serving the financial industry. Because of our continued growth and ongoing new software development activities, we are currently expanding our documentation department.

Metro Data Systems is looking for highly motivated individuals with the following:

- Strong writing skills
- Knowledge of computers or a strong sense of curiosity
- Effective interpersonal skills
- Excellent analytical skills

**Metro Data Systems (Box 12A)**
**The Daily News — Metro, CA 90020**

***Figure 1.2***
Typical Ad for Technical Writers

province of graphic designers and artists. Obviously, you need to become familiar with electronic publishing and computers because they will continue to be important tools for technical writers (we return to this in Chapters 13 and 15).

As a technical writer, you must also be able to write under the pressure of deadlines (Chapter 9). You must give careful attention to detail without getting bogged down—that is, you must know when to stop revising (Chapter 10). When instructors are asked to recommend students for jobs, employers often ask, "Does this person submit work on schedule?" Submitting assignments on time in the classroom demonstrates your ability to meet deadlines in other settings as well.

As a professional writer working for an organization, you must accept that you are not the owner of your document. Your name will seldom appear on your writing, and your talent may not be widely recognized outside your department. For that reason, technical writing could be called "egoless writing." Further, the ability to accept criticism and to dissociate personal feelings about your writing style and content is crucial—you can neither improve in the classroom nor survive on the job without it. To help students cultivate this ability, instructors often impose rigorous standards of their own, as well as encourage peer editing and collaborative work in the classroom. On the job, professional writers seek reviews of their writing because reviewers (by finding problems with technical accuracy or clarity, for example) can help to improve documents—which means that the writer will receive more credit for a job well done (see the discussion of reviewing in Chapter 11).

## Technical Knowledge

In the past, only applicants with in-depth knowledge of a specific technical field would be hired for technical writing positions. Such writers wrote primarily to experienced operators of equipment and highly trained technical personnel. In some organizations, such as scientific laboratories, a technical education remains an important qualification for employment.

However, many of today's technical writers prepare materials for the general public and for inexperienced users of equipment and systems. Such professional writers and their managers say that writers benefit by learning the relevant technology on the job, because having done so gives them more empathy with their readers (who must also learn that technology). For these writers, writing ability and analytical thinking are often more important than technical education.

Even though specific technical knowledge is not always crucial, to be an effective technical writer you must have intellectual curiosity. Furthermore, a broad base of general technical knowledge will make you much more adaptable—and employable. At the very least, you must have an interest in technology. If you are not curious about how and why things work, you will probably not enjoy or succeed in technical writing.

## Interpersonal Skills

As the want ad in Figure 1.2 suggests, technical writers must have the ability to work smoothly with people, especially technical specialists. Often technical specialists will be the primary sources of information; as a technical writer, you must be able to get along with scientists, engineers, and technicians. You must be able to look at things from their perspectives while still maintaining your own. You must also work collaboratively with publications professionals, such as artists, technical librarians, graphics specialists, and production personnel.

To work effectively with others, you must earn their respect—and you earn respect by being prepared, working with integrity, and taking pride in your work. You must learn to accept your own authority and expertise as a writer, just as you accept the technical expertise of engineers, scientists, and other professionals (we return to this in Chapters 3, 5, and 12).

## Critical Thinking

The job ad in Figure 1.2 states that the company is seeking an individual with "strong analytical skills." Having such skills means being able to think critically—to determine what is and isn't important, to weigh the merits of conflicting opinions, and to weed out useless or inaccurate information. Of course, writing itself requires critical thinking: the thinking it takes to analyze the rhetorical situation and your readers' needs and to adapt your strategies to meet those needs.

### Research Skills

As a technical writer, you will need solid research skills—the ability to locate the right people and documents, effectively interview technical personnel and potential users, evaluate source documents, and accurately summarize material and take notes. At times, however, you will have to go beyond standard resources and think like a master detective to locate a piece of crucial information or find the right person to interview (discussed in Chapter 5). Often technical writers need great creativity to solve research and writing problems.

On the job, learning must become a habit. As a newly employed writer, you may need to learn a technology that is not familiar to you. To become familiar with the technology, you might want to obtain a glossary of the technical terms, acronyms, and abbreviations, along with standard handbooks in the technology. It will also be important for you to become familiar with your organization's structure and informal protocol, as well as its stated policies (Reitman 103).

### Management Skills

As a technical writer, you must be able to manage your time; you must have enough self-discipline and initiative to work independently and produce writing in time to meet deadlines. You must be able to adapt quickly to new situations, because you will often work on several writing projects in a single day. As you gain experience, you may also develop the skill of leading others; one way to advance professionally is to become a manager of other writers (discussed further in Chapter 3).

## ACADEMIC PREPARATION

In view of the skills required of successful professional technical writers, a technical writing student must possess a commitment similar to that of students in, for example, law or medicine. Unlike some career areas, however, technical writers often have diverse academic backgrounds. Many successful technical writers today have English or journalism degrees. Most of these writers, however, have some background or interest in science or technology. Consider the following histories of students who have become successful technical writers.

> In high school, Louis decided that he wanted to be a professional writer—but he also enjoyed building electronic equipment. Based on a guidance counselor's advice, he enrolled in a midwestern university's technical writing program. He is now a senior writer at a computer company in the Northeast.

> Jim switched from mechnical engineering technology to technical writing in his final year of college. He is now a writer for an aerospace company in the Northwest.

> Mary returned to school after rearing a child, and she was an English major who enjoyed technical writing courses. She became a writer at a software firm in Texas, where she is now the manager of the technical writing department.

Bill was a math major in college who worked one year as a technical writer. He then enrolled in a master's degree program in technical communication. He is now an editor at a scientific research institute on the West Coast.

Pat, who had a B.S. in nursing, earned an M.A. in English. She now writes home therapy instructions for patients and teaches technical writing at a community college in Iowa.

## Developing an Appropriate Background

Regardless of your formal academic preparation, it is prudent to cultivate a background in *both* writing and technology. To continue to hone your writing skills, for instance, take courses in as many types of writing as possible, including creative writing. Such diversity will help you adapt to future demands. To learn some of the fine points of grammar (and thereby enhance your confidence as a writer), take language or linguistics courses.

To develop a technical background, take a few basic science and engineering courses; doing so will also help you discover how interested you are in technology. At the very least, you should learn the basics of computer systems and the principles of programming (Antoine 16). In addition, since technical writers often work in nonprint media, film and video courses are useful in helping you understand the principles of scriptwriting and the preparation of audiovisual materials.

Technical writers who have moved up in their organizations recommend that students take management courses, since one path of advancement is through management. Moreover, writers need to understand how to manage the whole document process (covered in Chapters 2 and 3). Management skills useful for both managers and writers (especially freelancers) include scheduling and budget preparation (Shultz 219); therefore, you should consider secondary courses in management information systems and general accounting.

Finally, if you plan to teach someday, either in a corporate education department or at a college, you will need to pursue an advanced degree (an M.A. or a Ph.D.) in a graduate program.

## Technical Writing Programs

At one time, only a handful of colleges offered formal training in professional technical writing. Most writers had to learn by doing. Today, however, many colleges and universities offer full degree programs at the undergraduate and graduate levels; many more schools have certificate programs or cores of advanced and professional technical writing courses. In addition to the standard advanced writing courses, such curricula often include electives in graphic design, layout, production, and printing processes, including desktop publishing (we return to this in Chapter 13). The diversity in such programs is important because as a professional technical writer, you may at some point work in a department where you must perform multiple tasks (write, prepare illustrations, and design pages) using a desktop publishing system (Farkas, Haselkorn, and Ramey).

## FINDING A JOB

To learn about job opportunities, seek your instructor's advice, talk to former students now employed as technical writers, and join professional associations. Pursue traditional sources as well, such as guidance counselors and "Help Wanted" ads in newspapers and trade magazines. When looking for positions, remember that job titles for both writers and editors vary greatly from organization to organization; they may be referred to as "document developers," "communication analysts," or "product support specialists," among other titles.

While you are in school, develop a *portfolio* (containing samples) of your writing. Save the materials you write for your classes; then select the best for presenting to potential employers.

When you begin your job search, remember that your letter of application and résumé are samples of your ability to organize information and use words and page layout effectively. Thus your writing should be flawless and your pages well designed.

## PROFESSIONAL ORGANIZATIONS

One way to find employment, continue to learn, and enhance your professional reputation is to join professional societies. The largest organization for professional technical writers is the Society for Technical Communication (STC), which accepts student members at reduced rates. You can find membership information in the STC quarterly publication, *Technical Communication*, or by writing to the society headquarters:

> Society for Technical Communication
> 901 North Stuart Street, Suite 304
> Arlington, VA 22203

Many areas of the country have local groups of writers, such as the Wisconsin Organization of Documentation Specialists (WORDS). Learn about the associations in your area. Technical writers in specific industries have formed other specialized groups. The following are typical:

> National Association of Government Communicators
> 80 South Early Street
> Alexandria, VA 22304

> American Medical Writers' Association
> 9650 Rockville Pike
> Bethesda, MD 20814

National Association of Science Writers
P.O. Box 294
Greenlawn, NY 11740

Participating in professional organizations is also an important way for you to learn how others are solving the problems that you may one day face on the job.

## WORKS CITED

Antoine, Valerie. "The Software Documenter: A New Specialist." *Technical Communication* 32.3 (1985): 16–18.

Dobrin, David N. "What's Technical about Technical Writing?" *New Essays in Technical and Scientific Communication: Research, Theory, Practice.* Ed. Paul V. Anderson, John R. Brockman, and Carolyn R. Miller. Farmingdale, NY: Baywood, 1983.

Farkas, David K., Mark Haselkorn, and Judith Ramey. "Desktop Publishing: The Author as Compositor." *IEEE Transactions on Professional Communication* 30.2 (1987): 87–90.

Kalmbach, James R., Jack W. Jobst, and George P. E. Meese. "Education and Practice: A Survey of Graduates of a Technical Communication Program." *Technical Communication* 33 (1986): 21–26.

Little, Sherry Burgus, and Margaret C. McLaren. "Profile of Technical Writers in San Diego County: Results of a Pilot Study." *Journal of Technical Writing and Communication* 17 (1987): 9–24.

Meyers, Richard, ed. "Special Issue on Freelancing." *Technical Communication* 33 (1986): 206–248.

Moran, Michael G. "The History of Technical and Scientific Writing." *Research in Technical Communication: A Bibliographic Sourcebook.* Ed. Michael G. Moran and Debra Journet. Westport, CT: Greenwood, 1985. 25–38.

Reitman, Paul. "The Trouble with Larry: A Look at the Problems Facing a New Writer." *Technical Communication* 34 (1987): 103–104.

Schmelzer, Richard W. "New Responsibilities for the Technical Writer." *Journal of Technical Writing and Communication* 11 (1981): 217–221.

Shultz, George E. "A Method of Estimating Publication Costs." *Technical Communication* 34 (1987): 219–224.

Society for Technical Communication. *Profile 88: STC Special Report.* Washington, DC: STC, 1988.

U.S. Dept. of Labor. Bureau of Labor Statistics. *Occupational Outlook Handbook.* 1988–1989 ed. Washington, DC: GPO, 1988.

Weber, Barbara. "Technical Writing Skills: A Question of Aptitude or Interest?" *Journal of Technical Writing and Communication* 15 (1985): 63–68.

Wicclair, Mark R., and David K. Farkas. "Ethical Reasoning in Technical Communication: A Practical Framework." *Technical Communication* 31.2 (1984): 15–19.

## FURTHER READING

Alred, Gerald J., Diana C. Reep, and Mohan R. Limaye. *Business and Technical Writing: An Annotated Bibliography of Books, 1880–1980.* Metuchen, NJ: Scarecrow, 1981.

Barabas, Christine. *Technical Writing in a Corporate Culture.* Norwood, NJ: Ablex, 1990.

Broadhead, Glen J., and Richard C. Freed. *The Variables of Composition: Process and Product in a Business Setting*. Carbondale: Southern Illinois UP, 1986.

Buchholz, William J. "The Boston Study: Analysis of a Major Metropolitan Business- and Technical-Communication Market." *Journal of Business and Technical Communication* 3.1 (1989): 5–35.

Burke, Michael. "Career Opportunities in Computer Documentation." *Technical Communication* 33 (1986): 13–15.

Firman, Anthony H. "Four Characters: Reconciling Publications Personalities." *Technical Communication* 34 (1987): 132–134.

Frank, Darlene. "Selling Yourself through Your Skills." *Outstanding ITCC Paper Series*. Washington, DC: Society for Technical Communication, 1986.

Girill, T. R. "Technical Communication and Ethics." *Technical Communication* 34 (1987): 178–179.

Gould, Jay R., and Wayne A. Losano. *Opportunities in Technical Communications*. Lincolnwood, IL: VGM Career Horizons, 1984.

Guinn, Dorothy Margaret. "Sleuthing, Diplomacy, Scheduling, and Budgeting for Technical Writing." *Technical Writing Teacher* 13 (1986): 13–17.

Hildebrant, H. W., and Kathryn Rood. "Social Reporting for Technical Writers." *Journal of Technical Writing and Communication* 13 (1983): 275–285.

Huss, Carol. "Running toward Professional Development." *Technical Communication* 34 (1987): 142–145.

Keeler, Heather E. "Portrait of a Technical Writer: An Annotated Bibliography." *Journal of Technical Writing and Communication* 19 (1989): 297–303.

Kellner, Robert Scott. "Preparing Technical Writing Students to Write for the Government." *Journal of Technical Writing and Communication* 10 (1980): 177–182.

Killingsworth, M. Jimmie, and Scott P. Sanders. "Portfolios for Majors in Professional Communication." *Technical Writing Teacher* 14 (1987): 166–169.

Senf, Carol A. "Technical Writing as a Career." *Technical Writing Teacher* 14 (1987): 68–76.

Society for Technical Communication. *Proceedings*. Published yearly.

"Special Section on Ethics." *Technical Communication* 27.3 (1980): 4–12.

# The Case History

### Chapter 1: The Professional Writer

Mary Mullins was a writer in the Technical Writing Department at First Wisconsin National Bank in Milwaukee, Wisconsin, for 4½ years. At the time the information in this Case History was compiled, she had just completed a procedural manual for users of a new international banking software system. The finished manual, *Introduction to Using the International Banking System*, is reproduced in the Appendix (pages 387–417).

   In Case History interviews following each of the first 14 chapters of this book, Mary will describe the steps in the document process that led to this manual, as well as some aspects of her own history as a technical writer. Her comments will provide an inside look at the writing, editing, and production practices of a technical writer employed by a large banking institution.

INTERVIEWER: What is the name of your department at First Wisconsin, and where does it fit within the bank's overall organization?

MARY: My department is called Technical Writing. It's part of the broader organization of Systems Support and Administrative Services within the bank's Information Services Division. [See Figure C.1.]

INTERVIEWER: How many other writers work in your department?

MARY: Ten writers work under the information services officer—that's my manager's title.

INTERVIEWER: I notice that a distinction is made between senior writer and technical writer on the chart. Is the senior writer a manager of sorts?

MARY: Only on a project basis. The senior writer might head up a project, with two technical writers reporting to him or her on an informal basis. The senior writer is the primary project planner and acts more as an adviser than as a manager of the other writers. Senior writers generally handle more complex projects—producing several manuals or using new documentation techniques, for example.

INTERVIEWER: What kinds of writing do you produce? Who are the end users?

MARY: We write procedural manuals or reference manuals. The end users are typically bank clerks, tellers, or officers, such as personal bankers, who use computer systems to do some kind of work.

   This project, the International Banking manual, involved a big software system that was being installed in the International Banking Department. The system was very complex, with many different features, and each feature was going to be used by a different cluster of employees in the department. For example, the traders, the foreign exchange people, the people who write letters of credit—each group would use a different feature of the system. Our purpose was to educate them—to introduce them to the sys-

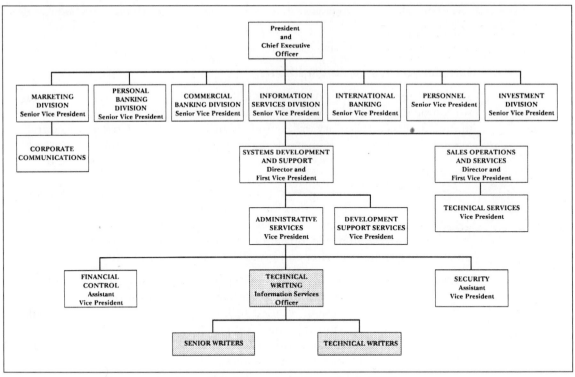

*Figure C.1* Partial Organization Chart, First Wisconsin National Bank (1988)

tem's features and show them how to use it. The *Introduction to Using the International Banking System* manual is actually the first in a series of nine manuals designed to document the whole system.

INTERVIEWER: What can you tell us about your own role as a technical writer at First Wisconsin? Is there anything about the job that surprises you or that you didn't expect?

MARY: One thing that surprised me about the job is how much I have to negotiate, do strategic planning, play politics, learn to handle how I talk to people—all things that I guess I didn't expect when I thought of the label "writer." A lot of my job involves dealing with people and getting information from them, trying to get the most current information and keep up with the many changes happening as a project is developed. And I've found that sometimes you have to speak up if you're not getting the information you need.

INTERVIEWER: What is your greatest challenge as a technical writer?

MARY: Managing all the details of a project. I sometimes envision myself as a charioteer holding lots and lots of reins (reins between every finger!). Writers have to keep track of a multitude of concerns, ranging from learning complex information to keeping a project within its budget to watching their spelling.

**INTERVIEWER:** What did you do before your present job at First Wisconsin?

**MARY:** I started out at another bank in another department, updating a database of personnel records. I'd taken a seven-month computer course because I tested well in logic and because computers were known to be a booming area in industry. From that training, I got the job updating the database.

Then I heard about an opening in that bank's Technical Writing Department. They wanted an assistant that they could bring up to the professional level. At that point I didn't know if I had the right skills, but they asked for a writing sample. I wrote a sample, and they hired me based on that. From there I moved on to my present job at First Wisconsin.

So in a way, you really see me "backing into" this profession. I don't think that could happen so easily today. More and more beginning technical writers are starting out with an educational background in technical writing.

**INTERVIEWER:** How much technical knowledge did you bring with you to your work?

**MARY:** I brought only the seven-month certificate in data processing. I'd never taken courses in technical areas, but I had always done well in English and Spanish, and I had a strong interest in language and literature. Plus I enjoyed reading very much; apparently, all the reading I've done has really built up my vocabulary.

**INTERVIEWER:** What kind of technology have you needed to learn in your present job?

**MARY:** Well, we document banking software.

**INTERVIEWER:** So you've probably had to learn something about finance and something about computers, and then the software that combines those two.

**MARY:** Right. For the finance part, I've taken several courses in banking, including some informal classes sponsored by the American Bankers' Association.

**INTERVIEWER:** What kinds of technology do you think technical writing students should study in school—assuming they don't know exactly what kind of technical writing they want to do?

**MARY:** Well, I'd say *any* technical writing student should know how to use a word processor. A mechanical drawing course or a document design course would probably be good, too, since graphics and design are usually part of the basics of a technical writer's job. And then maybe a sampling of courses in engineering or computers.

**INTERVIEWER:** What other skills do you see as important for the prospective technical writer?

**MARY:** Interviewing and other "people" skills. Also organizational skills—time management, organizing work, that kind of thing. Negotiating skills are important, too.

**INTERVIEWER:** What advice would you give a student for finding employment as a technical writer?

**MARY:** Present your résumé to the personnel departments of the companies you want to work for. Don't just mail your résumé—that doesn't have enough impact. Also, talk to some people who do what you want to do and ask them for advice.

I also think it's a good idea to become a student member of local technical writing organizations. Go to meetings and talk up your career goals with people you meet. Organizations like the Society for Technical Communication are really good for establishing a network. I went to a lot of STC meetings the first year I was a writer; I think that's especially important when you're first starting out.

**INTERVIEWER:** How important is continuing education for a technical writer?

**MARY:** As someone who's still working toward a degree, it's very important for me personally. At the moment, I'm not qualified for certain jobs because I lack a bachelor's degree. But I also think it's good to continue even if you already have a writing or English degree. A graduate degree may not be critical for employment, but it might be critical for promotion. And of course if you're going to be a professional, why stop learning about your profession?*

---

*Since the material for this case history was compiled, Mary was awarded a Society for Technical Communication scholarship and has completed her B.A. in English with a technical writing emphasis.

# 2

# The Document Process

The Document Process Defined
Principles of the Document Process
The Writer's Role in the Process

This chapter provides a general discussion of the process whereby a document is produced in an organization, whether that document is a proposal, a technical manual, a policy handbook, an employee publication, or a public information brochure. The chapter focuses particularly on the reasons why a document is created, the initial meeting that starts the document process, and the professional writer's role in the process.

The impetus for producing a document grows out of an organizational need or purpose. For example, an aircraft parts manufacturer must compete with other manufacturers for a NASA contract to supply spare parts for the space shuttle program. As a result, a writer is assigned to write and coordinate the preparation of a major government proposal for submission to NASA. In another case, a nationwide service association is swamped with complaints from its clients that its programs are inconsistent from state to state and even within some states. A writer is assigned to collect information and prepare a manual for regional directors describing standard procedures for their staffs to follow. The manual traced in the Case History that appears throughout this book grew out of a bank's need to train its staff to use a complex computer system to process international financial transactions. The bank assigned a professional writer to prepare the manual, which appears in the Appendix.

Generally, organizations need documents for any of the following reasons:

- To market a product or a service
- To support users, customers, or clients after a sale
- To manage more efficiently and effectively
- To reinforce positive images and build goodwill
- To win public support for a position or an issue
- To meet legal or ethical obligations

A single document may need to achieve more than one goal. For example, a computer manufacturer must not only teach customers to use its new personal computer but also convince them that the company stands behind the product. The company therefore needs a computer manual that will both convince buyers that the system is easy to use, maintain, and upgrade and reinforce a positive attitude about the company. Well-written, high-quality documents can help achieve many of the organizational goals that are key to the success of any enterprise.

## THE DOCUMENT PROCESS DEFINED

Documents are generally produced within an organization through a relatively systematic process. Such a process helps ensure that documents are finished in an efficient and timely manner. Although this process varies from one organization to another, its stages in most organizations are well defined and relatively identifiable (Broadhead and Freed 122–134). To help you visualize the process, the flowchart in Figure 2.1 presents the seven stages through which a document is typically produced. The solid lines in the flowchart indicate a direct movement of the document from stage to stage; the dotted lines illustrate either points at which stages may recur or points at which the document returns to the writer for further drafting before moving to the next stage. (The chapters in this book, which are referenced in brackets in the illustration, generally follow the order of these stages.)

### 1. Document Need Assessment

Because of specific organizational needs like those described above, the management of an organization calls an initial meeting to determine *if* and *how* a document may meet these needs or purposes. For example, a medical electronics manufacturer is interested in winning a contract to supply a 50-hospital medical corporation with sophisticated diagnostic equipment. As a result, someone in a position of authority calls a meeting. This first meeting may determine if the manufacturer is capable of responding to the medical corporation's request for proposals, or "RFP" (Brusaw, Alred, and Oliu 265–273). If an organization is not in a position to produce the item or service requested by the agency or client, writing a proposal will be a waste of time.

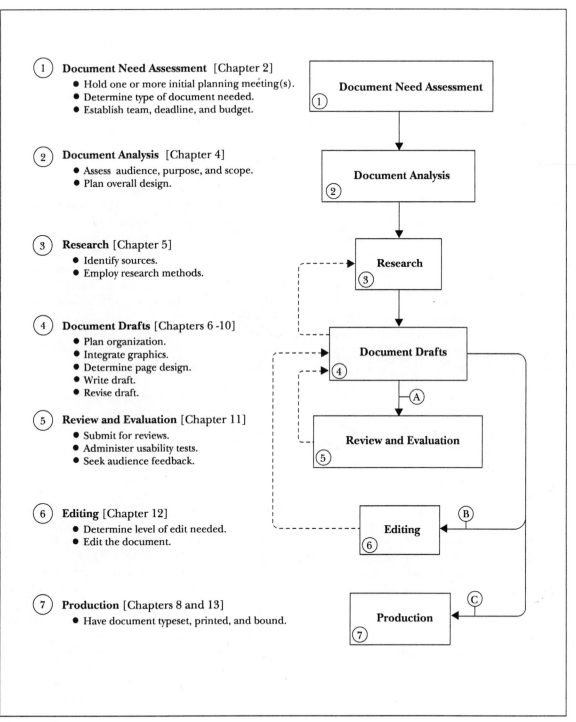

**1** **Document Need Assessment** [Chapter 2]
- Hold one or more initial planning meeting(s).
- Determine type of document needed.
- Establish team, deadline, and budget.

**2** **Document Analysis** [Chapter 4]
- Assess audience, purpose, and scope.
- Plan overall design.

**3** **Research** [Chapter 5]
- Identify sources.
- Employ research methods.

**4** **Document Drafts** [Chapters 6-10]
- Plan organization.
- Integrate graphics.
- Determine page design.
- Write draft.
- Revise draft.

**5** **Review and Evaluation** [Chapter 11]
- Submit for reviews.
- Administer usability tests.
- Seek audience feedback.

**6** **Editing** [Chapter 12]
- Determine level of edit needed.
- Edit the document.

**7** **Production** [Chapters 8 and 13]
- Have document typeset, printed, and bound.

*Figure 2.1* Flowchart of the Document Process

The participants in these initial meetings vary, depending on the circumstances and structure of the organization. However, the participants at such meetings could include representatives of senior management, the marketing division, the technical division, the graphics areas, the writing department, and editorial services. Often the writing department is represented by its manager; however, the writer assigned to the project may also attend, depending on the management practices of the organization and the meeting agenda.

Once the meeting participants have agreed that a document should be prepared, they usually decide on the general type of document (proposal, manual, brochure, etc.) that will fill the need. This decision is based on the organizational need or the form of document expected by a client. Another essential task is to consider a budget for the project. Based on the budget, further decisions, such as the sophistication of the overall design and even the binding, can be considered. These features are discussed in Chapters 4 and 8.

During the meeting, the ultimate deadline or publication date will be established. Depending on the organization's practices, a writer may be assigned to the project at the meeting, a writing team may be set up (see Chapter 3), or the writing department manager may be charged with assigning a suitable writer (or team of writers) to the project. This meeting also provides the writer or the manager with an excellent opportunity to identify key information sources and contacts as well as potential reviewers.

## 2. Document Analysis

Following the initial meeting, the writer or team should assess the audience, purpose, and scope of the document. As discussed in Chapter 4, the writer or team must also plan the overall design of the document on the basis of audience, purpose, and document type.

## 3. Research

The writer or writing team must next use awareness of audience, purpose, and scope to focus on gathering information that will produce a solid draft of the document (as described in Chapter 5). To conduct effective research, the writer must first identify the sources of information available: people, documents, online services, and so on. Although some of these sources may be identified during the initial meeting, the writer often goes on to establish a network of people who either know the relevant information themselves or can refer the writer to appropriate sources of information. As these sources are identified, the writer's next step is to determine and use all appropriate methods for gathering information, such as interviewing experts, observing processes, or searching online databases.

## 4. Document Drafts

During this stage, the writer (or writing team) must first organize the information gathered (discussed in Chapter 6). In addition, at some point during the organization of a draft, the writer should determine what visuals may need to be included and where they should be integrated (Chapter 7). During this stage, the writer may also consider page design (Chapter 8), either working with graphics specialists or preparing the illustrations and layout personally. When a satisfactory organization has been achieved, the writer will be able to produce a draft of the document (Chapter 9). After writing a first draft, the writer must revise it (Chapter 10).

Ideally, when the writer finishes revising the document, it should be ready for review and evaluation (stage 5 in the flowchart in Figure 2.1). However, during organization or any other step in producing a draft, the writer may discover gaps in the information gathered so far. If this happens, the writer must return to stage 3, research, as indicated by the dotted line from stage 4 to stage 3 in the flowchart. When the writer has gathered the missing information, he or she then incorporates it into subsequent revisions of the draft (stage 4).

## 5. Review and Evaluation

When the draft is as complete, accurate, and polished as possible within the constraints of the deadline, it is ready for review and evalution (path A, stage 5).

As described in Chapter 11, reviews are critiques of the draft by technical experts, peer writers, legal staff, management, or others. These reviewers typically search for inaccuracies, incomplete statements, inappropriate information, and other elements affecting the clarity or effectiveness of the document.

Evaluations also seek to identify weak areas in a document; however, they involve the actual use of a document to determine whether or not it enables readers to perform a task effectively and efficiently. Thus evaluations are more commonly used in checking documents (such as computer manuals) that are designed to *instruct* their readers in some way. One kind of evaluation, the usability test, employs subjects who are similar to the document's intended audience to perform the process the document describes. Based on an evaluation of these subjects' successes and failures in completing tasks, the document can be revised to make it more effective.

Another kind of evaluation, audience feedback, occurs after a document has been produced and is in the hands of the customers or clients who will use it. Audience feedback involves both customer validation tests and customer feedback. Customer validation tests are similar to usability tests but involve actual customers who use and test the document at their own workplace. Customer or client feedback includes written and oral responses from the buyers of a product after its purchase and use in their workplace. Such feedback could include direct responses from the audience in the form of written comments on questionnaires, comment cards, or other forms. It may also involve indirect responses based on the observations of company representatives who meet with customers. Audience feedback is particularly useful for postpublication updates of documents and new editions of publications.

Based on the findings during the review and evaluation stage, the writer revises the draft again (indicated by the dotted line from stage 5 to stage 4 in the flowchart). After the reviews have been incorporated, the document may then proceed to editing (via path B, stage 6). If an organization does not have an editing department or function, however, the document moves directly to production (path C, stage 7), where it is reproduced in its final form.

## 6. Editing

When a document is to be edited, the editor assigned must first determine the amount and kind of editing to be performed (the "level of edit," as described in Chapter 12). Based on the level of edit required, the editor marks the manuscript or the online version. When this work is complete, the editor returns the document (or "edited copy") to the writer for approval of the changes (indicated by the dotted line from stage 6 to stage 4 in the flowchart). As illustrated in the Case History section following Chapter 12, the writer and editor often communicate with each other and may even hold meetings during this loop in the process.

## 7. Production

When the writer is satisfied with the editing (or has simply incorporated the reviewers' suggestions when not working with an editor), the document is ready for final production (stage 7). In some cases, this stage may involve simply reproducing the final, proofread version of the document on photoduplicating equipment (as was done with the user manual discussed in the Case History and reproduced in the Appendix). In other cases, a manuscript (or disk) is submitted to a fully equipped, in-house production department by the writer, who often has very little to do with the document after that point. Similarly, a manuscript or disk could be submitted to an outside commercial graphic arts company for production.

The sophistication of the production process used may depend on the budget involved, which is often based on the importance or visibility of the document inside or outside the organization. It may also depend on the sophistication of the equipment to which the writer has access. Available and appropriate options and how to use them are described in Chapters 4, 8, and 13.

## PRINCIPLES OF THE DOCUMENT PROCESS

As you review the flowchart in Figure 2.1, keep in mind four principles concerning the stages of the document process:

1. Their order may vary.
2. They may overlap and merge.
3. They are cumulative.
4. They are recursive.

First, although the process shown in the flowchart is typical, the order of the components of the document process may vary. Depending on the practices of the organization, for example, document layout and design could occur before, during, or after the draft stage. You could work for an organization with a standard format for their documents (described in Chapter 14), or you might need to follow a set form that is always used for the type of document you are producing or that has been prescribed by the client (as in a grant proposal). In such cases, you must fit your draft into a layout determined well *before* the draft stage. Conversely, if the design has not been established and you are working with computer equipment that is capable of designing pages, you might create the design *during* the drafting process, using the principles we will discuss in Chapter 8. Finally, if you are working with an in-house or a commercial graphic designer, page layout and design could occur well *after* you have produced a draft of the document.

Second, the stages sometimes overlap and even merge. In some circumstances, for example, a document may be reviewed while it is still in the organizational or outline stage, before the draft is complete. In that case, stage 4, document drafts, would overlap or merge with stage 5, review and evaluation.

Third, the stages are cumulative. That is, the work of one stage does not necessarily end at the next stage; rather, the effect of each individual stage grows and accumulates through successive additions. For example, even though audience assessment in stage 2 should be as thorough as possible, you must continue to develop your audience awareness—your knowledge and sense of the reader—throughout the stages of the process (Dobrin 104–109).

Fourth, the stages are recursive; that is, they may occur repeatedly in the process. As discussed earlier, when the writer is preparing a draft, gaps in information may become evident. In that case, the writer must return to research (stage 3) and then apply the new information to the draft (stage 4). As noted, the dotted lines in Figure 2.1 indicate several points at which stages can recur. Further, within the document drafts stage (stage 4) itself, an individual writer's process may be recursive, moving back and forth between the steps of organization, page design, drafting, and revision.

As these principles demonstrate, the document process may not always occur in discrete steps or in a single, ideal sequence. However, conceptualizing stages in the document process and understanding how the stages generally occur are important in task organization. It facilitates planning, keeps the writer or writing team on a clear track, and helps ensure that the document will be produced by the deadline. Defining the stages in the document process is also important to your understanding of your role as a professional technical writer.

## THE WRITER'S ROLE IN THE PROCESS

The writer's role in the process is determined by the place of the writing function in the structure of an institution or organization. For example, in many small organizations without separate editing departments, graphics specialists, or elaborate production facilities, a writer is typically involved directly in most of the stages of the document process. By contrast, in large firms with well-defined departments and elaborate facilities, a manager may assign a writer to produce the text portion of a document only. After that point, the document might be handled by a staff of specialists who will review and edit, integrate graphics, design pages, and supervise the physical production of the document. In other words, the writer is only peripherally involved with many of the stages in the process—if at all.

The size of an organization alone does not determine the writer's role, however. At one major electronics corporation, for example, the writing function is undertaken by individual writers all around the United States. Each writer works with a desktop publishing system to produce final, "camera-ready" copy that is then duplicated and distributed from his or her location. Obviously, writers at this large firm are involved in most of the stages outlined in this book.

In any organization, new writers must learn quickly how the document process is organized, even though at first they may be involved only in updating and revising existing documents. Familiarity with that process is important because as a professional technical writer, your own writing process must fit the pattern of the document process in your organization. At a company where review occurs before a draft is completed, for example, you might have to submit outlines for review. In that case, you would have to be able to work from detailed outlines. Both Chapter 1 and the 14 Case History sections in this book contain information that can be helpful in learning about a typical document process in an organization.

*Checklist 2.1*

## Document Need Assessment

- ☐ What is the problem or situation stimulating the need for a document?
- ☐ Who should be involved in the initial meeting?
  - • Senior management
  - • Technical division
  - • Writing department
  - • Marketing division
  - • Graphics area
  - • Editorial services
- ☐ Will the document solve the problem or meet the need?
- ☐ What are an appropriate budget and time schedule for producing the document?
- ☐ What is the final deadline for the document?
- ☐ What form should the document take—manual, proposal, brochure, etc.—or should information be online?
- ☐ How might the form of the document be affected by use, document life span, or distribution?

☐ What management policy requirements or limitations will affect the scope?

☐ Who should write the document?

☐ What are the main sources of information?

☐ Who are possible reviewers for the document?

---

## WORKS CITED

Broadhead, Glen J., and Richard C. Freed. *The Variables of Composition: Process and Product in a Business Setting*. Carbondale: Southern Illinois UP, 1986.

Brusaw, Charles T., Gerald J. Alred, and Walter E. Oliu. *Handbook of Technical Writing*. 3rd ed. New York: St. Martin's, 1987.

Dobrin, David N. *Writing and Technique*. Urbana, IL: NCTE, 1989.

## FURTHER READING

Goswami, Dixie, et al. *Writing in the Professions*. Washington, DC: American Institutes for Research, 1981.

Monroe, Judson. *Effective Research and Report Writing in Government*. New York: McGraw, 1980.

Pakin, Sandra, and Associates. *Documentation Development Methodology*. Englewood Cliffs, NJ: Prentice, 1984.

# The Case History

### Chapter 2: The Document Process

INTERVIEWER: When does your department first become involved with a writing project?

MARY: When our manager is given a work request for the project. The request is usually mailed in by the system developers.

INTERVIEWER: After your writing manager gets that form, when do you become involved in the document process?

MARY: When I've finished a project and have time to take on something new, I let my manager know. Eventually she'll call me into her office and describe a work request for a project, then ask if I will take it on. Sometimes I'm asked to prepare a tentative scope for the project, but I won't be assigned for sure as the writer until we know how big the project is.

My manager will generally give me the request form and a few notes from her preliminary research on the project. Often she will have already done things like finding out the installation date or informing the programmers that our department doesn't expect to be able to meet their proposed deadline. [Figure C.2 shows the technical writing request form for the International Banking System manuals.]

INTERVIEWER: When, ideally, should a writer at First Wisconsin enter the process of developing a new product or system? It sounds as though the development might be completed by the time you first learn about the project. Would it be better for a technical writer to become involved a little earlier, for example, during some early meetings?

MARY: As a department, we've asked to be involved at the stage when a team of potential users have approved a preliminary design for the proposed system or system change.

But it depends on the project. On big projects, we'd especially like to be involved at those earlier stages, even if it results in some wasted time (if, for instance, we learn how the system works and then it changes). That way we can pick up on the actual meaning of the project. Smart developers will get us involved early because they want to get the documentation written on time.

INTERVIEWER: What was First Wisconsin's need for the International Banking System documents?

MARY: A big international banking software system was being installed in the International Banking Department. The documents were needed to educate the various users of the new system.

INTERVIEWER: Could you describe the normal document process at your firm?

MARY: For new manuals, we ask who's writing what, if our department will

# Technical Writing Request Form

Project name: _I BS – International Banking System_  Project number: _VTW045563_
Project team: _Trust / International_  Spectrum phase: _preliminary design_
Contact person: _Ralph Tatman_  ext: _3324_
User rep: _Donna Witegood_  ext: _1815_

Today's date: _10/12/86_
Installation date: _3/01/89_

Project description:
- ☐ MAPP  ☑ New system _IBS – International Banking System_
- ☐ Small  ☐ Single system impact _____
- ☐ Mnt  ☐ Cross-application _____
- ☐ BPR  ☐ Modified vendor package

- ☐ Screens (est. #new) _60_  (est. #revised) _____
- ☐ Reports (est. #new) _45_  (est. #revised) _____

TW requirement:
- ☐ Manual revision
- ☑ New manuals

Manuals affected: _____
_____
_____

Elaboration of requirements: _Please review existing documentation and develop a plan for completing documentation for the entire system. See Dave Hohenfeldt for a demonstration of the system._

Forward to: **Gail Atkins,** Manager of Technical Writing, 9th floor, as soon as the project has been approved.

*Figure C.2* Technical Writing Request Form

be doing training manuals or reference manuals, and other large-scale questions that help in setting the scope for the project.

Projects generally start with a work request. I do some preliminary research in which I look into the larger issues of how much information is needed. For example, I find out the audience and purpose for the documentation and quantify how many new screens and new reports are involved, how many processes need to be explained, and so on. I look for a rough idea of the size of the project. We mostly document screens and reports at my company, and we have rough formulas that say how many pages are needed for each typical component of the documentation we write.

Some tasks overlap because they are not done in a linear fashion. You don't do all the research and then write, for example. A lot of times I work up a draft that's full of questions and gaps, then go to the system developer and say, "This is what I have so far for this task, but I'm wondering where the reports fit in," or something like that. Writing, proofreading, and revising are overlapping tasks. Often I write something, look at what I wrote, and go back to the screen and change it.

INTERVIEWER: Generally, how do you manage your time when you have to handle numerous writing projects?

MARY: Once I know the amount of time estimated for each project and know the installation dates for the projects, I use a calendar to plot specific hours when I'll be working on each phase of each project. Before plotting time on the calendar, though, I note holidays, planned absences of reviewers and systems contacts, and the dates when new information is expected to become available. Then I assign specific days to work on particular tasks for the projects. That's my personal schedule.

On a larger scale, within our department, we are able to devise something called a Gantt chart on our text processing system. The Gantt chart allows us to plot the various stages of the document process on an overall project schedule. [The Gantt chart for the *Introduction to Using the International Banking System* manual is reproduced as Figure 3.2 in Chapter 3.]

INTERVIEWER: How did you estimate the number of pages for the International Banking System manual and the time you'd need to produce them?

MARY: Well, I took some pages from source documents that needed to be turned into user documents and added 30 percent, estimating that I'd need to expand the source document by that much. (I came up with the 30 percent figure by rewriting one source document as a test and finding that I had expanded it by about 30 percent.) For some tasks, I did some initial research, learned how to perform a task, and added 30 percent to the number of pages of notes I had for that task. Then I multiplied that number by the total number of tasks that needed documenting.

The time-estimating formula we use is based on classifying each page as a new page, a major-change page, or a minor-change page. We estimate that it takes 1 ½ hours to create a new page, ¾ of an hour to make a major revision to a page, and so on. Then we figure in factors that could influence that basic time. For example, it takes 1 ½ hours to create one new page and I've got eight pages to write, but since I'm not very experienced on the system I'll be documenting, I add 20 percent to the unadjusted hours. So that makes 12 hours plus 20 percent, which is 2 ½, for a total of 14 ½ hours.

INTERVIEWER: Then you use these estimates to set the schedule.

MARY: Yes.

# 3

# The Writing Team

The Functional Writing Team
The Peer Writing Team
The Supervised Writing Team
Cost Estimates
Project Schedules

Whether you work for a university writing grant proposals, for the government writing technical reports, or for a computer manufacturer writing software user manuals, you will work with other people and therefore become part of a team.

The term *writing team* is used to describe three kinds of collaboration. One is a group of professionals and specialists (writer, subject-matter specialist, editor, typist, artist, and printer) who work together to produce a printed document; this group is referred to in this chapter as a *functional* writing team. A second is a group of writers who work together as peers to produce a single document; that is referred to in this chapter as a *peer* writing team. A third is a group of writers who collectively report to a manager; this is referred to in this chapter as a *supervised* writing team. The group dynamics of the three writing teams differ because the teams are structured very differently. Writing teams are used to produce all kinds of business and technical documents, including proposals, annual reports, and technical manuals.

Some people also use the term *collaboration* to refer to one writer reviewing and criticizing the work of another; however, in our discussion of team writing, we will call that activity *peer review* rather than collaboration.

## THE FUNCTIONAL WRITING TEAM

Writers cannot work in a vacuum. They must use the knowledge and skills of others to achieve their own goals. Being able to identify these people—and then to establish and maintain good relations with them—is a critical skill for writers. Of course, the writer must make all decisions and accept all responsibility for the document that is produced by the functional writing team.

The functional writing team, which is probably the team most commonly thought of when the terms *collaboration* and *writing team* are used, includes the writer and the other professionals and specialists with whom the writer works to produce a printed document: subject-matter expert, editor, typist, layout artist, and production specialist or printer. The functional writing team is a team only in the loosest sense of the word, since the members never actually come together in one place to meet as a team. The writer works closely with each member of the functional writing team, however, at different stages of the document cycle (see Chapter 2).

The writer works with the subject-matter specialist throughout most of the document cycle. Only after review and evaluation does this person normally step out of the picture.

The writer works with the editor after review and evaluation and after revisions resulting from the review have been made, although the writer may solicit "unofficial" feedback from an editor prior to review and evaluation. The editor could be any of three different types of professionals. Sometimes a professional editor (someone who has been a writer but has proved to be especially good at improving the writing of others) is part of the functional writing team. At other times a peer writer may be designated to function as the editor on a particular project, or the project leader or manager may serve as the editor.

The writer works with the layout artist both when writing the draft (to get specific illustrations started) and after revising it (to achieve final page layout design and production of camera-ready art).

The writer works with the production specialist at the end of the document cycle to get the document printed.

## THE PEER WRITING TEAM

Allen et al. refer to the concept we are calling the peer writing team as *shared document collaboration* and define it thus: "shared document collaboration involves collaborators producing a shared document, engaging in substantive interaction about that document, and sharing decision-making power and responsibility for it" (70). They indicate that this type of team effort is distinguished from other types by a shared set of features.

- *Production of a shared document.* The goal of the writing team is to produce a single document or a series of related documents, which provides the team with unity of focus.

- *Substantive interaction among members.* Team members make approximately equal contributions at all levels, from stylistic details to standards to organization to content. Communication is always two-way and interactive—it is never superior to subordinate.
- *Shared decision-making power.* All members of the peer writing team share decision-making power as well as responsibility for the resulting document. Individual team members are not free to make decisions affecting the team without consulting the team (Allen et al. 84–85).

Allen et al. (85) further suggest three basic reasons for forming a peer writing team:

- The large size of the project or the time constraints imposed on it requires collaboration.
- The project needs more than one area of expertise or specialization.
- The project requires the melding of divergent views to form a unified perspective that is acceptable to the whole team or to another group.

The coauthors of this book, who have produced three college textbooks, two handbooks, and six additional editions of textbooks as a peer writing team, can substantiate the findings of Allen et al. with their own experience.

## Composition of the Peer Writing Team

Although the team is composed of peers, it recognizes and takes advantage of individual expertise. Team members must respect the professional capabilities of each member and be compatible enough to work together harmoniously (some conflict, however, is a natural part of any group project).

Although the team must designate one person as its leader, that person does not have decision-making authority for the team—just the extra responsibility to coordinate the activities of team members and organize the project. Leadership can be by mutual agreement of the team members, or it can be by rotation if the group produces multiple documents. Being team leader bestows no particular advantages; the team leader just gets to do a considerable amount of work in addition to his or her normal writing and research duties.

## Functions of the Peer Writing Team

The team ordinarily has four functions: planning the document, researching and writing the draft, reviewing the drafts of other team members, and revising the drafts on the basis of the reviewers' comments.

### Planning

The team plans, as a group, to the lowest level that is practical. Beyond a certain level, of course, the group does not have sufficient command of details to plan realistically and must leave any further planning to the individual team member who has the assigned responsibility for researching each section of the document being produced.

The team, collectively, should identify the audience, purpose, and scope of the project, as well as its goals and the most effective organization for the whole document. The team analyzes the overall project, conceptualizes the publication to be produced, creates a broad outline of the publication, divides the publication into segments, and assigns different segments to individual team members (often on the basis of expertise).

In the planning stage, the team should also produce a projected schedule and set any writing style standards that team members will be expected to adhere to. The agreed-on schedule should include the due dates for finishing drafts, for reviews of the drafts by other team members, and for revisions. It is important that these deadlines be met, even if the drafts are not quite as polished as the individual author would like, because a missed deadline by one team member holds up the work of the entire team. All team members should be familiar with the schedule and make every effort to meet it. The subject of scheduling is discussed in greater detail later in this chapter under "Project Schedules."

The team should not insist that the broad outline agreed on by the team as a whole be followed slavishly by individual team members. Once into the assignment, an individual writer may—and often will—find that the broad outline for a specific segment was based on insufficient knowledge and presents a plan that, in fact, is not possible or not desirable. The individual team member must be allowed the freedom to alter the broad outline on the basis of what is possible, more appropriate, or clearly more desirable.

### Research and Writing

Planning is followed by research and writing, a period of intense independent activity by the individual members of the team. Each member of the team researches his or her assigned segment of the document, fleshes out the broad outline with greater detail, and produces a draft from the detailed outline. The writers revise their drafts until they are as good as the individual writers can make them. Then, by the deadline established for drafts, the individual writers submit copies of their drafts to all other members of the team for review.

### Reviewing

During the review stage, team members assume the role of the reading audience and try to clear up in advance any problems that might arise for the reader. Each team member reviews the work of the other team members carefully and critically, but diplomatically. Team members review for things both large and small. They evaluate the organization of each segment, as well as each sentence and paragraph. They offer any advice or help that will enable the individual writer to improve his or her segment of the document.

The review stage may lead to renewed planning. If, for example, review makes it obvious that the original planning for a section was inadequate or incorrect or if new information becomes available, the team must revert to the planning stage for that segment of the document on the basis of the newer understanding and knowledge.

### Revising

The individual writers evaluate the reviews of all other team members and accept or reject their suggestions. Writers must be careful not to let ego get in the way of good judgment. They must evaluate each suggestion objectively, on the basis of its merit, rather than emotionally. The ability to accept criticism and use it to produce a better end product is one of the critical differences between a professional writer and an amateur.

## Conflict in a Peer Writing Team

Members of a group never have exactly the same perspective on any subject, and differing perspectives can easily lead to conflict. A group that can tolerate some disharmony and can work through conflicting opinions to reach consensus produces better results. Although mutual respect among team members is necessary, too much deference can inhibit challenge—and that actually reduces the team's creativity. Writers must be willing to challenge one another—tactfully and diplomatically.

It is critically important to the quality of the document being produced that all viewpoints be considered. Under such circumstances, conflicts *will* occur, ranging from relatively mild differences over minor points to major conflicts over the basic approach. Regardless of the severity of the conflict, it must be worked through to a conclusion or compromise that all team members can accept, even though all might not entirely agree.

Attitudes toward conflict among writers who have been members of peer writing teams range from "the least satisfying aspect of collaborating" to "painful but necessary" to "exhilarating" (Allen et al. 80). Conflict is indeed exhilarating when it breaks a logjam, leads to the solution of a problem, or opens a new avenue of approach that had not been pursued previously.

The dissonance created by conflict almost always generates more innovative and creative work. Although the end result of conflict in a peer writing team is usually positive, it can in the interim cause self-doubt as well as doubt about one's fellow team members. The important thing to remember is that conflict is a natural part of group work (Morgan et al. 23). We have come to call it *creative conflict*. Learn to harness it and turn it into a positive force.

## Advantages and Disadvantages of the Peer Writing Team

Many heads are better than one. The work a peer writing team produces can be considerably better than any one member could have produced alone. This is partly because members of a team lead each other down different paths than they would have thought of individually.

Getting feedback, immediate and sometimes contested and debated, is one of the great advantages of the peer writing team. Team members easily detect lack of organization, clarity, logic, and substance, and they point these things out during reviews. The fact that the writer receives multiple responses also makes criticism easier to accept; if all peer reviewers offer the same criticism, the writer can hardly debate the issue. It is like having a private panel of critics who have a personal stake in helping the writer do a good job. It is better to be criticized by friends and peers than by paying customers!

Another advantage is that creative conflict produces better and clearer thinking. Team members play devil's advocate for each other, taking contrary points of view to try to make certain that all bases are covered and that all potential problems have been exposed and resolved.

Team members become more aware of, and involved in, the planning stage than they would working alone because the planning process is externalized through team discussion. The same is true of reviewing and revising. Team members also develop a tolerance for the opinions of others (Morgan et al. 25).

Another advantage of the peer writing team is that team members help each other past the frustrations and the milestones. The individual writer is not alone; if help is needed in making a decision, the writer has someone to talk things over with.

Let us summarize the advantages of the peer writing team:

- Many heads are better than one.
- Feedback from team members leads to a better end product.
- Creative conflict often forces clearer thinking.
- Having coauthors improves planning, reviewing, and revising.
- The writer is not alone.

The primary disadvantages of the peer writing team are the demands that it places on the writer's time, energy, and ego. It does take more time and more energy than working alone. And writers have to suppress both ego and sensitivity—they must be confident of their abilities and yet able to accept criticism. This may be the hardest part of team writing.

## Leading the Peer Writing Team

If you are selected to lead a peer writing team, you will have the extra duties involved in coordinating the project, such as scheduling meetings, writing and distributing minutes of meetings, and keeping the master copy of the document being produced.

As a team leader, you should also prepare several crucial handouts for the other team members, based on decisions made in the early planning meetings. Three of these are project style guidelines, a schedule, and a review transmittal sheet.

### Project Style Guidelines

This document should establish guidelines for uniformity and consistency in the writing. This is especially important when a group of contributors write separate sections of the same document. It should cover these matters:

- Levels of headings and the style of each
- Capitalization
- Date format (day-month-year or month-day-year)

- Reference format (if references are used)
- Abbreviations, acronyms, and symbols
- Spacing and margins
- Preferred form of key terms that have several acceptable variants (e.g., *on-site* or *onsite*, *data base* or *database*)
- Proprietary, confidential, or classified information
- Disclaimers to satisfy legal or policy requirements
- Distinction between research sources that must be cited and those that need not be
- Advice to use the active voice, the present tense, and the imperative mood in most situations

### Schedule

All team members must know what is expected of them and when. That is the function of the schedule. Schedules can take different formats. Figure 3.1 shows a schedule used for part of the work on this book. Another format for a schedule is shown in Figure 3.2, which depicts the Gantt chart used by the Case History writer, Mary Mullins, and her colleagues (see Chapter 2 Case History, pages 28–31). Regardless of the format, the schedule must state explicitly who is responsible for what and when it is due.

### Review Transmittal Sheet

This is a routing sheet that is attached to the front of each draft. It presents at a glance the status of the project during the review cycle. The routing sheet lists in order everyone who must review the draft, as shown in Figure 3.3. Each reviewer initials and dates the sheet after review and returns the draft to the

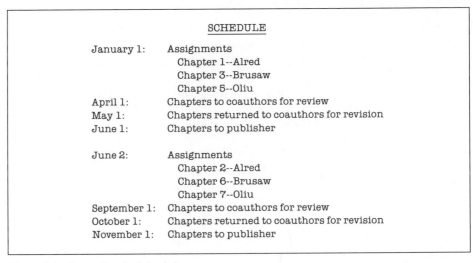

SCHEDULE

| January 1: | Assignments |
| | Chapter 1--Alred |
| | Chapter 3--Brusaw |
| | Chapter 5--Oliu |
| April 1: | Chapters to coauthors for review |
| May 1: | Chapters returned to coauthors for revision |
| June 1: | Chapters to publisher |
| June 2: | Assignments |
| | Chapter 2--Alred |
| | Chapter 6--Brusaw |
| | Chapter 7--Oliu |
| September 1: | Chapters to coauthors for review |
| October 1: | Chapters returned to coauthors for revision |
| November 1: | Chapters to publisher |

***Figure 3.1*** Example of a Schedule

## Writing/Review Schedule: Introduction to the International Banking System

**Legend:**
- ■ Writer tasks
- ▨ Reviewer tasks
- ☐ Tasks for other areas

**Timeline columns:** October 1 / October 15 | November 1 / November 15 | December 1 / December 15 | January 1 / January 15 | February 1 / February 15

**Tasks:**

- Research and planning
- Writing — Basic Skills Chapters; Overview Chapters
- Revising — Nov. 9 to Jan. 9
- Review Basic Skills chapters (30 pages) — Oct. 25 to Nov. 7
- Review Overview chapters and reference card (30 pages) — Nov. 9 to Nov. 25
- Revise after review — Nov. 28 to Nov. 30
- Marketing edits entire manual — Dec. 1 to Dec. 15
- Revise after special edit — Dec. 22 to Dec. 29
- Review entire manual (50 pages) — Dec. 22 to Dec. 29
- Prepare master for printing — Dec. 30 to Jan. 5
- Quality control check by TW manager — Jan. 6
- Print copies of master — Jan. 10 to Jan. 24
- Assemble and distribute manuals — Jan. 25 to Jan. 27

*Figure 3.2*  Sample Schedule

```
┌─────────────────────────────────────────────────────────────┐
│                  REVIEW TRANSMITTAL SHEET                     │
│                  Project: [Project Name]                      │
│                  Author: [Author's Name]                      │
│                                                               │
│                        Initial              Date             │
│                                                               │
│        Reviewer 1    _____         _____         │
│                                                               │
│        Reviewer 2    _____         _____         │
│                                                               │
│        Reviewer 3    _____         _____         │
│                                                               │
│        Reviewer 4    _____         _____         │
│                                                               │
└─────────────────────────────────────────────────────────────┘
```

*Figure 3.3*    Review Transmittal Sheet

author, who considers the comments and suggestions made by that reviewer and then forwards the draft to the next reviewer.

The team leader may also need to meet with the word processing staff to discuss various matters:

- The style guide
- Any special format requirements, such as margins and special tables
- The schedule: Will overtime be required? Two shifts? More operators or equipment?
- Any equipment compatibility concerns, especially for operators working in different locations
- When and under what circumstances disks can and cannot be duplicated

The team leader should also meet with the graphic arts staff to plan for certain items:

- Graphs, charts, drawings, maps, and other visuals
- The document's cover
- Any photographs necessary

Finally, the team leader should meet with the production staff to discuss these points:

- Scheduling
- Any special printing requirements, such as the use of color, special binding, document size, or foldout pages

## THE SUPERVISED WRITING TEAM

The supervised writing team is structured in the conventional hierarchical manner: several writers reporting to a project leader and several project leaders reporting to a manager. This mirrors the traditional superior-subordinate relationship that is used universally, regardless of occupation or profession. As a professional technical writer, you are almost certain to work on a supervised writing team at some point.

The description that follows is based on a model used by major multidivision manufacturers, which is appropriate since such corporations employ most of the professional technical writers working today (Gould 49). Smaller companies would have fewer writers and probably no project leaders; the project leader's function would be assumed by the manager. The process described is not limited to technical manuals; it works with any large project, such as a proposal.

In the large corporations, the supervised writing team consists of three levels: manager, project leader, and writer. Generally, the manager is responsible for the technical library for a family of products, the project leader is responsible for a set of technical manuals that document a subset of the family of products, and the writer is responsible for a manual within the subset.

## Manager

The Publications Department is part of a much larger group of departments that must cooperate closely to produce a major new product: Hardware Engineering, Software Engineering, Applications Programming, Technical Publications, Technical Education, Marketing, Systems Services, Quality Assurance, and Customer Services.

Representatives from all of these departments form a planning team for the overall project of developing, marketing, supporting, and servicing the new product. All these departments must coordinate their work and cooperate. If any one of them is out of synchronization with the others, all the pieces will not fit together at the end. The manager is a member of this team. The team, under the guidance of a group leader who is usually from one of the engineering departments, draws up a plan for the entire development process, including milestones that lead to a completion date for the whole project.

The manager then draws up a document plan that is part of the overall plan. The milestones in the document plan should include the projected dates for a rough draft of each publication, for reviews of each publication, for a draft of each publication for test use at a pilot site, for finished documents for product release, and for a customer verification test for the documents following release of the product.

Managers usually have between 10 and 15 writers reporting to them, although the number can go considerably higher. They usually also have three project leaders, although the number can vary from two to six. The primary function of the manager is to manage the resources of the department: personnel, equipment, facilities, and budget.

A key function of managers is planning, which consists of establishing objectives and determining the activities to be performed to accomplish those objectives. Planning is deciding what to do, when to do it, how to do it, and who will do it. By establishing objectives, priorities, and completion dates, planning enables the manager to make the most effective use of the department's resources. It includes scheduling and budgeting, as well as the knowledge that schedules and budgets must be revised and adjusted as results are achieved and conditions change. One of the primary benefits of planning is that it allows managers to guard against overcommitting or undercommitting their resources. Although planning takes a considerable amount of time, it saves managers much more time than it takes. As Dart Peterson puts it, "Hours you spend in planning will save you days of work and nights of worry" (66). Planning can also identify potential problems, enabling managers to develop the necessary action to prevent the problem from occurring or the contingent action to minimize the effect of the problem if it should occur.

Managers must determine the most effective organization for their departments. Organizing a department means defining the structure and the personal relationships that are necessary to meet the department's objectives. Managers must determine the number of writers needed in each supervised writing team, as well as the relationships among the various functions and jobs. Since the purpose of organization is to ensure efficiency, managers should base the organization of their departments on the work requirements and then modify the organization as necessary to accommodate the limitations of individuals or teams.

## Project Leader

The project leader functions as a supervisor of writers and a manager of projects. One project leader usually supervises three to five writers, although the number can go higher. Project leaders might or might not carry a writing workload, depending on how many writers report to them, how experienced and self-sufficient these writers are, and the extent of the leader's administrative duties.

The project leader's function has a narrower scope than the manager's. The manager determines the desired result and provides the resources necessary to accomplish it. The project leader then uses those resources to achieve the desired result.

Project leaders must make sure that each document fits into the document plan drawn up by the manager. The document plan is a high-level outline that identifies such things as the documents needed, the audience for each, the level of detail required for each, and the type of binding. Project leaders implement their part of the plan, schedule the work, and supervise writers to produce a total set of documentation for a new product. They allocate the resources provided them in the most effective way. They schedule their projects and assign writers to them, and they prepare regular reports to their managers indicating the current status of each project.

Project leaders must know which projects are going smoothly and which are not—and when to intervene. At some point in the life of a project, the project leader may have to call a halt and permit no further changes to a document in order to meet the projected deadline for the document (any further changes will have to be issued as revisions to the published document).

Project leaders must ensure that all parts of a manual read as if they had been produced by the same company. They must make certain that the terminology used across the whole system is consistent.

Project leaders must also be leaders, motivating their writers by providing challenging and satisfying assignments to them, according to their ability. They develop their writers in many ways:

- By setting goals for them
- By giving them needed guidance
- By teaming novices with experienced writers whenever possible
- By encouraging innovation and not punishing failure
- By providing needed training
- By giving writers free rein when they can handle it
- By teaching them to stop when they have produced a document that communicates effectively (any further polishing is wasted time)

Project leaders also perform some editing duties. They check the writer's document to make sure that the writer has met the established goals. They check the document for effective organization (usually by breaking the document back down to a topic outline and evaluating the outline for logical development). They check the document for completeness and accuracy to the best of their ability (they may not have the technical knowledge to do this and may have to rely on the technical reviewers). They check that the writing is at the appropriate level for the designated reader. They should also check for the appropriate use of headings and illustrations.

In doing all this checking, project leaders must master the art of criticism. If they approach it wisely, they can use criticism as a teaching device. But they should insert their criticism between layers of praise or recognition, honestly stated. They should start with a deserved compliment on some aspect of a person's past performance, then explain the problem, calmly and temperately, and how it can be overcome, and close, if possible, with a reassuring word so that the writer's self-esteem is restored.

## Writer

Just as the manager is responsible for the technical library and the project leader is responsible for a set of manuals, the writer is responsible for a publication or a manual.

For the writers, the end product is the distributed document. Writers are responsible for the following aspects of the documents they produce:

- Completeness and accuracy
- Writing quality
- Illustrations
- Mechanics
- Conformance to standards
- Reviews

- Layout and design
- Production
- Appearance
- Distribution

Although reviewers and the project leader may check for these things, they are all the writer's ultimate responsibility.

Writers must identify their sources of information, research their topics, create an outline of their documents that conforms to the document plan, and write the draft (including rough illustrations). They then submit the draft for technical review, at the same time submitting the rough art to the graphic artist for professional rendering. They revise the draft on the basis of the technical review and then submit the revised draft to *all* reviewers, including their project leader. On the basis of the reviews, writers make a final revision of the draft. Then they submit the final draft, along with the finished graphics, to the Production Department. The writers supervise the production and distribution of the document while preparing the rough draft of their next assignment.

## COST ESTIMATES

If you lead a peer writing team or manage a supervised writing team, you must know how to estimate costs and project schedules. You must have a cost estimate for any writing project in order to be able to establish a budget for the project. Here is a general rule of thumb for cost breakdown (Sachs 21):

Writing and editing: 50–80 percent
Illustrating: 5–30 percent
Production: 10 percent
Printing: 5–20 percent

To calculate actual costs, of course, you must have an hourly rate for each of these activities. An hourly rate must include direct labor, overhead, general and administrative costs, and profit. *Direct labor* includes the cost of hours spent writing, editing, illustrating, and making up the page layout. *Overhead* includes indirect labor (training, supervision, etc.), employee benefits, taxes (including social security, state disability, unemployment insurance, etc.), rent, utilities, and equipment. Overhead is usually expressed as a percentage of direct labor, anywhere from 50 percent to 200 percent (Sachs 21). *General and administrative costs* include salaries, payroll expenses, telephone, and insurance (some companies combine all this with overhead). Such costs can run from 10 to 25 percent of labor and overhead (Sachs 21). *Profit* is usually calculated at 5 to 20 percent of total costs, depending on the degree of risk involved in the project (Sachs 21).

For an hourly rate to be meaningful, you must assign a number of hours per page for writing, for editing, for illustrations, for copy composition, and for production. Then you must estimate the number of pages required for the document; to do so, you must be able to visualize the document in its final form, which requires the experience of having done estimates in the past. The number of pages needed for each section will depend largely on the amount of detail required, which must in turn be based on the readers' level of knowledge, the document's objective, and the complexity of the subject. Remember that documents that teach novice readers new concepts can be as time-consuming as those that explain complex concepts to more knowledgeable readers. To assess costs, you must also determine the quality of production you want; for example, an advertising brochure requires more quality than a set of engineering specifications intended for internal use.

The more stable the product environment, the fewer exceptions there will be to the traditional approach to cost estimating, and the better the traditional approach will work. In a world of new products and systems, however, things are more complex and difficult, especially when the document is being developed concurrently with—and as an integral part of—the product. Research is the most variable factor in the time estimate, and it is not easily quantified. For example, you cannot quantify "product design modifications," but you can know from experience the likelihood that they will be needed and alter the traditional formula to take them into account.

Whether a product is new or established and whether it is simple or complex will have a large influence on estimating hours. Simple and established products are least likely to change, and new and complex products are most likely to change; such change has a great influence on estimating the hours required to document a product. Covering an established or simple product in great depth, for example, can require less effort than a shallow treatment of a new and complex product. A careful analysis of such details can produce a weighting factor that you can use to estimate time accurately (Herrstrom 25).

Different segments of a document may also require a different amount of writing time, and the same can be true of various types of illustrations and the illustrator's time. Be sure that you have also included enough time for meetings, interviews, and review sessions.

Once you have calculated the specific hours per page required for the writer, the editor, the illustrator, and all others involved, you must work out the per-hour rate for each type of person in order to get your direct labor costs. This is a matter of simple arithmetic:

$$\text{Hourly rate} = \frac{\text{annual salary}}{\text{hours per year}}$$

Then multiply this per-hour rate times the number of hours calculated for each type of person to get direct labor costs. The accuracy of your final cost estimate, of course, will be only as accurate as your time estimates were.

We must make one final point. It is always wise to add a contingency amount of 10 to 20 percent to your final figures to cover the unexpected. Unexpected problems always develop during a project that will throw your estimates off if not planned for.

## PROJECT SCHEDULES

A schedule estimate is your best guess as to how much work is required to complete a document, expressed as milestones and expected deadlines. A good schedule is also accompanied by a list of the assumptions reflected in it. Five basic steps are involved in creating a schedule.

1. Identify the factors that affect writing time. Figure out the scope of treatment needed for the intended audience, the quality required to meet the document's objective, and the complexity of the product being documented.
2. Calculate the number of pages required, as described under "Cost Estimates."
3. Estimate the number of hours required to produce the pages, as described under "Cost Estimates." This is really just your best guess based on audience, purpose, and complexity, plus experience and historical data (Morison 101).
4. Identify your resources (number of writers who will be working on the project, number of artists, etc.).
5. Identify significant milestones in the project—completion of the draft, dates for reviews, dates for revisions, and so on.

The simplest way to create your schedule is to take an 8½-by-11-inch page and use the top for dates, breaking the year down by months and weeks, as shown in Figure 3.2. Down the left side of the page, list the projects that are scheduled (or the sections of the document, if it is a single document). Then enter the different phases of the whole document process, as shown in Figure 3.2. Computer programs designed to aid in project scheduling are available. If you have access to them, you might want to consider using them.

If your schedule should slip, take action immediately to rectify the situation so that subsequent milestones don't slip. Don't make the mistake of thinking that you can make up the time later—it never works. Go into overtime immediately if possible; if it is not possible, rework *all* the milestones in the schedule to reflect the change—and then be sure to notify *everyone* concerned about the change.

At the end of the project, take some time to evaluate how well you did with your schedule. Note the things that went well and the things that didn't so you will be able to do better next time.

*Checklist 3.1*

# Identifying the Appropriate Writing Team

Consider the following questions when considering what kind of writing team is needed for a project.

### Functional Writing Team

☐ Does the project require only one writer, working with other professionals such as a technical expert, an editor, a typist, a layout artist, and a printer?

### Peer Writing Team

☐ Does the project require a group of peers who share equally in decision making and responsibility for producing a document?

☐ Does the project require specialists with different areas of expertise?

☐ Does the large size or tight schedule of the project call for multiple writers?

☐ Does the project require the blending of different views?

### Supervised Writing Team

☐ Does the project require the traditional hierarchical structure of department manager, project leader, and writer?

---

## WORKS CITED

Allen, Nancy, et al. "What Experienced Collaborators Say about Collaborative Writing." *Journal of Business and Technical Communication* 1 (1987): 70–90.

Gould, Jay R. *Opportunities in Technical Communications.* Lincolnwood, IL: VGM Career Horizons, 1984.

Herrstrom, David S. "An Approach to Estimating the Cost of Production Documentation, with Some Hypotheses." *Proceedings of the 34th International Technical Communications Conference* (1987): MPD-24–MPD-27.

Morgan, Meg, et al. "Collaborative Writing in the Classroom." *The Bulletin* Sept. 1987: 20–26.

Morison, Susan B. "Estimating Resources." *Proceedings of the 34th International Technical Communications Conference* (1987): MPD-100–MPD-101.

Peterson, Dart G., Jr. "How to Produce That 'Impossible' Big Job." *Proceedings of the 35th International Technical Communications Conference* (1988): WE-66–WE-69.

Sachs, Irwin M. "Estimating Publication Costs." *Proceedings of the 34th International Technical Communications Conference* (1987): MPD-21–MPD-23.

## FURTHER READING

Couture, Barbara, and Jone Rymer. "Interactive Writing on the Job." *Writing in the Business Professions.* Ed. Myra Kogen. Urbana, IL: Association for Business Communication, 1989.

Prekeges, James G. "Accurate Estimating and Scheduling." *Proceedings of the 35th International Technical Communications Conference* (1988): MPD-47–MPD-49.

Taylor, Carol. "Estimating and Preparing Proposals for Contract Writing Projects." *Proceedings of the 35th International Technical Communications Conference* (1988): MPD-41–MPD-43.

Van Pelt, William. "Collaborative Writing with Computers: Software Documentation Projects for Technical Writing Classes." *Collaborative Technical Writing: Theory and Practice*. Ed. Richard Louth and Ann Martin Scott. Minneapolis: ATTW, 1989. 193–203.

Van Pelt, William, and Alice Gillam. "Peer Collaboration and the Computer-Assisted Classroom: Bridging the Gap between Academia and the Workplace." *Collaborative Writing in Industry: Investigations in Theory and Practice*. Ed. Mary M. Lay and William M. Karis. Amityville, NY: Baywood, 1991.

# The Case History

### Chapter 3: The Writing Team

INTERVIEWER: Some people think of professional writers as solitary people working in isolation. Is that true for you in your organization?

MARY: If you just looked into my cubicle for a moment, you might at first think I was working in isolation. But if you actually watched me work for a whole day, you'd see that I interview, consult with, and get reviews from many people—often six or more per project. In addition, I'm part of a team of about ten writers. We talk; share documents, graphics, and ideas; and critique each other's work. So my work is more interactive or collaborative than you might think.

INTERVIEWER: How about the idea that the document a writer produces is the result of only that person's work. Is that an accurate picture of what you do?

MARY: No, because I observe and document a real system that does something. I have to talk to the people who made it and to the people who use it, and then I write documentation to match how they say it works and how they say they use it.

As I research a project, I ask a lot of technical questions, but I also ask people their views on what they'd like to see in the document. I incorporate most of their suggestions. Then, when they review my documentation, the system designers and users (and my copyeditor) let me know whether my documentation clearly, concisely, and accurately describes the tasks I'm documenting. If it isn't satisfactory, they let me know exactly what changes they'd like to see, and I incorporate most of those suggestions. By the time the draft gets to my boss for her final check, it's been revised many times, based on comments from many people.

So the final document isn't something I've just made up out of my head—it's not fiction writing! It involves collaborating with a lot of people.

INTERVIEWER: When do these collaborations begin?

MARY: At the start of a project, I usually collaborate with my boss, the user representative, and the system designer to figure out a rough scope for the project.

Then, when I have a pretty good idea of the rough scope and I'm starting to plan the documents that I'll write, I usually interview other writers who've worked on that manual before I estimate the cost and schedule for the project. Experienced writers can advise me on any adjustments I might need to make to these estimates. For example, if the user rep has generally been inaccessible during portions of past projects, I will add time in the schedule to allow for his or her inaccessibility.

*49*

INTERVIEWER: What forms do these collaborations take?

MARY: I might meet informally with a writer who has worked on the system to find out that writer's impression of the system (or perhaps what to expect from the developers who are responsible for the system)—basically just to share some notes.

At some point during planning, I also determine the peer writer for the project—the writer who will critique my project plans or preliminary outline during the planning phase. By the way, the peer writer is usually the same person who will later copyedit the document.

I might send an electronic mail message to all writers in the department, asking if anyone has any text or graphics related to the feature I'm planning to document. That way, I can save time by copying and editing existing text and graphics instead of developing them from scratch.

INTERVIEWER: What are the benefits of working as part of a writing team?

MARY: Working as a team of writers, we get more than the sum of each individual's efforts. We can be a lot more productive.

I'm thinking of a particular situation when a fellow writer met with me regarding her project. She was very anxious about this writing problem she was facing. She talked it over with me, and then she said later how much of a relief it was just to talk about it instead of sitting alone thinking about it.

If you're working alone and you get stuck with a big writing problem, you can go crazy trying to solve it on your own. You tend to spend a lot of time puzzling over it, and you can really lose courage. But because this particular writer had someone to talk to, she could work through the problem, and she was able to relieve that pressure of trying to figure it all out by herself.

I'm more productive when I'm part of a team because I can get a quick answer from other writers instead of having to think everything through myself. For example, I've sent writers electronic mail notes asking for suggestions on rewriting a single sentence that is giving me problems. I usually get several good ideas.

INTERVIEWER: What problems have you encountered working closely with others to produce a "collaborative" document?

MARY: I was once asked to reformat some documents for another writer who was managing a project. I misunderstood her instructions to me, and I reformatted some documents that she wanted me to leave intact. The moral is, you have to spend more time on communication in a collaborative project than on a solo project.

You might also disagree on something like design decisions. If all writers on the team are at the same level within the organization, there might be compromises or tough decisions to be made.

Collaborating can also take more time than writing for solo projects. I mean, it could take more time than if half the project had been assigned to one writer and the other half to another writer.

INTERVIEWER: How have you solved those problems?

MARY: If I had that first project to do over again, I'd be more meticulous about echoing the oral instructions from the project manager—or perhaps confirming them in a memo to make sure I understood them thoroughly. I remember I took the instructions by just scribbling them on a note pad. That's not a good way to do it.

I'd minimize decision-making time by deciding design issues early or by developing a prototype and critiquing it before writing any drafts.

In terms of the extra time, it would probably work better if each writer had isolated responsibilities (say, one writer for screens, another for reports), and writers collaborated only on issues that would affect the work of all (such as definitions for terms that appeared in both the section on screens and the one on reports). It might be good to have someone acting as a tiebreaker for decision making or to have formal conflict management techniques in place for employees to follow.

INTERVIEWER: In general, what have you learned about getting along with other members of a writing team?

MARY: You have to be polite, always. You can never get so comfortable with someone that you forget courtesy. But you also shouldn't take politeness to the point where you're telling someone an idea is great when you really think it's lousy. If you find yourself getting into a tangle with another writer, you should call in an independent third party to help settle the dispute.

INTERVIEWER: What kinds of things is a writing team leader responsible for when managing a project?

MARY: For the International Banking project, the project leader carefully managed the schedule, making sure that interim deadlines were met and watching out for delays or other problems. She also did all the planning and decision making for the project, overseeing all the interactions between Technical Writing and other departments.

INTERVIEWER: What about consistency (in style and other matters) in the documents? Who looked after these details?

MARY: All writers on the project contributed to a set of writing conventions for the project that were developed during the planning stage. If the writers disagreed about something, the project leader had the tiebreaking vote. She could also choose to administer the conventions in some other way if she wanted.

# 4

# Audience, Purpose, Scope, and Overall Design

You should begin assessing the audience, purpose, scope and overall design of a document the moment you receive an assignment. Even though you need to refine your assessment throughout all the stages of the document process described in Chapter 2, you must define these elements as carefully as possible at the outset because they affect both the process and the products of writing.

The separation of audience, purpose, and scope is largely artificial. You cannot realistically exclude one from your mind as you contemplate the others. When considering the audience's needs, for example, you will inevitably weigh the purpose of the document, and when defining its purpose, you will establish part of the document's scope. Nevertheless, you must study these elements separately to understand fully how each helps to determine the content, form, organization, and language you use.

## ANALYZING YOUR AUDIENCE

The STC Code for Communicators shown in Figure 1.1 in Chapter 1 suggests the importance of audience analysis by stating that technical writers must hold themselves responsible for how well their audience understands their message and that they must place the readers' needs for information above their own need for self-expression. That the readers' needs are primary is a consistent theme throughout the history of technical writing. In *A Guide to Technical Writing* (1908), T. A. Rickard states, "Above all, put yourself in the other fellow's place. Remember the reader" (13); In *How to Write Computer Manuals for Users* (1982), Susan Grimm states, "The most important ingredient in the process of developing manuals is the user" (1). Not only is keeping the readers' needs at the forefront of your thinking a professional commitment, but it is often a practical business necessity as well. The reader of a manual who is frustrated may simply throw the manual aside, complete the task by instinct, with far less than successful results, and never use or buy one of that company's products again.

The challenge of audience analysis is to create for yourself the clearest possible image and understanding of the audience:[1] the ways readers will read your document; their familiarity with the subject; their role in the working environment; and their attitudes toward your organization, the subject, and the document itself. With such knowledge, you can reduce the barriers between yourself and the reader and reach a momentary common ground. This goal is challenging, especially for professional writers who must adapt their skills and experience to widely different audiences during their careers.

### Reading Theory and Its Implications

A key element in communicating with readers is an understanding of how they will read a "text" (written and visual material). Fortunately, over the past 20 years, a considerable body of knowledge has emerged to help us understand the reading process—how readers decode messages. By understanding readers' decoding process, writers can employ strategies to make the readers' work easier.

Researchers have discovered that readers do much more than passively absorb writing, sequentially decoding sentences and pages as if they were machines. Readers are much more creative. They draw inferences as they read, and they form their own concepts; they make meanings as they read, and the mean-

---

[1]Although some debate exists over the use of *audience* to refer collectively to groups of readers (Suchan and Dulek), Walter J. Ong points out:

"Audience" is a collective noun. There is no such collective noun for readers, nor, so far as I am able to puzzle out, can there be. "Readers" is a plural. Readers do not form a collectivity, acting here and now on one another and on the speaker as members of an audience do. We can devise a singularized concept for them, it is true, such as "readership." . . . But "readership" is not a collective noun. It is an abstraction in a way that "audience" is not. (11)

ings they make are often surprising leaps of imagination. In other words, readers remember not only what *we* tell them but also what they tell *themselves* (Flower 132). Readers also choose the information as they read, based on their needs (this will be illustrated later in this chapter).

Readers try to fit new information to what they already know or can recall. Much of the prior information that they recall is not held in a formal, explicit way. Instead, it is held as attitudes or images—loose clusters of associations. Cognitive psychologists call these clusters *schemata* (Huckin 92). Each individual cluster, called a *schema*, is held deep in our minds in what is known as long-term memory. Scientists view a schema as a placeholder containing "slots" for the components. A schema for the human face, for example, would include slots for mouth, eyes, nose, ears (Anderson 67). So if viewers look at a partially drawn face, their minds pull a "face schema" from long-term memory and fill the slots with the missing pieces as needed (see Figure 4.1).

A schema may also be abstract. If a reader had difficulty with mathematics in school, hearing that something is "mathematical" evokes in that person the cluster of negative images associated with the term (failed tests, the hated textbook, the stern teacher, etc.). We all store an enormous number of such schemata in our long-term memories.

Short-term memory works the opposite way: it processes words, phrases, and other "chunks" of information in a linear fashion. In other words, the mind adds each small piece of language until a larger structure is built. If the mind cannot perceive a larger structure, short-term memory adds chunks until the memory can no longer accommodate new ones. Ordinarily, *short-term memory can store only about seven chunks* of information before "memory decay" occurs (Huckin 96).

This research on how we process information has important implications for technical writers who must teach readers new material and enable them to follow procedures. For example, since all readers try to fit new information into old frameworks (or schemata), writers must learn what frameworks are familiar

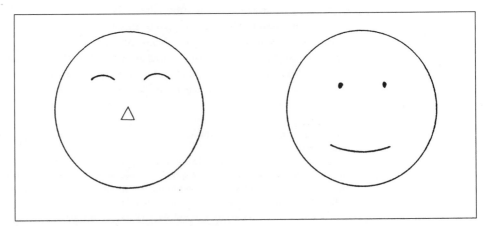

***Figure 4.1***   Simple Face Schemata (Incomplete)

to their readers and will help them anticipate the details that will follow. If a reader is told that a *numeric pad* is a set of keys arranged like the keys on a pocket calculator, the statement is effective only if that reader's mind contains a schema for *pocket calculator* (in other words, if the reader has seen or used one). When readers are not familiar with a subject, the writer should help them acquire the knowledge they need to create new frameworks (and thus contexts) and then fill in those frameworks with details that make their newly acquired knowledge useful and relevant.[2]

## Methods for Determining Specific Audience Needs

As reading theory makes clear, a writer must understand not only the reading processes for all readers but also the specific needs of each audience. You will learn some of those needs during meetings when you receive an assignment. Further, some organizations, such as government agencies and large corporations, hold long-established assumptions about their client audiences. In other cases, you will not receive detailed information about your readers, and you must aggressively seek it yourself.

The first step in identifying the needs of your audience is simply determining who will read and act on the document you are to produce. Beyond this simple identification, you must learn enough about various readers in your audience to develop an intuitive sense of their needs (Dobrin 104–109). To understand your readers that well, you should use as many techniques for analyzing the audience as possible. Traditionally, there are two approaches to audience analysis: *categorical* classifications and *heuristic* techniques (discovery through questions and particulars). Although both have limitations, together they can provide enough information about your specific readers to enable you to "sense" what they need.

### Categorical Methods

Categorical methods of audience analysis make generalizations about readers by type or classification. For example, one basic textbook views readers partly by level of subject knowledge (Lannon 2):

- Highly technical
- Semitechnical
- Nontechnical

Another approach categorizes readers primarily by function (Pearsall x):

- Layperson (who buys and uses)
- Executive (who makes decisions)
- Expert (who designs)
- Technician (who builds and maintains)
- Operator (who runs and makes simple repairs)

---

[2]Organizing strategies is discussed further in Chapter 6, and suggestions for using visual devices to build frameworks are made in Chapters 7 and 8.

Such generalizations are useful. However, by definition, they tend to blur the complexities of readers who cross such boundaries or combine features of several categories. Such categories can also tempt writers to accept inaccurate and imprecise generalizations about their audiences. For the professional technical writer, who must assess complex audiences with as much precision as possible, assumptions based on categories should serve only as a starting point. They must then focus and refine their generalizations with the heuristic techniques (described later) that are appropriate and that time and resources permit.

All categorical methods define readers using assumptions about four audience features: subject familiarity, functional roles, psychological factors, and special considerations.

*Subject Familiarity*    As a technical writer who has completed a thorough research process, you must remind yourself that *your readers are always less familiar with the subject than you are*. Be on guard: you can easily become so blinded by your familiarity with your subject that you begin to assume that your reader knows as much about it as you do. To be safe, write for a slightly lower level of knowledge than you assume they possess by providing knowledge they may lack; the more knowledgeable readers can simply ignore the few details they do not need.

Readers range in their degree of subject-matter knowledge between two extremes: experts and novices. An expert (or specialist) is someone completely familiar with a subject; a novice (or nonspecialist) is someone with little or no knowledge of a subject. Seldom will you write for complete novices or complete experts; generally, you will write for readers who fall somewhere in between. By understanding the characteristics of readers at each extreme, you can adjust to readers at various points on the continuum.

*Expert Audience.* For experts in a subject, research has shown that a writer may use appropriate technical terms, detailed illustrations, scientific notations, and equations with no loss of readability (Huckin 101). Such devices, in fact, condense material for experts in a handy visual shorthand. Further, when writing to a uniformly expert audience, you do not need to overexplain with lengthy examples, operational definitions, and analogies. However, you should not overcomplicate your writing for experts; you should apply the principles of clear and accessible writing. One researcher finds that it is "a popular intellectual myth that high-level technical audiences can be insulted by simple writing, simple expressions, and direct answers to problems that are anything but simple" (Caernarven-Smith 122). Robert Day, a former editor of a scientific journal and past president of the Council of Biological Editors, admonishes experts writing for other experts to keep clarity as a goal and not to lose sight of less knowledgeable "experts":

> I believe that the key to scientific writing is clarity. Successful scientific experimentation is the result of a clear mind attacking a clearly stated problem and producing clearly stated conclusions.
>
> . . .
>
> Of course, you will have to use specialized terminology on occasion. If such terminology is readily understandable to practitioners and students in the field, there is

no problem. If the terminology is *not* recognizable to any portion of your potential audience, you should (i) use simpler terminology or (ii) carefully define the esoteric terms (jargon) that you are using. In short, you should not write for the half-dozen or so people who are doing exactly your kind of work. You should write for the hundreds of people whose work is only slightly related to yours but who may want or need to know some particular aspect of your work. (1, 168)

*Novice Audience.* The novice (or nonspecialist) does not have enough subject-area knowledge to establish context or a framework for the information and thus know what's important. The novice must rely heavily—perhaps solely—on the information presented in the text. Research indicates that if key, general information is structurally buried in a text, the nonspecialist reader will not recognize it as being important and will thus "pay less attention to it, infer few details about it, comprehend it poorly, and recall it poorly" (Huckin 95). Therefore, always introduce details after general information that provides a framework for them.

To make writing more readable for novices, make use of existing frameworks (or schemata) stored in their long-term memory. That is, start with concepts and ideas within the readers' experience and support them with clearly connected metaphors, analogies, examples, operational definitions, and the like. For example, when describing an arm on a manufacturing robot welder, you might refer to its "elbow" to a novice, even though an engineer would call it an "articulating hydraulic connection." Making such anthropomorphic (human-feature) comparisons may mean employing the principle of "communicative accuracy," the principle of presenting technical concepts somewhat incompletely or inaccurately for the sake of communication. Consider, as another example, a patient going into heart bypass surgery. That patient needs simple line drawings, even if somewhat incomplete, to understand the process; the heart surgeon, by contrast, needs fully detailed, accurate medical artwork—perhaps actual surgical photos or a videotape.

For the novice, avoid jargon and technical terminology if at all possible, and carefully define terms used in a special sense (such as *boot* to mean "start up" in computer terminology). If possible, find simple substitutes, such as "a surgical instrument sterilizer" rather than its technical name, *autoclave*. Cushion the introduction of any unfamiliar terms when they cannot be avoided. Avoid mathematics, equations, chemical formulas, and statistical or scientific notation. Replace exponents like $10^6$ with 1,000,000 or one million. Give comparisons when using estimates. You might write "Chances are 1 in 1 million, or about the same as hitting the jackpot in your state lottery twice in the same year." Keep in mind as well the inability of short-term memory to retain more than about seven chunks of information, as mentioned earlier. For all readers—but especially for novices, who must struggle with new concepts—group details, facts, instructions, and so on in sets or clusters that do not overload short-term memory.

When writing for novices, however, respect their intelligence. Perhaps a *Life* magazine statement to new staffers is the best advice: "Never underestimate the intelligence of your readers and never overestimate their knowledge" (Houp and Pearsall 72).

*Intermediate Audience.* Of course, many readers are not simply experts or complete novices, so you will need to apply these generalizations prudently. Your readers might also be sophisticated in some areas of a subject but relative novices even in related areas. For example, even a lab technician may be unfamiliar with the particular testing equipment you are documenting. Such readers need to know how the equipment is both similar to and different from other devices in the same class. Readers who have used word processors, for example, don't need the basic definitions of program disks or disk drives that first-time users do. However, such readers do need to be told how the features of a new word processor relate to existing systems and programs.

Executives and managers pose special problems when you analyze their level of familiarity with the subject. Depending on their education and career path through an organization, their technical knowledge can vary widely. You can assume safely that they have wide (but perhaps not deep) knowledge of their organization's technology—certainly more than a novice. But each case must be assessed individually using the pertinent heuristic techniques described later in this chapter.

Writing a document for audiences composed of readers with different levels of subject knowledge will be discussed in the section of this chapter devoted to overall design.

*Functional Roles*   In addition to their level of subject familiarity, audiences are often classified according to their role in a profession or an organization or as consumers or clients.

Each profession or vocation—engineering, accounting, computer science, marketing, medicine, education—has unique needs that you must meet to enable such readers to perform their work efficiently and effectively. Systems programmers, for example, need comprehensive descriptions of operating system functions and how to modify them. Computer service technicians often need exploded-view drawings and tables of precise measurements to repair hardware. Physicians need drug descriptions that include side effects and reactions with other drugs; pharmacists also need information about generic equivalents.

The organizational role of readers can also help you determine their needs. For example, executives and other high-level decision makers are interested in finance and personnel deployment; they often need to know the long-term planning implications of proposals. Decision makers in governmental bodies and agencies also need to be aware of the public policy implications and politics involved (see Monroe for a detailed account of analyzing the politics in government documents). Managers need to know how to make the best use of resources, how to motivate their employees, and how to solve problems.

Consumers and clients outside your organization need to know *what* and *how*. They need *why* only if it helps them understand how to accomplish a goal. You can assume that they want practical, not speculative or theoretical, information. Consumers and clients could include home users of products and services, industrial purchasers of equipment, clients for whom special services are developed, and buyers of products like mainframe computers. Because of their position as consumer or client, such readers often expect glossy, easy-to-read "how-to" documents for products and services they are being asked to buy.

*Psychological Factors*   Consider your audience's expectations and attitudes about the document itself. How willing are readers to invest effort in reading the document? One research study found that different users of the same manual often vary substantially in their willingness to use it:

> Generally, a user is willing to learn from a manual if the required effort appears worth the potential benefit. But various users weigh efforts and benefits differently. Intermittent users, such as managers who seek immediate solutions to pressing problems, will likely balk at the time required to understand lengthy or complex instructions, and they will seek other less demanding methods for gaining help. However, users such as project engineers who see the computer as a powerful tool for solving problems, will put forth much time and effort to initially learn a program, even if it means forgoing a speedy completion of a task. (Mirel 122)

Another study found that patients with a disease were both unable and unwilling to read fairly simple material when they were hospitalized. However, shortly after they were released, they willingly read difficult material, even medical textbooks, about that disease (Nagy 6). Their curiosity about their condition provided the motivation necessary for them to read through the most challenging technical material.

In any case, you will need to consider your readers' attitude toward your message before they read your document, determine the attitude you want them to have after reading your document, and then adjust your writing strategy accordingly. There are three broad categories of reader attitudes: supportive-interested, anxious, and skeptical.

*Supportive-Interested Audience.* A supportive-interested audience knows the general subject, accepts the source of information as credible, and approaches the document with a positive attitude. Readers who perform minor home fix-it jobs will not be intimidated by a task requiring the assembly of a device. Such readers may or may not have deep technical knowledge, yet they are generally quite *willing* to read and use, for example, assembly instructions for a new piece of equipment if they expect a positive outcome.

Don't take reader interest for granted. Work to make the document readable and usable—even the most supportive readers have probably seen confusing instructions or reports in the past, and they could be suspicious. Reinforce their positive associations by letting them know, for example, what steps or information they may skip.

However, such readers may be so eager that they skip important information. In general, you should slow overeager audiences to the point where they take the theory of operation, their own technical competence, and danger into consideration (Caernarven-Smith 195). You can slow overeager audiences with warnings at openings:

> STOP: Read the following section, "Safe Operating Practices," before attempting the procedures that are described in Chapter 1.

You can also slow overeager audiences by placing requirements on procedures—what the user must know or what tools are needed. For example, the assembly instructions for an electronic device warn:

Before attempting to operate the system, make sure that all the connections are fastened tightly.

*Anxious Audience.* You may often write for a technically insecure audience that is unfamiliar with your product and its technology. Further, readers of patient care information, first-time users of a computer system or homeowners dealing with environmental hazards may be nervous, embarrassed, under pressure, or fearful. Sometimes people inside an organization fear that a new technology may cost them their jobs, and their apprehension can make them uncooperative. They may also approach the document with anxiety because of lack of knowledge or because of bad experiences with similar documents.

Because such readers are timid about approaching the subject, they need reassurance at the outset. Reassure anxious readers through analogies that compare unfamiliar concepts with ones that are familiar and unintimidating. For example, a manual written for first-time users of word processors states, "Think of a word processor as a typewriter that produces text on a TV screen and stores it in an electronic filing cabinet." A document with a self-deprecating voice and noncondescending tone that demystifies a technology by comparing it to generally understood information helps anxious readers overcome their fears. Further reassure this audience that missteps on the first time through are normal and have no serious consequences, if that is the case.

*Skeptical Audience.* Sometimes you may write for a skeptical audience. Learn why these readers do not accept your organization's credibility or your view of a subject or product, using the heuristic methods we will discuss later. You must overcome preconceived opinions by adapting the message to the readers' experiences and perceptions. For example, use a role model or an authority figure whom you know the audience trusts. For example, for a booklet on exercise and the heart, messages from a well-known, recovered heart-attack victim and a renowned heart surgeon would be effective. Present both sides of an issue as you make your case *in language that respects the readers' opinions.* Personalize the text where possible—sound "human" and avoid the "institutional passive voice" where possible. (See Chapter 10 on using the active voice.)

*Special Considerations*   Some readers require additional consideration. Elderly readers, because of failing eyesight or mental ability, may require, for example, larger type sizes or special warnings to seek assistance before using a device. If some readers are semiliterate, you need to maintain a high ratio of visual devices to words. For such readers, the use of graphics and page design are crucial (discussed in Chapters 7 and 8).

Consider also possible cultural differences among your readers: the order of elements expected in prose and graphics, culturally loaded words, and so on. Be especially aware of American idioms if nonnative speakers of English may be in the audience, and avoid culturally insensitive terminology, such as master-slave jargon for computer networks (see Stevenson).

## Heuristic Approach

To refine and overcome the limitations of such categories, based as they are on prior assumptions, you can employ a heuristic approach to audience analysis.

With this approach, you gather information about the audience through systematic questioning and analytical observations. Depending on the time and resources available, use the following tactics:

- Accumulate quantifiable facts about the audience.
- Analyze the audience's organizational duties.
- Question and observe the audience directly.
- Analyze relevant documents.
- Stand in for the audience (gain "hands-on" experience).

First, accumulate as much factual or statistical information about an audience as possible, such as the following:

| | |
|---|---|
| Age range | Gender (if appropriate) |
| Spending habits | Recreational interests |
| Formal education | Audience size |
| Work experience | Language(s) |
| Location(s) | Physical limitations |

Such factual information can serve as a starting point in fleshing out the assumptions of the categories. A questionnaire is particularly useful in gathering such information (Brusaw, Alred, and Oliu 532; Caernarven-Smith 100).

Second, you can learn about the audience's functions and duties by examining company organization charts and job descriptions. You might also create diagrams that illustrate the readers' proximity to each other and to the subject or activity. For an illustration of this process for technical reports, see Mathes and Stevenson's "ego-centric organization chart" (17).

Third, consider interviewing select representatives from the target audience to enhance and verify your general knowledge and assumptions about that audience. (Chapter 5 provides interviewing tactics.) If you understand the methods of social science research—such as surveys (see Keeler), the "Delphi" technique (a series of questionnaires on the basis of which responses are ranked), and focus groups—use them to gather information about your readers (see Delbecq, Van De Ven, and Gustafson). Understand that these techniques are time-consuming and expensive. If your organization has a market research department, you may be able to use it to help you gather information about potential readers in terms of knowledge of your company and product, technical expertise, and circumstances under which the product will be used (see Ridgway). You can also directly observe potential readers at work to learn what they might need to complete a task.

Fourth, examine materials, like those in the following list, that reveal such reader needs and expectations as typical problems readers experience with documents or products, the design they find most comfortable, or topics they expect the document to cover.

- Notes from company instructors who train customers
- Reports from the sales staff
- Maintenance reports that indicate potential problems
- Publication standards for the document
- Past communications or documents revealing reader expectations

- Publications that target audiences regularly read[3]
- Journals' data on their subscribers

Finally, learn the audience's needs by using the product, system, or procedure yourself. If you are writing about a device or system that readers must assemble or operate, assembling or operating it yourself may, in fact, prove the most useful method of assessing audience needs. Standing in for your readers will also help you empathize with their situation.

Specific advice on using the information gathered through audience analysis as you write the draft will be given in Chapter 9.

## REFINING THE PURPOSE AND SCOPE OF THE DOCUMENT

As shown in Chapter 2, a document often grows out of a business or organizational need or purpose. Readers too have practical purposes for reading documents: to assemble a product, to locate information, to follow a procedure, to answer questions, to make a decision, to understand an issue. So, as suggested earlier, readers pay attention to information that they feel fulfills their purpose. Depending on readers' purposes, in fact, they can consciously choose a "reading style" (Huckin 98). Figure 4.2 presents a number of possible reading styles. In addition, some readers may casually glance at a document and decide to file it for future reference.

Although readers can easily switch from one style to another, they usually begin with the one they think will efficiently fulfill their purpose for reading. You must try to anticipate the reading styles your readers are likely to use so that you can determine the organization and design of a document, as shown later in this chapter and in Chapter 8.

### Purpose of the Document

To ensure that the document meets both the organization's needs and the readers' purposes, ask the following question: "What specifically do I want the readers to be able to do, know, accept, or believe as a result of reading or using the document?" When you can answer that question precisely, you can formulate a statement of the document's purpose.

Although stating a purpose at the outset prevents you from getting sidetracked, a purpose statement is especially useful throughout the document process for your management and for collaborators in cowritten documents. Such statements, of course, are intended primarily for the writer or team members and will seldom appear in the final document, although a modified version might

---

[3]Researchers can gather information about readers through *content analysis*, which statistically measures and analyzes the topics and views in publications that particular audiences usually read (see Krippendorff).

| Reading Style | Reader's Purpose | Example |
|---|---|---|
| Skimming | Reviewing to get the general drift of a document | A manager skimming a report for background prior to a meeting |
| Scanning | Reviewing to find specific items or pieces of information | A PC user scanning for ways to solve a problem |
| Search reading | Scanning and then carefully reading specific passages | An engineer reading journal articles, stopping where sections provide details to use in writing a proposal |
| Receptive reading | Concentrating on the entire document for thorough comprehension | A writer reading a specification full of information needed to write a manual |
| Critical reading | Evaluating both ideas and details | A grant officer reading the final two proposals for a $50,000 grant |

***Figure 4.2***    Reading Styles

appear in an introductory section. The following are typical purpose statements for documents:

> This proposal describes our company's capabilities and strengths in aircraft parts manufacturing so that NASA decision makers [readers] will be convinced to award us a contract.

> This manual shows buyers [readers] of our PCs how to use the software for locating information they need so that they can make the best use of the system for their needs and consider the purchase of our product a wise choice.

> This instructional manual provides step-by-step procedures for teaching wellness classes so that clinic staff members [readers] will follow consistent, standard practices.

To construct a clear statement of purpose, as in the sample statements above, use strong, action-oriented verbs such as *award, believe, convince, use, describe, enable, follow, learn, locate, provide, understand.*

Once you have established a clear purpose, guard against losing sight of it as you become involved with the other steps of the writing process. One experienced professional writer puts her purpose statement on an index card and posts it near her desk so that she can glance at it regularly.

As suggested in our sample purpose statements, a document may have several purposes. A single document may need to instruct, serve as a reference, and promote the value of a product. If a document has more than one purpose, you may need to decide which purpose is most important and emphasize it. Another method of dealing with multiple purposes (discussed later in this chapter) is to design parts of a document, or even create separate documents, to accommodate each purpose.

## Scope of the Document

Scope is the breadth and depth of detail to which you need to cover the subject in a document. Scope is determined primarily by your audience's needs and the document's purpose, and scope is often a delicate balance between brevity and clarity. You must provide readers with *only* the information they need to know to perform their jobs efficiently but *enough* to perform their tasks properly and safely.

### Requirements of Audience and Purpose

Readers, for example, need *appropriate* information, rather than just more information or irrelevant information. One study found that explaining how a computer worked in general did not enable users to infer the actual procedures needed to operate the system (Mirel 113). However, to achieve clarity, you may need to provide some redundant information for readers following processes or instructions—such redundancy (which has been called a "consistency heuristic") helps users check to see if they are taking the appropriate steps (Mirel 114). In other words, you may need to repeat information rather than referring readers to other pages or sections.

You should determine the general scope of a publication at the outset, but you must refine the scope throughout the document process as you sift the information and write the document. At the start of a project, list the possible content—the tentative scope—based on your purpose and what you know about your readers. Suppose that you are assigned to write a manual for operators of construction equipment. At the start of the project, you might make the following list of kinds of information the user may need:

- Basic instructions for operation
- Routine maintenance procedures
- Sophisticated service procedures
- Specifications and parts lists

As you work through the document process, you must refine the scope by adding appropriate details, subordinating others, and eliminating unnecessary ones. Many organizations and writers check the scope by having potential readers use an early draft of the document to check what details should be added and which should be cut. In the case of the manual for operators of construction equipment, such a process could reveal the need to add a chart with troubleshooting procedures. This is the process of refining scope. (Chapter 11 describes how to conduct "usability" tests.)

### External Constraints on Scope

A number of factors other than audience and purpose may also affect the scope of the publication. One constraint may be the physical size of a document. An 8½-by-11-inch reference card, for example, can hold only so much information and still remain legible. Other external constraints on scope may include these:

- Schedule and deadlines
- Company policy (some subjects and details cannot go outside the firm)
- Budget (including writer's time and printing costs)
- Distribution
- Availability of information
- Storage (filing)

The probable life span of a document may also help determine its scope: if it is short, the scope can be narrow; if it is long, the scope must be wider since the longer a document is in use, the more contexts and possible applications it will need to accommodate. Some experts suggest, for this reason and a number of others, that no computer manual's scope "should be any broader than explaining the use of a specific system" (Grimm 28).

Many government and grant proposals set out rigid content and length requirements. For some documents, such as military manuals, the content and even the order of presentation are prescribed. Figure 4.3 shows sections from a 47-page military specification, which instructs writers on the preparation of manuals for personnel who maintain supplies at military depots. Notice that the specification prescribes the scope as well as organization and even part of the language.

Both the general content and the length of journal articles are often constrained by publishers of trade and professional publications in their instructions to authors. For example, the statement shown in Figure 4.4 appears in each issue of *Technical Communication*.

## OVERALL DESIGN

Audience, purpose, and scope are the crucial factors in the design of a document, from large issues like organization to small ones like the typeface used. The goal of design is to make the readers' tasks easy and help them find the information they need quickly. Early in the document process, you need to consider overall issues of design; later you need to work on the details of design. (Chapter 8 discusses page layout and design in detail; Chapters 7, 13, 14, and 15 cover other design issues.)

The large issues that you must consider early in the process include (1) dividing the publication into modular units or parts for multiple audiences and

MIL-M-63041C (TM)

**Military Specification**

**Manuals, Technical**

**Preparation of
Depot Maintenance Work Requirements**

. . . . . . . . . . . . . . . . . . . . . . . . . . . . . . . . . . . . . . . . . . . . . . . . . . . . . . . .

3.10 *Order of presentation.* The DMWR shall be arranged in a sequence which is logical in terms of the functions to be performed. The following outline shall be followed:

Front Matter
Chapter 1    Introduction
Chapter 2    Technical Support Requirements
Chapter 3    Preshop Analysis Operations
Chapter 4    Overhaul Operations
Chapter 5    Quality Assurance Requirements
Chapter 6    Preservation, Packing and Marking
Appendix A    References
Appendix B    Repair Parts and Special Tools List
Appendix C    Expendable Supplies and Materials List
Appendix D    Depot Mobilization Requirements
(*Other Appendixes)
Glossary
Index

   *As specified by the procuring activity.

. . . . . . . . . . . . . . . . . . . . . . . . . . . . . . . . . . . . . . . . . . . . . . . . . . . . . . . .

3.12 *Chapter 1 Introduction.*

3.12.1 *Section 1 General.*

3.12.1.1 *Scope.* This paragraph shall contain the following statement: "These instructions are for use by depot/contractor personnel. They apply to the (insert name of equipment) and, in case of conflict, take precedence over all other documents pertinent to depot maintenance of the item. Conditions of overhauled (insert end item, assembly, subassembly or component) shall be that utility and performance is equal to that of a condition code A as defined in AR 725-50."

*Figure 4.3*    Samples from a Military Technical Manual Specification

> *Technical Communication* publishes articles of professional interest to technical communicators—including writers, editors, artists, teachers, managers, consultants, and others involved in preparing technical documents. Before writing an article, look over several recent issues of the journal to make sure your topic and approach are appropriate for our readers.
>
> Also, please look over the literature in the field and cite any relevant publications, so that your article builds on and extends previous work, if there is any.
>
> . . . . . . . . . . . . . . . . . . . . . . . . . . . . . . . . . . . . . . . . . . . . . . . . . . . . .
>
> ### Manuscript Preparation
>
> Articles should usually not exceed 5000 words (about 20 double-spaced pages). . . .

*Figure 4.4*    Excerpts from a Statement of Editorial Guidelines

purposes and (2) planning overall visual principles. Many of these principles may already be addressed by a company style manual or set of document design standards. If no such style guide exists within your organization, you may ensure consistency by developing a manual that sets forth these standards, as described in Chapter 14. If the document is large and involves numerous cowriters and collaborators, standards are essential and should be prepared early.

## Modular Design

Your analysis of audience and purpose may reveal that you must write several reports, manuals, or other documents—or leave some audiences for another time. You may determine that you need a package of documents related to the same subject. If that is the case, consider using a matrix similar to the one in Figure 4.5, which is based on one developed by Sandra Pakin and Associates. Such a matrix is useful in determining the needs of different audiences and helping you decide if you need to create, for example, three-ring-bound manuals for in-house programmers, reference cards for clerical staff, or professionally printed and bound booklets for customers (Pakin 43). As the matrix illustrates, a single project or system can include many documents (a "library" of documents) aimed at various audiences within an organization. Such libraries standardize information while accommodating the needs of distinct audiences.

For multiple audiences with widely varied backgrounds addressed in a single document, such as a report or a proposal, aim various elements and sections of the document at different sets of readers, as the following partial table of elements and typical readers illustrates.

| Document | Description | Primary Audience | Medium |
|---|---|---|---|
| Preinstallation brochure | Short introduction on the system's operational benefits and review of the installation schedule | General and line managers | Three-fold card |
| User's guide | Explanation of procedures for the operating system, including running and correcting reports and checking employee input | Managers who approve time reports for employees | Manual: three-ring binder |
| Reference summary card | Review of operational commands for running the system | Managers who operate the system | One-page card |
| Reference card for time reporting procedures | Graphic review of the punching-in and punching-out procedures, plus time report verification | All employees | Poster |
| Accounting procedures manual | Overview of the accounting portion of the system and explanation of operational procedures | Bookkeepers | Manual: three-ring binder |

**Figure 4.5**   Matrix for Documents Related to a Computer System

| *Section of Document* | *Target Readers* |
|---|---|
| Cover and binding | Clients and customers |
| Reference card | Regular and frequent users |
| Abstract | Decision makers and librarians |
| Table of contents | Occasional users (overview) |
| How-to-use-document section | All readers |
| Index | All users (detail) |
| Lists of abbreviations | Novices |
| Executive summary | Decision makers and experts |
| Body | Primary audience |
| Headings | Experts scanning for information |
| Summarizing lists | Experts (review); novices (study aid) |
| Recommendations | Decision makers |
| References | Experts and other researchers |
| Appendixes | Secondary expert audiences |
| Glossary | Novices |

Such segmenting of documents allows readers to focus on the information they need.

For separate audiences with different levels of expertise, consider also designing "two-track" documents. These involve lengthy explanations and simple procedures for novices and terse introductions and summarizing lists for experts. Of course, when writing for such multiple audiences, you may need to make a trade-off by satisfying the primary audience first and the secondary audience as necessary or possible.

For example, if your primary reader is an executive and your secondary reader is a technical specialist, you should not include a large amount of technical detail that would obscure the main point for the executive, even though the technician might find such detail interesting. But you could include a small section containing detailed technical information for the specialist without interfering seriously with your message to the executive, especially if you label that section "Technical Analysis" or in some other appropriate way.

## Visual Design

Desktop publishing has enabled professional writers, using well-established visual design and page layout techniques, to make the individual page an organizing unit within itself. (Chapters 7 and 8 provide details on graphics and layout.) Further, various elements of a document (type of binding, quality of production, headings, title, symbols, page design, corporate authorship) not only provide the context but also create an image for the document. The overall quality of design or image depends in part on the readers' expectations and the purpose of the document. For expensive consumer products or high-level consulting services, readers expect a high-quality, lavish design. For tax-supported organizations, by contrast, readers expect a modest but carefully designed publication reflecting a frugal, institutional image. Of course, budget and resources may be the final arbiters of design quality.

Visual elements—including graphics, boxed sections, and white space—also contribute by making the topic of each section visually prominent, helping to orient readers. For example, readers who must assemble a simple device or product need a clear picture of the fully assembled product. Any required warnings must stand out visually from the rest of the text. And for readers of lesser literacy, visual clustering and orientation are essential.

Because it clusters information and makes use of readers' visual frameworks (the schemata discussed earlier), visual design is a powerful tool for communicating with readers. Therefore, begin to consider the visual elements as you assess your audience, purpose, and scope and make them integral to your project as you move through the document process. Use the appropriate chapters in this book and any resources available to you—graphics and layout specialists, colleagues, publication standards, desktop publishing capabilities—to enhance your working knowledge of visual design and communication.

*Checklist 4.1*

## Basic Audience Features

☐ Identify readers' familiarity with the subject
- Expert
- Novice
- Intermediate

☐ Identify readers' functional roles
- Professional or vocational specialty
- Organizational role
- Customer or client

☐ Identify psychological factors related to the readers
- Supportive or interested
- Anxious
- Skeptical

☐ Identify special considerations
- Physical limitations
- Language barriers
- Cultural barriers

---

*Checklist 4.2*

## Methods for Refining Audience Analysis

Use as many of the following methods as possible within the time, resources, and budget available.

☐ Gather facts and statistics about the readers

☐ Examine organizational charts and job descriptions

☐ Interview and observe potential readers

☐ Analyze relevant documents
- Training materials
- Sales reports
- Publication standards
- Communications to readers
- Publications read by the target audience
- Subscriber data

☐ Operate, use, or assemble the device, system, or product yourself

---

*Checklist 4.3*

## Purpose of the Document

☐ Assess the organization's needs for the document (see Chapter 2)
- Sell a product or service
- Support users and customers after a sale
- Build goodwill or image

- Manage efficiently
- Meet legal or other requirements

☐ Consider the readers' purposes

- Assemble an item
- Follow procedures
- Operate a device or a system
- Understand a topic
- Repair something
- Answer a question
- Locate information
- Make a decision

☐ Anticipate the reading style of your readers

- Read for general sense
- Read carefully to understand
- Evaluate for logic and conclusions
- Read portions only
- Search for specific items
- Casually glance and file for reference

What should readers know, do, or believe when they have read or used the document? Formulate this into a statement of purpose.

---

*Checklist 4.4*

## Scope of the Document

☐ Determine the topics to cover, using audience and purpose

- List possible contents of the document
- Observe a typical user working with the product or system
- Note potential topics or details to cover

☐ Determine external constraints on scope

- Budget and time available
- Size of medium (8½-by-11-inch reference card, for example)
- Prescribed or recommended length
- Elements required by standards or specifications
- Standard elements required by the form (proposals, etc.)
- Order of parts required by the specifications
- Organizational policy (topics not to be discussed)
- Document life span, storage, or distribution
- Information availability

---

*Checklist 4.5*

## Overall Design

☐ Determine if the project requires separate documents

☐ Develop a matrix (see Figure 4.5)

☐ Determine if a single document should be divided into sections
- How-to-use section
- Abstract
- Table of contents
- Lists of abbreviations
- Executive summary
- Recommendations
- References
- Appendixes
- Glossary
- Index

☐ Determine if the design must accommodate two audiences
- Create two tracks with section headings and summary lists
- Satisfy the primary audience, accommodating the secondary audience as possible

☐ Develop a strategy for the initial visual design of the document
- Should standards or existing specifications be followed?
- Should standards or a style manual be developed?
- What quality or image does the audience expect?
- Should literacy level or physical or mental considerations affect the design?

☐ List special graphic or visual design needs based on reader needs, purpose, and scope

☐ Review the checklists in Chapters 7, 8, 13, 14, and 15

---

## WORKS CITED

Anderson, Richard C. "Schema-directed Processes in Language Comprehension." *Cognitive Psychology and Instruction*. Ed. A. M. Lesgold, et al. New York: Plenum, 1978. 67–82.

Brusaw, Charles T., Gerald J. Alred, and Walter E. Oliu. *Handbook of Technical Writing*. 3rd ed. New York: St. Martin's, 1987.

Caernarven-Smith, Patricia. *Audience Analysis and Response*. Pembroke, MA: Firman, 1983.

Day, Robert A. *How to Write and Publish a Scientific Paper*. Phoenix: Oryx, 1988.

Delbecq, A. L., A. H. Van De Ven, and D. H. Gustafson. *Group Techniques in Program Planning: A Guide to Nominal Group and Delphi Processes*. Glenview, IL: Scott, 1975.

Dobrin, David N. *Writing and Technique*. Urbana, IL: NCTE, 1989.

Flower, Linda. *Problem-solving Strategies for Writing*. Orlando, FL: Harcourt, 1981.

Grimm, Susan J. *How to Write Computer Manuals for Users*. Belmont, CA: Lifetime Learning, 1982.

Houp, Kenneth W., and Thomas E. Pearsall. *Reporting Technical Information*. 6th ed. New York: Macmillan, 1988.

Huckin, Thomas N. "A Cognitive Approach to Readability." *New Essays in Technical and Scientific Communication: Research, Theory, Practice*. Ed. Paul V. Anderson, John R. Brockman, and Carolyn R. Miller. Farmingdale, NY: Baywood, 1983. 90–108.

Keeler, Heather. "A Writer's Readers: Who Are They and What Do They Want?" *Technical Communication* 36 (1989): 8–12.

Krippendorff, Klaus. *Content Analysis: An Introduction to Its Methodology*. Newbury Park, CA: Sage, 1980.

Lannon, John M. *Technical Writing*. 4th ed. Glenview, IL: Scott, 1988.

Mathes, J. C., and Dwight W. Stevenson. *Designing Technical Reports*. Indianapolis: Bobbs, 1976.

Mirel, Barbara. "Cognitive Processing, Text Linguistics, and Documentation Writing." *Journal of Technical Writing and Communication* 18 (1988): 111–133.

Monroe, Judson. *Effective Research and Report Writing in Government*. New York: McGraw, 1980.

Nagy, Patricia M. "Writing Educational Material for Patients." M. A. thesis. U of Wisconsin-Milwaukee, 1985.

Ong, Walter J. "The Writer's Audience Is Always a Fiction." *PMLA* 90 (1975): 9–21.

Pakin, Sandra, and Associates. *Documentation Development Methodolgy*. Englewood Cliffs, NJ: Prentice, 1984.

Pearsall, Thomas E. *Audience Analysis for Technical Writing*. Encino, CA: Glencoe, 1969.

Rickard, T. A. *A Guide to Technical Writing*. San Francisco: Mining and Scientific Press, 1908.

Ridgway, Lee. "The Writer as Market Researcher." *Technical Communication* 32.1 (1985): 19–22.

Stevenson, Dwight. "Audience Analysis across Cultures." *Journal of Technical Writing and Communication* 13 (1982): 319–330.

Suchan, James, and Ron Dulek. "Toward a Better Understanding of Reader Analysis." *Journal of Business Communication* 25.2 (1988): 29–45.

## FURTHER READING

Keene, Michael, and Marilyn Barnes-Ostrander. "Audience Analysis and Adaptation." *Research in Technical Communication: A Bibliographic Sourcebook*. Ed. Michael G. Moran and Debra Journet. Westport, CT: Greenwood, 1985. 163–191.

Lay, Mary. "Procedure, Instructions, and Specifications: A Challenge in Audience Analysis." *Journal of Technical Writing and Communication* 12 (1982): 235–242.

Wagner, Carl G. "The Technical Writing Audience: A Recent Bibliography." *Technical Writing Teacher* 14 (1987): 243–264.

### Chapter 4: Audience, Purpose, Scope, and Overall Design

INTERVIEWER: Who were the primary readers of the *Introduction to Using the International Banking System* manual?

MARY: The primary readers were 150 people in the International Banking Department (IBD) at First Wisconsin. Most readers were clerks, some professional traders, and some bank officers in various areas within International Banking, such as commercial loan processing, letters of credit, foreign exchange, and investment. Managers would also use the material, but they would only need it occasionally.

INTERVIEWER: What did you find out about the audience for this manual? How did you determine your readers' level of technical knowledge?

MARY: The majority of the employees were clerical staff, primarily bilingual women for whom English was a second language. They performed tasks (such as recording payments and answering correspondence) that required their bilingual skills.

There were also several groups of professional employees who had specialized skills: finance officers with extensive knowledge of international finance, traders expert in foreign exchange trading. And there were two managers for the clerical staff.

I asked Marty Freier, the person in charge of training IBD users of the new system, how much IBD personnel knew about computers. He said there were a few personal computers in the area, but that was it. By going back and forth between Marty and several clerks in the area, I gradually learned the details of the audience's technical knowledge.

The handful of people who used the PCs understood the specific tasks they did on the computer and were familiar with the computer keyboard and with the things you have to do with a computer—log on, enter a password, and so on. These employees generally felt that they had mastered the application they worked with. However, some computer concepts were unfamiliar to them, so they stopped me if I let computer jargon slip into our conversations. Most employees, though, had jobs that did not expose them to a computer system.

INTERVIEWER: What was the attitude of the audience toward the new system you were documenting?

MARY: I would say it was skeptical. Not exactly hostile, but definitely skeptical; they had some reservations. They'd already seen some earlier documentation, written by the system developers, that confused them.

When I planned the manual, I made a point of keeping the tone as warm as possible to overcome the audience's skepticism. I knew that if I could resolve some of their confusion, I could win them over.

INTERVIEWER: Were there any cultural or language factors you needed to consider?

MARY: Since many of the employees were not native English speakers, I had to be careful not to include idioms. I don't normally use complex constructions in my sentences anyway, but for this audience I tried to simplify the text even more. However, I always tried to avoid talking down to people, both in my writing and in my contacts with employees.

I don't know that I did anything specific in the manual to address the diverse cultural backgrounds of the audience. I don't know that a technical document *could* do much to address that. I assumed that since they were part of the banking community, their participation in *that* culture was more important than their personal cultures (for the purposes of this manual).

INTERVIEWER: How did you visualize your reader while you were writing?

MARY: Once I got to know Jenny Acevedo (a clerical staff person who turned out to be one of the main user reviewers for the project), I visualized her. (I also asked her lots of questions as I was drafting the manual.) Before I became acquainted with her, though, I would just imagine the sea of desks in the International Banking area, and I would think about the open environment and what it might feel like to do bunches of paperwork there. I was fortunate to have direct contact with a reader like Jenny on this project—that's unusual.

INTERVIEWER: What, in a sentence or two, was the purpose of the *Introduction to Using the International Banking System* manual?

MARY: To provide the IBD people with the basic knowledge that everyone was going to need to use the system. You might say it was a first-time training aid, but it was not designed primarily as such. It was designed primarily to be there as a reference *after* training or when a new employee was hired. But the idea was also that the manual was going to be there when they were first learning to use the system, so they could get going on it.

INTERVIEWER: So you had more than one purpose in mind?

MARY: Yes, both training and reference.

INTERVIEWER: How did you determine the purpose of the document?

MARY: Purpose was decided from the top down, on a large scale; that is, we kind of worked our way from the more complex to the more basic documents that were going to be needed for the system. It turned out that the International Banking people wanted some marketing documents; some procedural manuals for things like business transactions, communications screens (sample computer screens showing how to transfer funds or how to send electronic mail), and special functions, such as maintaining customer records; a report manual; and some basic system documentation.

*I* decided that we needed the *Introduction* manual. The system designers originally wanted to slip basic information, such as data entry rules, into the procedural manual for the business transactions screens. But I realized that many more people would need the basic information than would need the documentation for the business transaction screens. For example, the traders and wire service clerks would need to log on but would not need to know anything about the business transaction screens.

In addition, because the system was so large, I felt that it also needed a coherent introduction. Overall, it would be cheaper and more reasonable to have that coherent introduction and the basic procedures in one place, rather than having them scattered or, worse, repeated in several manuals.

INTERVIEWER: What is the scope of this document, based on the audience's needs and the purposes?

MARY: The *Introduction* is one of nine end-user manuals designed to document the International Banking System. It introduces the system as a whole (what its basic components are and what it's supposed to do); then it shows how to perform basic tasks on the system, such as logging on and off and using menus, and provides other basic information, such as screen formats and data entry rules.

The manual is geared for daily users—especially first-time users, so it doesn't assume familiarity with the system. It includes a glossary of system codes and terms.

INTERVIEWER: Yes, but I notice the glossary doesn't include any banking terms.

MARY: Right. That's because built into the scope is the assumption that readers of the manual *do* understand the common terms and processes of international banking.

INTERVIEWER: How did the audience, purpose, and scope affect the overall design of the document?

MARY: Since it was an internal First Wisconsin document, I kept the binder and spine label very simple (and inexpensive). I organized the manual so that the simplest tasks came first, to make it as readable as possible. I tried to limit chapters to 10 to 15 pages, figuring that that was about as much as someone would want to hear about a topic without becoming overwhelmed.

The system designers had already set format conventions for all documentation, which is unusual for a project. The design included side margin captions (headings on the left, with corresponding text indented 25 spaces). The margin caption format was the format we used in most of our department's manuals, and it was especially recommended in this case because the audience was skeptical of the system. We felt that more white space would make the manual less intimidating to read.

# 5

# Gathering Information

This chapter introduces you to a range of information sources available to technical writers on the job and presents proven strategies for successfully obtaining the information needed from those sources. You can acquire background information for your writing assignments from a variety of sources:

- Publications
- Computer databases
- Technical experts
- Your own knowledge and hands-on experience
- Observations of people and processes
- Survey results

## GETTING THE ASSIGNMENT

As the first step in your information gathering, use the prewriting conference discussed in Chapter 2. Some of the questions raised in that conference will help you determine where to go for information and how much of it you need before beginning the first draft. The detailed analysis of audience, purpose, and scope in Chapter 4 is also central in focusing your information-gathering efforts. The following questions will help focus your thinking:

- What is the purpose of the document?
- Who is its primary audience?
- Will secondary audiences be targeted?
- How broad is the document's scope?
- Will written source material be provided?
- Who are the technical experts? Will they be available?
- Will the product or equipment being documented be made available?
- Will field testing of the publication be necessary?
- Can I observe the necessary people and processes?

## RESEARCHING DOCUMENTS

Before you begin researching written sources, be sure to call the relevant technical experts to ensure that your source documents, including their flowcharts, schematics, diagrams, and the like, are up to date.

### Similar or Related Documents

First, be aware that nothing you write will be completely new. Somewhere, someone, maybe even in your organization, has written a document similar to the one you are at work on. If you are writing about a product new to you, manuals may already exist for similar products. Find them and examine them attentively, looking at their scope, organization, page layout, point of view to the reader (direct or indirect), and use of figures, tables, and such aids as table of contents, margin tabs, and subject index. Review as many documents similar to the kind you are preparing as time permits, examining each for useful elements that apply to your document. If your company or organization has a library of such documents, search it, too. Other technical writers and your supervisor should prove to be good sources of this type of information.

Do not hesitate to borrow passages from in-house documents, either. As long as the information fits your context, is accurate and well written, and originated within your company or organization, use as much of it as you need. Professional writers refer to such passages as *boilerplate* and use them extensively in some settings. (Proposal writers, for example, use the same language to describe a product's attributes from one proposal to the next, as long as the context is appropriate.) Using boilerplate from in-house documents is not the same as incorporating passages from published works because your employer already owns the rights to in-house material. The Copyright Act protects any work from the moment of its creation, regardless of whether it is published or even contains a notice of copyright (©). Under this law, you are free to use works written by employees of your own organization, but you must obtain permission to use works written elsewhere.

When using in-house information, review it and, if necessary, revise it carefully for consistency of content, style, and format with the work into which it will be incorporated.

## Relevant In-House Documents

In addition to publications that provide boilerplate or that serve as models for format, scope, and point of view, review the in-house documents containing the raw materials needed for the contents of your publication, such as these:

- Design specifications
- Manuals for earlier models and for related equipment
- Memorandums and correspondence
- Reports
- Procedure documents
- Codes and standards

Review this information carefully, noting areas that are unclear or incomplete so that you can ask for clarification in subsequent discussions with the originators. Use caution before adapting the contents and organization of these documents to your own, however, because they are seldom prepared with your audience in mind, a point emphasized for technical writers at the Allen-Bradley Company:

> Most engineering documents are topic-oriented. They describe the physical components and properties of a product or system. Typically, they describe the way the product was designed and built, not the best way to use it. However, such a description may not tell the user how to assemble, install, use, and repair the product. (Allen-Bradley 2-2)

Be especially wary of technical specifications, which are often the primary source of written information for technical writers in certain fields. Intended for knowledgeable insiders, technical specifications ("tech specs") contain a spare narrative of particulars prescribing the design details, materials, dimensions, and so on, for a product. Specs are densely packed with information, like a dictionary, and typically adhere to a rigid format and section-numbering scheme for ease of internal cross-referencing. Although specs may contain information indispensable to your project, they provide a poor model for how the information should appear in your finished document.

The ease with which you can find various documents depends on how well they are organized and filed for retrieval. When you first join an organization, familiarize yourself with how it files the technical information created in-house and received from outside. If your organization provides training in how to locate such information, take it; if not, get to know the people who catalog it: in-house librarians, technical information specialists, or document technicians. Your willingness to learn from them will give you ready access to your organization's relevant information sources. (Computerized information-retrieval systems will be discussed later in this chapter.)

## Trade Documents

Vast amounts of information are also available in the trade documents in your field. Technical disciplines spawn a wide range of publications, including magazines, reports, newsletters, booklets, brochures, conference proceedings, and membership directories. The corporate library or information resource center will be the most convenient place for locating this information.

As you seek documentation, keep the following guidelines in mind. The more recent the information, the better. Journal articles are essential sources for current information because books take longer to write, publish, and distribute than journal articles. Conference proceedings can be an even better source of up-to-date information. They contain papers presented at meetings of trade, industrial, and professional societies about recent research results or work in progress. Much of this information either will not be published elsewhere or will not appear for a year or more in a journal. When studying an article, consider the author's reputation before accepting the information uncritically. Is the author considered an authority in the field? Has the author written other, well-respected articles or books on the subject? Your professional colleagues can help you answer these questions until you become familiar enough with the field to hold informed opinions.

## Copyright Permission

A word of caution about using copyrighted sources: scrupulously avoid plagiarism. The temptation to incorporate long passages, tables, photographs, figures, and the like from outside sources without first seeking permission from the copyright holder may be especially great for documents you write that are intended solely for in-house use. Doing so, however, could subject your firm to an expensive and embarrassing lawsuit. You could also lose your job. So identify all quoted and reprinted matter clearly. Bear in mind that plagiarism has practical as well as legal and ethical ramifications. Failure to document sources deprives readers (and you) of important information about where your ideas came from. For this reason, it is wise to cite even sources in the public domain, such as publications of the federal government. When you seek permission to use copyrighted material, adapt the sample form letter in Figure 5.1 to your needs.

# COMPUTERIZED INFORMATION RETRIEVAL

Computerized information-retrieval systems put a wealth of information at your disposal. You can use such systems to check references, verify facts, conduct literature searches, locate experts, learn the location of specific publications or whole classes of publications, and, in some cases, gain online access to the full

[YOUR ORGANIZATION'S LETTERHEAD]

Subject: Request for Permission to Reprint Copyrighted Material

Dear Publications Manager:

I am preparing a [brochure, product manual, report, journal article] titled
"[_____]" for [name of your firm or organization] to be
published in [Title of Publication]. I would like your permission to reproduce the
following material:

Text Excerpt:   [title, author, date, and other publication data; first and last
                words of passage]
                From page ___ to page ___.
                Approximate number of words or pages: ___.

Table:          [title, number, page number, and data about the publication from
                which the table was obtained]

Figure:         [title, number, page number, and data about the publication from
                which the figure was obtained]

Full credit will be given to the author and publisher. Please indicate any special
wording you may require in a credit line. You may use the release form, which is
provided for your convenience at the end of this letter, for responding to this
request. A copy of this form is enclosed for your files. Should you have any
questions about this request, please call me at [(area code) phone number]. I
would greatly appreciate your prompt consideration of this request.

                                  Sincerely,

                                  Name, Title, etc.

_____

You have my permission to use the material stipulated in this letter.

Signature: _____

Title: _____ Date: _____

Preferred Credit Line: _____

_____

_____

*Figure 5.1*   Sample Permission Form Letter

text of publications. Some systems are dedicated exclusively to in-house information (the material produced exclusively by and for a specific company, governmental agency, or other organization). You can also use systems available in academic, research, and corporate libraries to compile bibliographies of materials in published literature.

## In-House Systems

Dedicated in-house systems typically contain databases that store user manuals, specialized reports, policy and procedure documents, specifications, codes and standards, letters and memorandums, engineering drawings, and the like. Depending on their degree of sophistication, these systems permit access to their databases through a variety of search modes. Seated at a terminal, you may enter the title of a document, its author, a code for the type of document (letter, memorandum, report, manual), its date (or a range of dates if you don't know the exact date or if you want everything for a specific period), a document identification code (such as a report number), or some combination of these. After the system locates the information in the database, it may display a synopsis of it on the screen and direct you to where a paper or microfiche copy of the document can be located. The paper copy may be filed in a central repository elsewhere in your organization, or it may be stored in microfiche form in a special file. Some systems even permit full text retrieval, so that you can call up text and illustrations on your screen and "page" through them at the terminal.

Figure 5.2 shows a "search" screen from an in-house computerized database of documents. The record shown concisely describes a one-page preliminary report of February 1, 1991 (**020191**), that was to be followed in 30 days by a final report. The researcher seeking a copy of the report can consult the "address" shown at the upper right corner of the screen that indicates the microfiche card number (**FICHE START: 44420**) and the frame on that card (**342**) where it can be located. The screen also shows the author of the report (**C. R. Dietz**), an abbreviation of the author's affiliation (**EUTCPL**), the recipient (**J. N. Grace**), an abbreviation of the recipient's affiliation (**NE R2**), and a variety of other coded information that the researcher may need. Screens for in-house databases tend to look forbidding to the uninitiated, which is why organizations usually provide on-the-job training in the use and interpretation of these systems.

Other in-house systems make it possible for you to conduct searches of the database using subject terms and phrases, like those found in the card catalog of a library, rather than limiting you to searching by author's name, type of document, or date of issue.

These systems can be valuable sources of information if you understand their capabilities well enough to exploit their full potential. Accordingly, take as much training in their use as your firm provides, and practice on the system until you become comfortable using it.

## Bibliographic Databases

The other common databases available for computer searching contain indexes, abstracts, and other bibliographic reference works. Searches of these "electronic

```
      ACCESSION NO: 8802170308 SN:   AVAIL:  PDR   FICHE  START: 44420 FRAME1: 342
      DATE ISSUED: 020191     NO. OF PAGES:     1p.           END: 44420 FRAME2: 342

      DOCKET1: 05000324 DOCKET 2:    DOCKET 3:     LPDR: Y SSN:*
      AUTHOR:  NAMES       ORG       RECIP:  NAMES       ORG
            DIETZ, C.R.       EUTPL        GRACE, J.N.      NE R2


      TITLE:  Submits supplemental response to IE Bulletin 80-13 re insps
            of core spray piping. Corrective actions & final results of
            insps will be provided in final rept within 30 days.

      DOCUMENT TYPE: CLUTN  DOCUMENT  ID NO.:


      FECN: PDR  FIL1: ADOCK FIL2: 05000324 FIL3: Q       880201


      FILE PACKAGE NO.: 8802170308* DOCKET DATE:       DATE INDEXED:  000000
                                                       SUPPLEMENTS?:

            SALES:      SIZE: A4 LANGUAGE: A    PRICE:

      Type P to print, V to view the next record or Q to quit
```

***Figure 5.2***   Search Screen from an In-House Computerized Database

libraries," the majority of which are leased from commercial vendors, are usually conducted by reference librarians or other trained search analysts. These analysts can access hundreds of databases for pertinent published articles on a given subject. Figure 5.3 shows a printout of a bibliographic citation and abstract retrieved from the LISA (Library and Information Science Abstracts) database of Dialog Information Services, Inc. It shows the LISA access number, title, month and year of publication, abstract, and printed location of the summary of a report written by a committee of the National Science Foundation.

The bibliographic sources in these databases exist in printed form and can be searched manually; however, the advantages of computer-searching these sources are numerous and significant. First, computer searches are comprehensive. They provide access to hundreds of databases of published sources. Such a wealth of sources could be matched by only the largest university or public library. They also contain the most up-to-date information. Printed indexes typically lag from six weeks to one year behind the literature they cite; most are three to six months behind. Computer searches are also fast. A skilled searcher can locate in an hour what a manual search could take weeks or months to find. They are thorough, allowing searchers to look for key subject terms in the titles and abstracts of reports and articles as well as the standard subject headings. They permit searches for combinations of information, like an author and a

171696     85-5909     Library and Information Science Abstracts (LISA)
**Reliance upon scientific publications and face-to-face contacts called impediments of progress and communication of scientific information**
Information Hotline
SOURCE: 16 (4) Apr 84, 1, 16–17
The USA National Science Foundation Committee on Science, Engineering, and Public Policy has prepared a report, The five-year outlook on science and technology, in which it criticizes scientists for relying upon scientific publications and face-to-face contacts rather than the new communications technologies. The report considers the former methods waste time and slow scientific development. The report identifies trends and opportunities in 8 selected areas of science and engineering. It also considers basic research and advanced technology critical to further economic growth and finds the traditional boundaries between areas of science and technology are dissolving.

**Figure 5.3**   Citation from a Commercial Bibliographic Database

subject ("William Zinser and word processing"), a subject and a subject ("fiber optics and automobile instrumentation"), and a subject for a given period ("text-editing software since 1985"). Finally, these databases allow searches for work by a specific author, whereas printed indexes provide only subject-term access to information.

## INTERVIEWING FOR INFORMATION

If you are working in a technology that is under development, little written material may be available to help answer your questions. You may need, instead, to talk to the technical experts who are developing a product or a system. Having recorded the names of technical experts during the prewriting conference, arrange to interview them. Membership directories of professional associations are also valuable resources for locating technical experts outside your organization.

Interviewing is a crucial information-gathering technique. A recent survey of a cross section of technical writers working for computer, medical, financial, public utility, and manufacturing firms revealed that they conduct three to five telephone and face-to-face interviews to gather information each week (McDowell, Mrozla, and Reppe 50). Other recent estimates put the amount of time technical writers spend researching (including interviewing and asking questions in some settings) at between 40 and 75 percent (Lamping 151; Levine 55).

### In-Person Interviews

Most interviews are conducted either face to face or over the telephone. The in-person interview requires a formal request that you can make by telephone or in a letter. (Send a letter only if the interviewee works in a different city and if you can schedule the interview far enough in advance.) When you make the request, identify yourself and the purpose of the interview, explaining briefly the type of information you need and why you think the interviewee can provide it. If a mutual professional acquaintance recommended the expert to you, mention the referral. Offer to arrange the interview at the other person's convenience within a time frame that will accommodate your schedule as well. Keep an eye on the calendar, too. Arranging to meet people near holidays and vacation periods is risky; if time permits, schedule around these periods.

Prepare for the interview on the principle that you'll get out of it what you put into it. Your research of in-house and published sources provided essential background information, but you need additional preparation. If you are new to the field, begin by studying the vocabulary that defines the concepts central to the project. This knowledge will enhance your understanding of what you read and of what you are told during the interview; it will also allow you to translate these ideas into clear English for your audience more easily. Furthermore, your grasp of such fundamentals will gain you the respect of the experts you interview. Also learn as much as you can about the interviewee and his or her department or organization. If your interviewee has published in the field, read all work related to your project.

Guided by information from the prewriting conference and from your other research, prepare a list of questions. Write your questions in random order as they occur to you, arranging them into a logical sequence later, before the interview. Remember your audience, too. Represent the readers' point of view by using what you have learned during audience analysis to ask questions that they would ask.

Arrive for the interview on time. After the introduction and brief initial small talk, record the name and title of the interviewee, as well as the date and place of the interview. Then be prepared to guide the discussion. Be pleasant but purposeful. You are there to get information, so don't be timid about asking difficult questions, beginning with those on your list. Allow for flexibility, too. Do not follow the list so rigidly during the interview that you cannot ask logical follow-up questions or pursue another line of questioning should the conversation take an important but unexpected turn. However, should the interviewee get off the subject, use your prepared questions to guide the discussion in the proper direction.

Keep the questions objective and friendly. You are there to interview, not to debate. Let the interviewee do most of the talking; encourage this by asking questions that cannot be answered by a simple yes or no. If the person is not very forthcoming, ask leading questions: "Can you tell me anything more?" or "Can you elaborate in more detail on that?" You must also exercise tact. Be attentive about when to interrupt with a follow-up question and when to remain silent as the interviewee formulates an answer to your question. Listening carefully usually pays dividends because it helps establish rapport with the speaker and takes advantage of the fact that most of us like to talk about our work to someone who's interested.

The interview logically ends when your prepared and follow-up questions have been answered to your satisfaction. Then ask a final open-ended question:

"Is there anything else you'd like to add?" This gives the interviewee a last chance to add anything missed during the interview. Leave the door open for follow-up interviews as well by asking if the interviewee will be available if you have further questions. Before leaving, remember to ask for any documents or illustrations that the interviewee has offered to provide.

### Interpersonal Skills

Technical writers must have the ability to work effectively with other people. This crucial trait may be sorely tested, however, during information-gathering interviews with technical experts. Writing about her experiences as a manual-writing intern at a large computer firm, a technical writing instructor points to some of the problems she faced.

> I . . . learned the importance of maintaining good interpersonal relations. Programmers often consider writers secretaries at best and nuisances at worst. They generally think their job is only to develop programs, not to explain their programs in person to the writer. Also, their perspective is narrow. Thoroughly familiar with their product, they are unable to see it from the user's point of view. Thus, when the writer asks the programmer to explain a particular bit of information, the programmer may say with a sweep of the hand, "Oh, the user will understand." ("You don't," they imply, "but the user will.") I learned to be persistent, tactful, and thick-skinned. (Fridie 317)

This attitude on the part of the technical staff, regardless of the setting, may arise because the expert is too busy or has not scheduled enough time to talk to you. One writer observed of programmers, "When they're under pressure to complete a system, they may feel they're being paid to program, not to talk to you" (Grimm 11).

An uncooperative source puts you in an awkward position. With an unwilling interviewee, establish an attitude of shared goals about the task at hand: "We're in this together." Practice this attitude when requesting the interview, during the interview, and in follow-up contacts after the interview. If the technical expert remains uncooperative or even hostile, don't return the attitude in kind. Be confident in what you know about the technology and in the special organizational and writing expertise you bring to the profession. Invoke your knowledge and experience when you must, but don't do so petulantly; advance your position courteously and objectively. Of course, what you advance must ring true. To echo the advice in Chapter 1, keep your professional knowledge fresh by formal education, continued reading, and active involvement in professional organizations. Listen to and read what your colleagues are saying. Your skills, backed by wide-ranging knowledge of the field, will inevitably earn you respect and cooperation. Use them to your advantage during interviews.

You must also phrase your questions tactfully, avoiding assertions that improperly challenge the interviewee's knowledge.

*Poor:* "What you just said contradicts what so-and-so just told me."
*Better:* "I may have misunderstood what you just said, but I recently heard that . . ."

Likewise, do not question the interviewee's competency.

*Poor:* "Why didn't you design the program/system/equipment to perform this extra function? It would have been easy!"

*Better:* "Would it have been possible to add this extra function?"

*Best:* Don't ask. The design is already either finished or nearly finished, so it's unlikely to be redesigned to accommodate your suggestion.

Such a question also carries the implicit assumption: "If you were as smart as I am, you would have done a better job." If the question "Would it have been possible . . ." seems logical in the context of the interview, ask it. Phrasing it this way gives the technical expert the opportunity to explain why the feature was not added or how it might be added later, without the danger of a wounded ego.

Finally, do not try to impress the interviewee with your knowledge of the subject about which he or she is the expert. Emphasizing what you know can get in the way of your questions and frustrate the interviewee.

Be aware of two other situations in which your interpersonal skills will be put to the test. Some interviewees may be either too terse or too rambling. Those who give truncated answers or skip essential steps in their explanations seldom intend to be discourteous or obtuse; they just assume that you know as much as they do and so won't need to worry about all those transitions from point to point. This interviewee must be drawn out and asked to repeat steps in more detail. Remember, if you can't follow what's being explained, you will be unable to explain it so that your readers can follow it either.

Conversely, ask people who ramble too much questions that can be answered more pointedly and succinctly. Sometimes all that you can do with someone who rambles along, getting hopelessly snarled in an explanation, is suppress the panicky feeling welling up within you that you'll never understand what's being said, look for a good opening, and then, leaning forward to emphasize your intent, ask the person to pick up the explanation at the point where you got lost. You must be tactful but persistent with both types of problem interviewees. If you leave without the information you need, you probably won't be able to obtain it anywhere else.

### Nonverbal Communication

Your demeanor, like your questions, communicates messages about how well you're listening. Sit facing the other person with not more than the distance of a desk between you. Keep your posture relaxed, leaning forward to show attentiveness as circumstances demand. Maintain as much eye contact as your note taking will allow, especially when you ask questions. If your posture is rigid or if you lean away from the interviewee or sit hunched over your notepad scribbling notes and avoiding eye contact, you communicate nervousness and detachment, not interest and attentiveness.

## Telephone Interviews

A telephone interview can be a quick, efficient, and relatively inexpensive way to gather information. Rather than having to travel across town or even across the country, you can sit in your office, freely taking notes with your research material spread out for quick reference where you sit.

The ease with which a telephone interview can be conducted depends, to some extent, on whether you already know the person and on your following a set of guidelines that communicates courtesy and professionalism.

1. Create a list of questions ahead of time, as for a personal interview. The list will focus the interview, making the most efficient use of the time available, and ensure that you get the information you need. Again, pursue relevant follow-up questions, regardless of whether they are on your list.
2. Identify yourself and the purpose of the call. If you do not know the person, introduce yourself, indicate where you work and your job title, and explain the purpose of the call. For persons outside your organization, refer to a mutual professional acquaintance, if you know one.
3. Be aware that a phone call to someone during working hours is likely to interrupt something the person is working on. If your call will be brief, say so, asking if the interviewee has a few moments to talk; if the call will take longer, give the person an honest estimate of the time you'll need. Nobody likes to be asked to answer "just a few questions" to discover 15 minutes later that you had planned all along to ask many more than a few.
4. Observe some elementary courtesies during the interview. Speak clearly and distinctly. When you are taking notes, don't let your silences last too long—an occasional "uh-huh" or "OK" provides the necessary feedback that you're attentive to what's being said. Also, be careful not to get sidetracked from the topic. Should the interviewee begin to wander, tactfully steer the conversation back on track.
5. Finish with a final, open-ended question, if it's appropriate, and ask if the interviewee will be available to answer follow-up questions in the future.

## Recording Information

How will you record the answers to your questions? What's the best way to capture the essential steps in an important technical process that you learn in a rapid-fire give-and-take? For all practical purposes, you have two options: take memory-jogging notes, or tape-record the interview.

### Note Taking

If you plan to take notes, consider the following suggestions:

- Develop a shorthand technique for recording information. Consider dropping vowels and use symbols for common terms. A sntnc mt lk lke ths w/ mst vwls mssng + symbs thru/o, lke ~ fr *approximately* or + fr *and*.
- Concentrate on recording major points and the specific details that support them: facts, figures, examples.
- Use common sense. If the interviewee mentions three steps in a process, make sure you get all three before going on.

- As quickly as you can after the interview, review and clarify your handwritten notes. Build this step into your schedule. No one takes flawless notes during an interview, and no matter how good your memory, you will forget nuances, and even important points, unless you review your notes for clarity soon after the interview.

### Tape Recording

Using a tape recorder has several distinct advantages over note taking. First, it permits you to be a better listener. Freed of note taking, you can concentrate on the details of what's being said and develop a better rapport with the interviewee, since you need not focus on your notes. You also have the confidence of knowing that you can review the entire interview verbatim at your convenience. Reviewing interview tapes serves a valuable self-teaching function, too. You can detect and correct flaws in your interviewing technique, such as interrupting the speaker too often or failing to follow up on important points at the times they're made.

Tape recorders have disadvantages, too, however. Some people have an aversion to having their conversations recorded verbatim. Help them overcome this anxiety by mentioning in your request for the interview that you plan to bring a tape recorder. When you arrive, set up the recorder quickly and unobtrusively, preferably to one side where it's inconspicuous. Interviewees tend to take cues from the interviewer, so the less attention you pay to the recorder, the better. Also, do not take a new or otherwise unfamiliar recorder with you. The possibilities for mistakes are too great. If the machine is new to you, practice with it until you become completely familiar with operating it. The ease with which you set up the machine will help minimize the impact of its presence. Finally, check beforehand that the equipment works properly so you can concentrate on the exchange of information during the interview rather than on the means by which the information will be captured.

## Confidentiality

During a telephone or face-to-face interview, you may find yourself the recipient of a personal or professional confidence that you are asked not to betray. You may also be made privy to confidential or proprietary information about the inner workings of your organization. Unless these confidences or disclosures involve criminal wrongdoing, keep them to yourself. If you become known as a person who betrays a confidence, regardless of whether you were explicitly asked not to do so, you will find yourself cut off from otherwise essential sources of information. In addition to making you less effective at gathering information, such a perception could get you removed from sensitive projects or even fired.

## TAPPING YOUR OWN KNOWLEDGE AS A SOURCE

As your job experience grows, so will your mastery of the principles and vocabulary of your field. For example, if you've written one service manual for a laser printer and you're assigned to write a manual for a newer model of the printer, you already possess a wealth of knowledge that you can apply to the new project. This knowledge and experience make *you* an important source of information as you conduct research for the new assignment. Fortunately, systematic techniques exist to tap this valuable source: brainstorming (manual and computer-aided), mind mapping, and freewriting.

## Brainstorming

The purpose of brainstorming is to "get the juices flowing" as you generate ideas about how to solve a problem or about how to organize a mass of information you or your team already possesses. Brainstorming can be done by individuals or by groups on a notepad, on 3-by-5-inch index cards (one idea to a card), on a chalkboard, or on a personal computer.

### Handwritten Brainstorming

Beginning with the pertinent information learned in the prewriting conference, randomly jot down as many ideas as you can think of about the subject. Where you jot your ideas is a matter of personal style, but 3-by-5-inch note cards are convenient if you are working alone because they can be shuffled and rearranged easily when you begin to organize your preliminary outline. If you are working in a group, a chalkboard works best. (Assign someone to record the ideas for subsequent analysis.) After you exhaust your initial store of ideas, ask, for each idea, *what, when, who, where, how,* and *why,* and list the details that these additional questions bring to mind. When no further thoughts come to mind, analyze each one you have recorded, discard redundant notions, and group the items in the most logical order based on your purpose and your reader's needs. Your audience and purpose will provide the basis for arranging these ideas and the information from your other sources.

The result of this grouping will be a tentative outline of your project—be it a pamphlet, a brochure, a newsletter article, an operator's manual, or a video script. As an outline, however, it will be sketchy or incomplete. These gaps will point up where further research is needed. The tentative outline will also provide a framework for integrating any new details that the additional research yields.

### Computer-Aided Brainstorming

Software for personal computers can assist in the prewriting tasks of recording, sorting, evaluating, and organizing ideas. Some of these programs are used in conjunction with word processing software; others operate independently. Such software can help you (1) put thoughts randomly into a flexible framework (or listing) for expansion and rearrangement (the classic brainstorming technique); (2) clarify, test, and expand ideas about a topic; and (3) organize the ideas listed into a final working outline.

The first type of software allows you to enter what you already know about the topic in a random-order listing. After the list has been compiled, you can evaluate it, moving and inserting single ideas or clusters of ideas anywhere in the list. This process mirrors the method of asking *what, when, who, where, how*, and *why* about each idea during traditional brainstorming sessions. Such software accommodates the thinking process, which moves by fits and starts, frequently doubling back on itself. These programs can be used by individuals or by groups. Groups can work at one terminal simultaneously or separately at different times and places, in which case the work files are subsequently merged.

The second type of prewriting software goes beyond the type just described by actively engaging you in a kind of dialogue. You respond to a series of questions about a topic that are posed by the software with the goal of stimulating and focusing your thinking. The initial questions ask for information about the subject and purpose of the writing. Phrases from these and subsequent answers are reformatted into a series of follow-up questions that help you explore and record information you already know, whether consciously or subconsciously.

This question-posing software does not comprehend what you enter, of course. Instead it responds to specific syntactic information in your answers to frame a series of follow-up questions. Thus such programs may be well suited to technical writers who need a structured technique for drawing out their knowledge of past work as they embark on a new assignment. Such software also has the advantage of being open-ended. It places no limits on the size of the response as it leads you through your self-analysis. Equally important, it makes you aware of areas that you know nothing about. This awareness helps you recognize where additional information is needed.

The third type of software allows you to make the transition from brainstorming to writing by helping organize your thoughts into outlines. This software also uses language from your responses to frame questions and organize the answers into a preliminary outline.

For any of the software discussed, you can print out the results at any point in the process for review and analysis. You can examine the printout by yourself or circulate it for comment to others. How useful are such software programs? The only valid test for usefulness is to try such a program yourself. One reviewer of such programs notes that the software tends to be most useful to writers "who regularly plan with outlines, who tend to write things they are sure about, and who rarely update entries in paper and pencil outlines" (Dobrin 105).

## Mind Mapping

Some writers find a technique called *mind mapping* conducive to recording and organizing ideas created during brainstorming sessions. (It is also sometimes called *branching, ballooning, clustering,* and *cognitive mapping*.) Begin with a blank sheet of paper, typically the size of a desk pad. Think of a key word or phrase that best characterizes your topic, and put it in a circle in the center of the paper. Then put any pertinent subtopics that the main topic calls to mind in circles or boxes around the main topic, connecting each to the center circle as though adding spokes to a wheel hub. Each subtopic in turn should stimulate additional subtopics, which in turn you add in circles connected to the parent subtopic. Continue the process until no further ideas come. The technique is more effec-

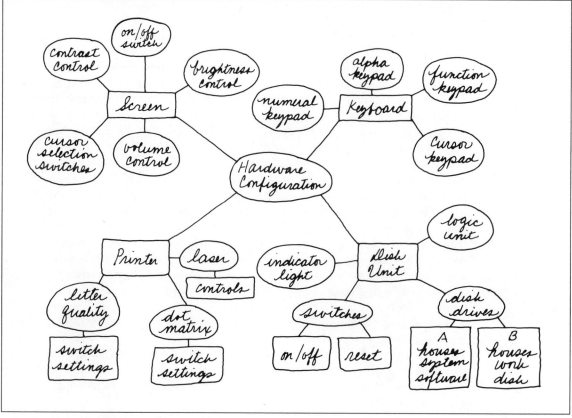

***Figure 5.4***   Mind Map of a Personal Computer Hardware Configuration

tive when ideas are expressed in one or two key words. The resulting "map" will show clusters of terms grouped around the central concept. Figure 5.4 depicts a complete grouping of ideas used to introduce the hardware components of a personal computer.

After creating the map, write out the topics in a linear outline based on your purpose, audience, and scope. A tentative outline for the sample mind map would look as follows:

PERSONAL COMPUTER HARDWARE CONFIGURATION

I.  Disk Unit
   A.  Logic unit
   B.  Indicator light
   C.  Switches
      1.  On/off
      2.  Reset

    D. Disk drives
       1. A—system software
       2. B—work disk
  II. Keyboard
    A. Alphabetic keypad
    B. Function keypad
    C. Numerical keypad
    D. Cursor keypad
 III. Screen
    A. On/off switch
    B. Cursor selection switches
    C. Contrast control
    D. Brightness control
    E. Volume control
 IV. Printer
    A. Letter-quality—switch settings
    B. Dot-matrix—switch settings
    C. Laser—controls

## OPERATING A PRODUCT OR A DEVICE

When you write a procedure or an instructional manual for a piece of equipment, an office or shop procedure, or new software, one way to gather information is to perform the procedure or operate the equipment yourself. By doing so, you become both a researcher and a participant in the project; you conduct a self-interview, in effect. By conducting hands-on research, you act as a stand-in for all future users.

Before you begin, obtain permission to operate the equipment or carry out the procedure. Find an expert to give you an overview of the task and detailed oral instructions about step-by-step procedures. Having someone talk you through the process is particularly important if any steps are potentially dangerous to you, to others, or to the equipment. Adhere to all applicable safety rules. Later, you will incorporate them into the procedures you write.

Take preliminary overview notes of the task, and make increasingly detailed notes as you become familiar with each step. Organize the information, and write a preliminary draft. Because you are so close to your draft at this point, have someone unfamiliar with the equipment or procedure follow your instructions while operating the equipment or performing the procedure as you watch. Carefully note when the person encounters difficulty, and rewrite the pertinent passages until they are clear and easy to follow. Then have someone else equally unfamiliar repeat the procedure with your revised instructions.

If no one else is available, follow your own written instructions, word for word, as though you had never performed the task. Again, make sure that experts are available to point out how to avoid potential hazards, incorporate these guidelines into the document, and have the first and all subsequent drafts reviewed by these experts. Finally, if you discover a flaw in the system, product, or equipment during your research, bring it to the attention of someone in a position of responsibility as soon as possible.

## OBSERVING PEOPLE AND PROCESSES

The only way to obtain some kinds of information is through firsthand observation. Anthropologists, biologists, psychologists, and others spend careers planning, conducting, assessing, and publishing their observations. Technical writers, too, often gather information about how people perform specific tasks, how certain manufacturing processes occur, or how specific types of equipment and systems operate.

Plan for your observations as carefully as you would for an interview. Request permission ahead of time. Observe the "chain of command" that exists in every organization when you seek permission in-house. Direct your request, or have your supervisor direct your request, to the person with the authority to grant permission. Seeking permission in advance is also important because such observations, particularly at an outside organization, could be against company policy or even illegal.

Once that hurdle has been cleared, use your analysis of audience, purpose, and scope to determine the kind and amount of information needed. If you are writing a manual for use by an equipment repair technician, for example, you must gather detailed information about each step in a repair procedure and for every contingency situation as well. Thus the more detail you gather, the better. If you are writing the script for a videotape presentation on a manufacturing process for potential corporate customers, however, you would likely gather detailed information but use it selectively in the script. After all, the audience can watch the process unfold on the screen rather than having to rely solely on a detailed narrative.

Look for the unexpected as you gather information. If the technicians performing a step-by-step task habitually take a shortcut around one step that the equipment designer did not anticipate, note this practice. Either the step is unnecessary or it is essential for reasons the technicians aren't aware of at that point in the process. If the step is essential, you must build in a special notice at that step when you write the manual, forewarning users of the consequences if they do not follow each step properly. If the shortcut harms neither the technicians nor the equipment, also note that and bring it to the attention of the designer.

Be aware that gathering information from direct observation has disadvantages, too. It is time-consuming and may be complicated and expensive. Recognize also that some people do not communicate well and that others may have a point of view that is too narrow for your purposes. Good planning, including preliminary interviews with them or their supervisors, can help you avoid such people.

## QUESTIONNAIRES

A questionnaire—a series of questions on a particular topic sent out to a number of people—is a sort of interview on paper. It has several advantages over personal interviews and several disadvantages. A questionnaire allows you to test the thinking of many more people than personal interviews would. It enables you to obtain responses from people who are geographically widespread, too. Respondents to a questionnaire do not face the pressure of someone jotting down (or recording) their every word—a fact that may result in more thoughtful answers from some respondents. Finally, questionnaires cost less than numerous personal interviews. (The discussion of customer feedback in Chapter 11 covers other ways of obtaining information from customers and clients.)

Questionnaires have drawbacks, too. People holding strong opinions are more likely to respond than those who do not, thus biasing the results. An interviewer can follow up on an answer with a pertinent question; at best, a questionnaire can be designed to let one question lead logically to another. Furthermore, mailing a batch of questionnaires and waiting for replies take considerably longer than a personal interview does. (For guidelines on creating questionnaires, see Brusaw, Alred, and Oliu 539.)

*Checklist 5.1*

### In-Person Interviews

☐ Request the interview, allowing adequate lead time.

☐ Complete background research on the topic and on the interviewee.

☐ Prepare a list of questions.

☐ Begin by asking the prepared questions.

☐ Let your body language communicate attentiveness.

☐ Be an active listener, and ask appropriate follow-up questions.

☐ Record major points and specific details.

☐ Close with an open-ended question about the topic.

☐ Ask permission for additional interviews, if necessary.

☐ Review and clarify your notes immediately after the interview.

*Checklist 5.2*

## Telephone Interviews

- ☐ Complete background research on the topic and the interviewee.
- ☐ Prepare a list of questions.
- ☐ Telephone, identify yourself, and state the purpose of your call.
- ☐ Give the interviewee an honest estimate of the amount of time you'll need.
- ☐ Keep the interview focused with your list of questions.
- ☐ Ask appropriate follow-up questions.
- ☐ Record major points and specific details.
- ☐ Close with an open-ended question about the topic.
- ☐ Ask permission to call for additional information, if necessary.
- ☐ Clarify your notes immediately after the call.

---

*Checklist 5.3*

## Interpersonal Skills for Interviews

- ☐ Promote an attitude of shared goals.
- ☐ Be confident of your knowledge and skills.
- ☐ Do not spend time talking about your knowledge when you are there to learn from the interviewee.
- ☐ Do not challenge the interviewee's knowledge.
- ☐ Do not question the interviewee's competency.
- ☐ Draw out terse interviewees.
- ☐ Have unfocused interviewees rephrase points you're unsure of, or paraphrase them yourself, asking for verification of your paraphrasing.

---

### WORKS CITED

Allen-Bradley Company. *Technical Writer's Style Manual: Guidelines for Preparing Technical Documentation*. Milwaukee, WI: Allen-Bradley Co., 1985.

Brusaw, Charles T., Gerald J. Alred, and Walter E. Oliu. *Handbook of Technical Writing*. 3rd ed. New York: St. Martin's, 1987. 539–544.

Dobrin, David. "Some Ideas about Idea Processors." *Writing at Century's End: Essays on Computer-assisted Composition*. Ed. Lisa Gerrard. New York: Random, 1987. 95–107.

Fridie, Pamela. "Interpersonal Skills for Technical Writers." *Technical Writing Teacher* 13 (1986): 316–317.

Grimm, Susan. *How to Write Computer Manuals for Users*. Belmont, CA: Lifetime Learning, 1982.

Lamping, Marilyn. "The Compleat Writer." *Proceedings of the 34th International Technical Communications Conference* (1987): RET-149–RET-153.

Levine, Leslie. "Interviewing for Information." *Journal of Technical Writing and Communication* 14 (1984): 55–58.

McDowell, Earl E., Bridget A. Mrozla, and Emmy Reppe. "Information-gathering Interviewing in the Technical Writer's World of Work." *Technical Communication* 33 (1986): 49–50.

## FURTHER READING

Anderson, Antony. "Computer-aided Lateral Thinking." *New Scientist* Nov. 1985: 64–65.

Broadhead, Glen J., and Richard C. Freed. *The Variables of Composition: Process and Product in a Business Setting*. Carbondale: Southern Illinois UP, 1986.

Dolan, Donna R. "Computer Searching and the Technical Writer." *Journal of Technical Writing and Communication* 10 (1980): 183–188.

Dragga, Sam. "Technical Writing and Library Research: Pairing Objectives." *Technical Writing Teacher* 12 (1985): 107–110.

Glossbrenner, Alfred. *The Complete Handbook of Personal Computer Communications*. New York: St. Martin's, 1983.

Hersey, Paul, and Ken Blanchard. *Management of Organizational Behavior*. 4th ed. Englewood Cliffs, NJ: Prentice, 1982.

Holcombe, Marya, and Judith Stein. *Writing for Decision Makers*. 2nd ed. New York: Van Nostrand Reinhold, 1987.

Kemp, Fred. "The User-friendly Fallacy." *College Composition and Communication* 38 (1987): 32–39.

Krull, Robert, and Jeanne Hurford. "Can Computers Increase Writing Productivity?" *Technical Communication* 34 (1987): 243–249.

Lesko, Matthew. *Lesko's New Tech Sourcebook: A Directory to Finding Answers to Today's Technology-oriented World*. New York: Harper, 1986.

McCormick, Mona. *The New York Times Guide to Reference Materials*. New York: Random, 1985.

Metzler, Ken. *Creative Interviewing*. Englewood Cliffs, NJ: Prentice, 1977.

Mills, Carol B., and Kenneth Dye. "Usability Testing: User Reviews." *Technical Communication* 32.4 (1985): 40–44.

Nellis, Marilyn K. "Using Interdisciplinary Faculty to Teach Library Research." *Technical Communication* 33 (1986): 47.

*New York Public Library Desk Reference*. New York: Webster's New World, 1989.

Oestriech, Linda. "Mind-Mapping: The Road to Well-organized Technical Publications." *Proceedings of the 34th International Technical Communications Conference* (1987): RET-66–RET-69.

Rodrigues, Raymond, and Dawn Wilson Rodrigues. "Computer-based Invention: Its Place and Potential." *College Composition and Communication* 35 (1984): 78–87.

Schwartz, Helen. "Teaching Writing with Computer Aids." *College English* 43 (1981): 239–247.

Wallia, C. J., ed. "Technology Reviews." *Technical Communication* 34 (1987): 175–176.

Wresch, William. "Computers in English: Finally beyond Grammar and Spelling Drills." *College English* 44 (1982): 483–490.

### *Chapter 5: Gathering Information*

INTERVIEWER: What source material did you use in preparing the *Introduction to Using the International Banking System* manual?

MARY: One of my best sources was the project design proposal: an outline of the new system written by the system designer for the approval of end users. It's essentially the basic content of the new system in condensed form. Users sign off on the design proposal to approve it. It's useful to me as the technical writer because it lets me see what the *users* think is important.

I also had many conversations with the assigned user reviewer. For instance, when I was having trouble fixing on the audience for the manual and the level of detail readers would need, I called the user reviewer and we discussed it until I got "unstuck." I also interviewed managers, auditors, and accountants in the International Banking Department (to learn more about the audience and set the scope).

I referred to an article on how to write manuals for non-English-speaking readers from the journal *Technical Communication*. It turned out to be only indirectly related to my task, but it was comforting to see that the article mentioned some techniques I was already using. I also referred to an accounting paper written by the system designer for potential purchasers of the system. It was useful just to look it over to find out what I didn't know.

I took a few terms from a company glossary; reading the glossary got me a bit more familiar with the business of international banking. I also went back to the section on that topic in a textbook I had from a class on banking. And I got a lot out of the *Dictionary of Banking and Financial Services*. It reminded me of what I already knew about banking and exposed me to some new terms—a good background source.

One of my most useful sources was the edited version of what the systems people had written as documentation. I circled the technical jargon in their version, then rephrased it into something that would make sense to a novice. For example, I circled the phrase "when customers and accounts are first loaded onto the system" because I wasn't sure whether readers would be familiar with the verb *loaded*.

INTERVIEWER: Are you ever your own source of information?

MARY: Yes. But I would not say that I was much of a resource in the first year of my job, because it takes time to get oriented to a new company or department.

INTERVIEWER: But after the first year, did you become a technical resource in and of yourself, by virtue of the projects you had worked on?

MARY: That would be a fair statement, but I would qualify it by noting that I can't always be sure my knowledge is current. I might know how a system worked the last time I documented it, but I would never rely solely on my own knowledge because it might be out of date.

INTERVIEWER: Do you ever borrow the organization or even passages from publications within your organization?

MARY: Yes.

INTERVIEWER: In the classroom, such copying would be considered plagiarism, and in general, using the work of other organizations and authors would be a violation of copyright. Why is that not true within an organization?

MARY: Within the organization, all information can be shared because the company already owns the rights to it, and reusing it serves our goal of presenting a unified message to our public. Also, since many of our readers use several of our manuals, it contributes to the continuity of the manuals for them to repeat duplicate information exactly instead of providing the same content in different forms.

INTERVIEWER: How important is gathering information from technical experts in your organization?

MARY: For this project, that was the only way to get the current information for a lot of aspects of the system. Even the system documentation was not current, so I relied heavily on interviews with key people.

INTERVIEWER: Could you describe briefly the process you go through to find the right person to interview?

MARY: I start with my first contact. In this case, my manager recommended a first contact, whom I asked to recommend others I should talk to. Our departmental files of existing manuals also list past contacts for those manuals.

INTERVIEWER: Do you have any tips for a new writer who is trying to get in touch with the right person?

MARY: Know your company's organization chart.

INTERVIEWER: How do you get respect from technical experts? Any advice for students?

MARY: I'm reminded of the Society for Technical Communication publication called *Respect: How to Get Some*. Do your homework; don't ask the experts to answer a question that you could find the answer to yourself; set up the meeting so that both of you know what to expect; and generally be pleasant, calm, and controlled. Also, asking for preparation materials that you can review before meeting with a technical expert is a good idea. In other words, come prepared.

INTERVIEWER: Could you describe briefly how you conduct an interview?

MARY: I set it up beforehand. I tend to go overtime in interviews, so I set a time limit with the interviewee. I write a memo presenting the agenda and give it to the interviewee prior to the interview. In technical writing, one of the cardinal rules is "Don't waste your technical expert's time." An agenda, a planned location, and a time limit all help in meeting that obligation.

INTERVIEWER: How do you take notes when interviewing?

MARY: If I expect the interview to be dense, with a lot of technical explanations, I tape the interview. Surprisingly, interviewees often feel flattered if I ask to tape them.

INTERVIEWER: What note-taking technique do you use when you're not taping?

MARY: I always use note cards. In early planning meetings, I usually take handwritten notes that I later transfer to note cards. When I ask interview questions from note cards, I usually write the answers right on the cards, in a different color ink than I used in writing the questions—just for my own ease of reading.

Often a drawing will become a source of information. For instance, if I ask a question such as "How far back can you backdate transactions?" the interviewee is likely to take a notepad and start drawing a time line for me to answer the question. I always save drawings like that—they're often an important source of information. I also find that I retain the understanding of a drawing long after I first saw or drew it.

If an interview is taking place later in the project, I may have questions embedded in my draft. In that case, I bring along the draft, with highlighted questions, and put my notes directly on the draft. Since the questions are there already, I can easily jot simple answers ("yes" or "500 feet" or something like that). Then when I'm working with the draft a few days later, I can still associate the answers with the questions.

INTERVIEWER: Do you ever interview by phone? Any tips?

MARY: It's harder, but I have done it. If you have a copy of whatever written documentation you're referring to in front of the person being interviewed, that helps. Then, as you ask questions, you can both refer to the XYZ description on page 5 as you talk instead of having to give the interviewee all the background for every question before you ask it.

INTERVIEWER: How do you tap your *own* knowledge as you gather information? Do you have some technique for getting your own ideas down on paper?

MARY: I might write down a description of the system I'm documenting *as I understand it*, outline a particular procedure, or even draw a picture of a system—any way to convey what's in my head. The end result will undoubtedly have holes in it, so at that point I can say, "I know this much, but what's missing?" and can then formulate questions based on my observations.

As I'm writing a draft, I might insert facts about the system that occur to me as I write. These facts can remind me of how the thing works. For example, I might put in a note such as "Hmm, this system works like the XYZ system." That reminds me to look at the manual for the XYZ system.

INTERVIEWER: Do you observe people or processes as you gather information?

MARY: Yes. For this particular project, I watched some International Banking clerks learn a new task on the test system. That was great—a great way to become familiar with the users, listen to typical users' questions, and so on.

# 6

# Organizing Information for a Document

The research process described in Chapter 5 produces a mass of bits and pieces of information. This information is meaningless in such a chaotic state, and it will remain meaningless until you find a way to present it to your readers that will enable them to understand it quickly and easily. In other words, it will remain meaningless until you have found an effective way to *organize* it. You may write the greatest sentences and paragraphs ever penned, but if your readers do not get the information they need because the material is not well organized, your perfect sentences and paragraphs are worthless (Huston and Southard 179).

By organizing your information well, not only do you bring all the pertinent information about your subject together, but you also present the information to your readers in a pattern that they will recognize. Such recognition makes it easier for readers to understand the subject; it also means that they are more likely to read your document (Huckin 91). You make it possible for your readers to learn in an hour or two what might have taken you a week or two to learn—the difference being that when you learned it, the information was not all in one convenient place and presented to you in a recognizable pattern, as it is for your readers after you have organized it for them. Different parts of it might have been in a report, in a specification, in someone else's mind, in a trade journal article, or in your own background, training, and experience. You had to collect all the bits and pieces of information and bring them together; yet you yourself probably did not fully understand the subject until you found a way to put all the information together in a pattern that you found recognizable and logical—a pattern that placed the most important points in superior hierarchical positions (Huckin 92, 95).

The preceding paragraph could constitute a definition of the technical writing process: the technical writer collects all of the pertinent bits and pieces of information from whatever sources may yield them and then shapes them into an organizational pattern that readers will readily recognize. When your readers recognize how you are developing your subject, they fall into step with you. And if your document is well organized, your readers *will* recognize your organizational strategy—if not consciously, certainly subconsciously (as described in Chapter 4). Your readers will instinctively look for a recognizable pattern of organization when they read your document, and they will not be comfortable until they find it (Flower 135). If they do not find it, they will be very uneasy after reading your document, unsure that they really understood it.

The purpose of organization, then, is to bring order and shape to your writing to make it easy for your readers to understand your subject. You must decide what order and shape your writing will take *before* you begin to write the draft because if you try to write the draft without having first worked out your organizational approach, you will be attempting too many tasks at the same time. You will be trying to (1) select an organizational approach, (2) structure your document, and (3) write the draft—and it is difficult, if not impossible, to perform these three steps simultaneously. You should select your organizational approach first, structure your document second, and write your draft third (Broadhead and Freed 128). If you try to do all three things at the same time, you will bog down in confusion like the many technical professionals (engineers, systems analysts, etc.) who write draft after draft as they try to sort it all out, understanding only dimly what they are trying to do. As a professional writer who must meet tight deadlines, you will not have time for such a trial-and-error approach to writing.

Organizing your information effectively before you begin to write has many advantages:

- You provide structure to your writing by assuring that it has a logical beginning, middle, and end.
- You give proper balance to your writing by making sure that one part flows smoothly to the next, without omitting anything important.
- You emphasize your key points by placing them in the positions of greatest importance.
- You make larger and more difficult subjects easier to handle by breaking them into more manageable parts.

- You concentrate exclusively on telling your readers exactly what you want them to know when you write your rough draft (in other words, you do not try to develop your subject, structure your document, and write your draft simultaneously).
- You facilitate the reader's ability to recall the information you are presenting (Huckin 91).

## ORGANIZING YOUR SUBJECT

So how should you organize your document? What organizing strategies and influencing factors should determine your organization?

### Organizing Strategies

Selecting an organizational approach is a matter of determining how you can "unfold" your subject to your readers in a way that will make it easy for them to understand it. Following are the most common organizational strategies, most of which you are no doubt familiar with already.

- General to specific
- Specific to general
- Sequential or chronological
  - Step by step
  - Question and answer
  - Myth versus fact
  - Numeric/alphabetic
- Most used to least used
- Simplest to most complex
- Cause and effect
- Problem and solution
- Division and classification
  - By task
  - By function
  - By component
- Increasing order of importance
- Decreasing order of importance
- Comparison and contrast
- Spatial

Organizational strategies are very often used in combination. Rarely does a writer rely on only one organizational strategy throughout an entire document. For example, you might divide the large subject into component parts and then arrange the parts by order of importance—but your primary organizational strategy would be division and classification. The important thing is to base your overall outline on your primary, or dominant, organizational strategy.

## Influencing Factors

Basically, three factors should influence your selection of an organizational strategy: (1) the internal logic of your subject, (2) your readers' needs, and (3) the purpose of your document (Huston and Southard 180). Each of these influencing factors has an individual impact, but they also have a collective impact on your document. Any one of them might be the primary influencing factor in selecting the most appropriate organizational strategy in different situations (although the remaining two will also influence organization).

To instruct medical or nursing students about the human heart, for example, you would present the organ as a pump made up of chambers, each of which has a certain structure and function. In this case, the natural *internal logic* of your subject would suggest division and classification as the most appropriate organizational strategy.

But to inform patients who have just had a heart attack about the structure and functions of the heart, you would elect to use an organizational strategy that sympathetically addresses the *readers' needs* and anxieties. A clinical description of the heart as a mere organ for pumping blood would not be suitable; instead, you would likely choose a sequential organizational structure that would respond directly to the questions most commonly asked by survivors of heart attacks: a myth versus fact approach or a question and answer approach, for example.

By contrast, if you were writing a sales brochure to persuade heart surgeons to purchase a cardiac catheter (an instrument used to inject dye into the heart to make an X-ray possible), your *purpose* of convincing your readers that your product is superior to its competitors would take precedence, and you would most likely use an order of importance organizational strategy that begins with the most important benefit to be realized by the surgeon selecting your product and continues to the least important benefit. Let's consider these three influencing factors individually.

### The Internal Logic of Your Subject

Most topics have a certain inherent internal logic. In the absence of any other overriding concern, such as audience anxiety, you should let this internal logic determine your organizational strategy.

For example, you could logically begin a troubleshooting manual with the problem and trace backward to its cause, or you could begin with the cause and show the sequence of events that leads to the problem. Either way, you would be using a cause and effect organizational strategy. For similar topics, you might logically use a problem-solution organizational strategy, beginning with a description of the problem and moving on to its solution.

To document a new topic that is similar to a more familiar topic, you would logically compare the new subject to the old one, considering the similarities between them and thereby enabling your readers to make certain broad assumptions about the new subject based on their understanding of the familiar one. Or you could contrast the new subject with a familiar one, considering the differences between them. For example, if you were documenting a new model of an office printer, you could explain how the new model differs from the old one and then point out ways in which the two are alike.

When describing a physical object, such as an electric motor, the internal logic of the topic would suggest that you divide it into its component parts and explain the function of each part and how all the parts work together, using division as an organizational strategy. If you were describing a group of objects, such as the tools required to repair an electric motor, you would logically group them according to common characteristics, using classification as an organizational strategy. Division and classification can also be used with abstract concepts; this book, for example, uses division and classification as a strategy for discussing the writing process.

If you were documenting software for a new computer system, the internal logic of the topic would suggest that you begin with a general statement of the function of the complete software package, then explain the functions of the larger routines in the software package, and finally deal with the functions of the various subroutines within the larger routines. In this case, you would be using a general to specific organizational strategy. In other situations you might use a specific to general organizational strategy. For example, if your subject were highway safety, you would logically begin with a specific highway accident and then go on to generalize about how certain details of that accident are common to many similar accidents. These generalizations could in turn prompt recommendations for preventing such accidents.

You would logically use a spatial organizational strategy to describe the physical appearance of a device from top to bottom (or the reverse), from inside to outside (or the reverse), or from front to back (or the reverse).

### Your Readers' Needs

Your readers' needs also influence organization. For example, consider readers who are apprehensive or fearful. If you were to write a corporate press release for people located in an area that had been selected for a chemical waste disposal site, your primary concern would be to ease their fears. You could use a question and answer or a myth versus fact organizational strategy that would raise and then respond to the issues you know to be uppermost in your readers' minds.

If your readers are skeptical or hostile, as readers of an annual report bearing bad financial news might be, you could use an increasing order of importance organizational strategy that carefully builds the case, point by point, for why the corporation is in financial trouble. Or you could use a variation on this strategy by presenting specific evidence first and then using it to build logically to a general conclusion that attempts to persuade your readers to accept your point of view.

If your readers are apathetic, as your department manager might be about a proposal you are writing, you could use a problem and solution organizational strategy, first to convince your manager that a problem exists and then to offer your solution to the problem. You could use a cause and effect organizational strategy to achieve the same end.

If your readers are uninformed novices, as readers of a training manual might be, you could begin with something they are familiar with as a basis of comparison and then move on to the unfamiliar. (With more expert readers, you would more likely use a simple step by step, or "cookbook," approach.) The less experienced and knowledgeable the reader, the more important the structure of the presentation of the information becomes (Huckin 95). You must keep in mind also that the same document might be used first as a tutorial and later as a reference work; your organization should take such a possibility into account, primarily through the use of appropriate headings, as discussed in Chapter 9 (Huston and Southard 179).

### The Purpose of Your Document

The final major influence in selecting your organizational strategy is the purpose of your document. If your purpose is to *instruct* your readers, as would be true for a software tutorial or a repair manual, you would most likely use a sequential organizational strategy that represents the steps of the procedure or operation in the sequence of performance. Sequential organization would also include question and answer sequences, myth versus fact sequences, and numeric and alphabetic sequences. If you wished to emphasize the time required for each step, as you would for instructions about developing film, for example, you would use a chronological organization.

If your purpose is to *inform* your readers, as would be true for a reference manual, you might use a division and classification, comparison, spatial, or general to specific organization, depending on the internal logic of the subject and the reader's experience with the product or situation. To inform your readers about the operation of a mechanical device, for example, you might use division and classification. To inform your readers about a new model of an existing product, you might use comparison and contrast. To inform your readers about the physical appearance of a product, you might use a spatial method of development.

If your purpose is to *persuade* your readers, as would be true for a sales brochure, you might use a problem and solution, cause and effect, order of importance, or specific to general organization. For example, if you were writing a proposal to convince your manager to fund a project that you consider worthwhile, you might begin with the most important reason why the project should be funded and work your way down to the least important reason. Or you could begin with the least important reason and build to the most important at the end, where it would be freshest in your manager's mind when he or she finished reading your proposal. Or you could use a specific to general organization, presenting the facts of your argument and building on them to reach a convincing general conclusion. The following example, from an electric utility company's annual report, uses an increasing order of importance strategy to convince stockholders that the declining profits experienced in the current year are not a cause for undue concern.

First, the decline in our earnings is a short-term problem. It has been caused by the lack of rate relief for the South Fork Project (SFP). SFP, a nuclear power generating station of which our subsidiary, Central Power and Light (CPL), owns 25.2 percent, is the largest capital investment ever made by our system. CPL has invested $2.3 billion in SFP and has annual operating expenses for the plant, including depreciation, of $140 million. CPL has earned no cash return on the majority of this investment and has had no cash recovery of its operating expenses.

Second, once CPL's rates reflect SFP, our earnings should improve. CPL filed in February 1989 for a rate increase to begin recovering its SFP costs.

Third, we are positioned for continued growth. Our kilowatt-hour sales to all our retail customers grew in 1989. The largest improvement was in our sales to industrial customers, but the growth was also accompanied by an increase in the number of residential customers our system serves.

Fourth, we have a strong record as a reliable supplier of electricity for our customers and as a reliable investment for our owners. For more than ten years, we have compared the performance of our four electric operating subsidiaries with that of other investor-owned electric utilities in the Southeast and found that our system has one of the best records for overall performance—and one of the most consistent.

Finally, we have experienced management in key positions and a program to develop our best future leaders.

### All Factors Combined

Differentiating the internal logic of your subject, your reader's needs, and your purpose in writing the document is helpful for teaching purposes. Working professional writers, however, rarely consciously separate these factors in their thinking. They intuitively understand that they may well need to consider a combination of all three influencing factors when organizing documents.

As you gain experience as a professional technical writer, you will begin to do many of the things discussed in this chapter instinctively. As you are doing your research and learning your subject, for example, you will begin subconsciously thinking, "OK, now I understand that. What organizational strategy can I use to explain it so that my readers will understand it quickly and easily?" You will think automatically about your readers' needs and the purpose of your document and what you must do to satisfy both requirements. While you are doing your research, in other words, your search for the most appropriate organizational approach will be "cooking on the back burner" of your mind, and you will often find that by the time you have finished your research, you will have already worked out the most effective organizational strategy to meet the needs of your subject, your reader, and your purpose. Until you learn to do this intuitively, however, allow yourself a period of careful and deliberate thinking about the logic of your subject, your readers' needs, and the purpose of your document in order to achieve the most effective organizational strategy.

## STRUCTURING YOUR DOCUMENT

Most of us are confronted constantly by deadlines. We are harried and harassed. It is not surprising, therefore, that when it comes to writing, people sometimes say, "I don't have *time* to outline!" As a professional technical writer, however, the truth is that you don't have time *not* to outline. Only by constructing some kind of well-planned outline—be it formal or informal—can you structure your document effectively. Writing about complex technical subjects is difficult and demanding even under the best conditions because the complexity of the subject often seems to defy a clear and simple explanation. That is why outlining is so important. The more complicated the subject is—the more difficult it is to explain—the more important a good outline becomes to the professional writer.

The McGraw-Hill Book Company even reminds its book authors of the importance of outlining.

> Before starting to write, some authors outline the chapter in minute detail and then revise the outline over and over. This is no doubt the best method for achieving good organization, proper coverage, and continuity. Some writers prepare only a broad list of topics they plan to cover and start immediately to write. In other words, they don't know what they want to say until they have tried to say it! Although this may not seem efficient, it works quite well for some people. Of course, these authors will rewrite their chapters more often than the careful organizer. (McGraw-Hill 7)

The following scenario depicts a common on-the-job writing experience shared by beginning writers. Mark, an engineer, has to write a specification. He sits down at his desk with a notepad and a pen or at his word processor, thinks for a moment, and then starts to write "off the top of his head." Before finishing, he realizes that his draft is rambling too much—but now he sees how he could have written it more logically and effectively. So he starts all over again, confident that he will do it right this time. If the specification is a sizable document on a complex technical subject, however, it is far more likely that the engineer will have the same experience again, but to a lesser degree; the second attempt will be better organized than the first, but it will still not be as well organized as it could be. So he will start yet again. And again he might get it right this time—or he might not. And so on, frustratingly, through draft after draft.

When amateur writers like our hypothetical engineer work this way, they spend far more time and effort on their writing than they can justify—because if they took the time to prepare a good outline before beginning to write the draft, they would not need to write more than one rough draft. That doesn't mean that they would write a final draft the first time; it would be a *rough* draft, and a rough draft, by definition, still needs a lot of work (tightening, refining, clarifying). The difference between a rough draft prepared from a good outline and a rough draft prepared without an outline is that the outlined draft will be better organized, needing only polishing to produce an effective final draft. A rough draft produced off the top of the writer's head, in contrast, will probably be disorganized and in need of a complete rewrite because the writer has not adequately developed the subject or structured the document. Professional writers understand that it is much easier and quicker to polish a well-organized rough draft into a final draft than it is to write draft after draft, trying not only to polish sentences and paragraphs but to arrive at a workable organization at the same time. Simply put, an effective outline is essential to an effective document.

A good outline provides many advantages.

- It exposes errors in your logic.
- It shows you where more research is necessary before you begin to write the draft.
- It indicates your starting point and keeps you moving logically so that you don't get lost in unrelated side paths before reaching your concluding point.
- It allows you to evaluate your range of material to make sure that you have covered all the critical issues.
- It may expose the fact that an aspect of your topic that you thought to be major is in fact minor (or vice versa).
- It may reveal new ideas and fresh perspectives—*before* you begin a first draft.
- It lets you move parts about so that you can experiment to see what arrangement of your ideas is most effective.
- It can be circulated for approval before you begin writing the draft (see Chapter 11).

Barabas reports the following results of outlining in a study of an organization:

> Not surprisingly, . . . over three times the number of good writers as compared with poor writers begin by constructing a written outline. Whereas none of the good writers denied using an outline or plan for their reports, 36 percent of the poor writers said they never use an outline or plan, either written or mental. (188)

Outlining is so critical to good writing—and so many writers seem to have trouble with it—that a quick review is in order.

## Creating Your Outline

The purpose of an outline is simply to group related things together, sequence them logically, and establish a hierarchy of importance (Flower 137). You should not try to create your outline until you have completed your research and recorded all your notes, of course, because until then you will not know enough about your subject to be able to create a good outline.

It is easiest, and very effective, to start with a *list* outline that simply presents the major divisions of your topic (or the main points you want to cover in your document) in a sequence that fits the organizational strategy you have decided to use. For example, if you were documenting the writing process, you might use the following list outline:

Preparation
Research
Organization
Writing
Revision

If you are ready to begin the outlining process, you should find it relatively easy to determine the major divisions of your topic; if you are unable to make such a list, you probably do not yet know enough about your subject and should return to the research stage of the writing process.

For the documents you will write as a professional technical writer, the list outline is just the starting point. The next step is to evaluate the major divisions of your subject to see if they break down further. (Again, you should be able to do this essentially off the top of your head if your research is complete; if you can't do it off the top of your head, you probably don't know enough about the subject yet to be able to outline it effectively and should return to the research stage and continue to study.) When you have broken down your major divisions and arranged the subdivisions to fit your organizational strategy, you have a *topic outline*. For example, the topic outline for a document on the writing process might look as follows. (Although roman numerals and capital letters are shown here, other methods of labeling divisions and subdivisions are also quite workable; outlining can vary from writer to writer in the particular form it takes.)

    I. Preparation
       A. Identifying your readers
       B. Identifying your purpose
       C. Identifying your scope
   II. Research
       A. Researching internal documents
       B. Interviewing technical experts
       C. Using your own background
       D. Using the product being documented
       E. Researching external documents
       F. Using the questionnaire
  III. Organization
       A. Developing your subject
       B. Selecting your organizational strategy
       C. Structuring your document
  IV. Writing
       A. Writing effective sentences
       B. Writing effective paragraphs
       C. Using effective writing techniques
           1. Emphasis and subordination
           2. Parallel structure
           3. Transition
   V. Revision
       A. Allowing a cooling period
       B. Revising in passes
       C. Checking for completeness and accuracy
       D. Checking for sentence problems
       E. Checking for mechanical problems

In creating the list outline and the topic outline, you are essentially working off the top of your head. You are able to do that because you know the subject well enough, as a result of your research, to divide it into major and minor divisions. But you also may have collected written notes during your research process, many of which you may *not* be able to recall immediately off the top of your head. To ensure that nothing important is overlooked, you may want to combine your topic outline and your notes into a *detailed outline*. To do so, gather your topic outline and your notes together. Read your first note to determine what it is about. Then decide where the subject of that particular note fits within your topic outline, and key the note to that point in the outline. Repeat this process for each note. For example, the following notes are keyed to the topic outline just given.

| | |
|---|---|
| IV.A | Sentence variety is very important to smooth writing. |
| IV.C.2 | Faulty parallel structure is a primary cause of awkwardness. |
| I.A | You must know three things about your readers: their knowledge of the subject you are writing about, what use they are going to make of the information you are giving them, and their general educational level. |
| V.D | Make one pass over your rough draft just to check for wordiness. |

When you have keyed all your notes to your topic outline, create a detailed outline by inserting the notes into your topic outline according to your keying system. When you have finished, you will have a rough detailed outline—rough because your notes will be in completely random order. Now you need to go over your rough outline and arrange your notes in the most logical sequence, indicating the sequence with Arabic numbers, as in the following example.

*Research*

| | |
|---|---|
| 2 | Can't be done if you don't understand subject. |
| 5 | Five sources of information: internal documents, technical experts, your own background, using the product, external documents, questionnaires, etc. |
| 4 | Create working outline from notes. |
| 1 | Purpose of most technical writing: to explain something. |
| 3 | Compile a complete set of notes during research. |

This may take several passes, especially if you have 10 or 12 notes under one heading, but it is worth the time it takes to be certain that everything is in the right place.

Note that when you have completed your detailed outline, it will probably be a mixture of brief phrases (the topics) and longer passages (the notes). This is fine—a good detailed outline is often a combination of briefer topic headings and longer, more detailed notes (often in sentence form).

It might help to think of the outlining process as a pyramid, with your subject at the top point of the pyramid and your detailed notes across the broad bottom of the pyramid, as shown in Figure 6.1. (Of course, *each* of the five major divisions of the writing process would need to be broken down, just as "Organization" is in the illustration.)

Your outline is a working document. Keep reviewing and improving it until you think it is right, and even then continue to be critical of it. You will find it

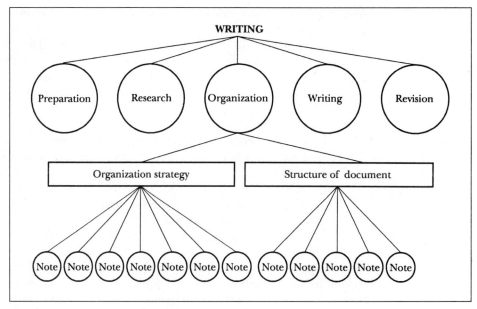

***Figure 6.1***  Stylization of the Outlining Process

much easier to see organizational logic—or the lack of it—in the outline than in the draft. Don't rush the outlining process. Keep going over your outline until you no longer find adjustments you need to make. Check and double-check it for completeness and accuracy. Make sure that it follows your organizational strategy and that it sticks to your subject, not straying into unrelated or loosely related side paths that might be of interest to you but not to your readers. Check to make certain that all major sections are in fact major and that all minor sections are in fact minor. Also check your outline for parallel structure. The less certain you are of your writing ability or of your command of your subject, the more important it is to work from an outline—and the more detailed that outline should be.

Treat illustrations as an integral part of your outline. Whenever you find that something is especially difficult to explain and could perhaps be conveyed more clearly in a graphic form, indicate in your outline that an illustration should accompany the explanation. If an idea for the illustration occurs to you, draw a rough pencil sketch into the outline; if not, just write the word *illustration* at the appropriate place in your outline and draw a box around it to indicate to yourself that you need to come up with an idea for that illustration. Then each time you go over your outline to improve it, give some thought to what that illustration might be. If you work your illustrations into your document in this way, your text and illustrations will work together harmoniously because you planned them from the start to work together. By planning your illustrations early, you might also ensure that they will be ready when you need them.

## Aids for Organizing Your Outline

When routine outlining techniques prove inadequate or do not seem appropriate, as sometimes happens, a number of techniques can help you overcome the problem.

### Brainstorming

For short documents for which the required information is all in your mind, brainstorming is often the most effective approach. Use the manual or computer-assisted brainstorming techniques discussed in Chapter 5. After you have recorded all your notes, read them over to look for common themes or topics, and list these. Then sequence these themes or topics in a logical order to create a list outline. Assign each a roman numeral or a capital letter. Then go back to your brainstorming notes and evaluate each note, either crossing it out or keying it to one of the headings in your list outline (by roman numeral or capital letter). Type each heading, followed by all the notes you have keyed with that roman numeral or capital letter. Then sequence the notes under each head in the best order. Now you have a well-organized outline.

### Flip Chart or Chalkboard

A flip chart or a chalkboard is an especially effective aid for creating an outline when you are working with a writing team because the whole team can see the information. Use the flip chart or chalkboard as you would use a notepad or a computer screen if you were working alone. If you use a flip chart, be prepared to tape pages to the wall as you work so they will be visible to everyone. Also consider using this technique when you can't seem to make a logical organization jell clearly in your mind by using more conventional techniques.

Using a flip chart or chalkboard within a group is a creative type of brainstorming: instead of recording everything you know about the subject, you are collectively applying your creative skills to arrive at a logical and effective way to present your topic to your readers. Put on the flip chart or chalkboard the best approach you can come up with. Then apply all your collective critical and creative faculties to try to improve what is there. Try different things, and don't give up if you don't immediately produce something you consider good. Be patient—even stubborn—and keep working until you have produced something you are happy with. And you will, if you don't give up.

### Experimenting with Your Topic Outline

Take a critical look at your topic outline to make sure that you have it right. See if any of your minor divisions could instead be major divisions, or vice versa. Try different sequences for your major divisions and then for your minor divisions. Try every arrangement that might *possibly* make sense, and then select the best arrangement of major divisions and the best arrangement of minor divisions.

### Computer Programs

A number of computer programs on the market are designed to help writers outline. This technology is still in its infancy, however, and most of these programs are more useful for analyzing the logic of an existing draft than for creating logic out of chaotic bits and pieces of information (Krull and Hurford 248). Succeeding versions of these computer programs might be more effective than those available today. Don't rule them out, but also don't accept them blindly, without testing them to be sure they are effective. Computer programs might be especially suitable to help in generating ideas for very large projects.

## WORKING WITH PRESCRIBED FORMATS

Some types of documents use prescribed formats that the writer must follow, although most allow some organizational discretion within the prescribed format. Use whatever degree of freedom the prescribed format offers you to make your document as well organized as possible.

*Trade journal articles* should adhere to the general organizational patterns evident in the articles regularly published in the journal. If you plan to write such an article, either send for a copy of the journal's publications standards or study the journal and adhere to the types of organizational patterns you see in articles published there.

Many *technical reports and technical specifications* follow a relatively rigid organization that has been specified by a company or a department in order to achieve consistency in the type of information included and its relative location. Within the required sections of these documents, however, you still have some organizational freedom. The example in Figure 6.2 is from one corporate department's guidelines for preparing a trip report.

---

Subject: Trip Report to: (destination of trip)

    Customer: (company name and address)

    Trip Dates: (dates you were on site)

Service Provided: (problem solving, sales assistance, installation, consulting, trade show, etc.)

Met with: (list the people you met with and their titles)

Purpose/Situation: (give a brief description of the events that led up to the trip)

Action Taken: (give a brief description of what was done to resolve the problem or what action must take place before the problem can be resolved)

Results: (give a brief description of what corrected the condition)

Conclusion: (give a brief description of anything that needs follow-up)

---

*Figure 6.2*    Sample Trip Report Guidelines

---

BULLETIN 90-1

Model 4819

Electronic Page Count Meter

The new model printers have an electronic page count meter rather than a mechanical meter. This bulletin explains how to save the page count after replacing the Operator Panel Board.

Problem

When the Operator Panel Board is replaced without transferring the D308 chip to the new board, the page count is lost. Chip D308 (nonvolatile RAM) contains the electronic page count. The page count billing can get very confusing when this chip is not transferred because the old page count needs to be recorded and the new count starts at zero.

Resolution

When you replace the Operator Panel Board, transfer chip D308 from the old board to the new one. This chip, the tallest chip on the board, will be stamped with either "nonvolatile RAM" or "D308." Place the chip from the new board back onto the old board prior to returning it to the Parts Center (failure to do this may be cause for the district to be charged the full price of the board).

---

*Figure 6.3*    Sample Field Service Bulletin

*Military manuals* must adhere to rigidly prescribed military specifications and permit very little freedom of organization. Figure 4.3 in Chapter 4 (page 66) shows sample portions of a military manual specification.

*Field service bulletins* often adhere to a general sequence of specific headings, but within that loose organization the writer has freedom of organizational expression. Figure 6.3 shows a complete sample bulletin.

*Checklist 6.1*

## Document Organization

☐ Consider all appropriate organizing strategies (see page 105).

☐ Consider all influencing factors (see pages 106–109).

☐ Select the best organizing strategy to satisfy these influencing factors.

☐ Create a logical outline to reflect the selected organizational strategy.

☐ Consider whether you must work within a prescribed format.

## WORKS CITED

Barabas, Christine. *Technical Writing in a Corporate Culture*. Norwood, NJ: Ablex, 1990.

Broadhead, Glen J., and Richard C. Freed. *The Variables of Composition: Process and Product in a Business Setting*. Carbondale: Southern Illinois UP, 1986.

Flower, Linda. *Problem-solving Strategies for Writing*. Orlando, FL: Harcourt, 1981.

Huckin, Thomas N. "A Cognitive Approach to Readability." *New Essays in Technical and Scientific Communication: Research, Theory, and Practice*. Ed. Paul V. Anderson, John R. Brockman, and Carolyn R. Miller. Farmingdale, NY: Baywood, 1983. 90–108.

Huston, Kathy, and Sherry Southard. "Organization: The Essential Element in Producing Usable Software Manuals." *Technical Communication* 35 (1988): 179–187.

Krull, Robert, and Jeanne M. Hurford. "Can Computers Increase Writing Productivity?" *Technical Communication* 34 (1987): 243–249.

*The McGraw-Hill Author's Book*. New York: McGraw, 1978.

## FURTHER READING

Anderson, Paul V. "Organization Is Not Enough." *Courses, Components, and Exercises in Technical Communication*. Ed. Dwight W. Stevenson. Urbana, IL: NCTE, 1981. 163–184.

Felker, Daniel B., et al. *Guidelines for Document Designers*. Washington, DC: American Institutes for Research, 1981.

Goswami, Dixie, et al. *Writing in the Professions*. Washington, DC: American Institutes for Research, 1981.

Harrington, Henry R., and Richard E. Walton. "The Warnier-Orr Diagram for Designing Essays." *Technical Writing and Communication* 14 (1984): 193–201.

Hays, Robert. "Model Outlines Can Make Routine Writing Easier." *Technical Communication* 31.1 (1982): 4–8.

Plung, Daniel L. "The Advantages of Sentence Outlining." *Technical Communication* 31.1 (1982): 8–11.

Seltzer, Jack. "Arranging Business Prose." *Writing in the Business Professions*. Ed. Myra Kogen. Urbana, IL: Association for Business Communication, 1989. 37–64.

### Chapter 6: Organizing Information for a Document

**INTERVIEWER:** When do you begin to think about the overall organization of a document?

**MARY:** When I'm planning the scope of the project. It's part of the whole estimating process—you have to know what the structure's going to be to get a realistic idea of what the project requires.

**INTERVIEWER:** What was your overall pattern or method of development for the International Banking System documentation?

**MARY:** After I knew, at least in a general way, what pieces needed to be in a specific manual, I evaluated the best organization for those pieces. For the *Introduction* manual, this organization turned out to be (1) overview, (2) simple tasks, (3) more complex tasks, and (4) supplemental items, such as glossaries. In other words, basics first, then more complex things.

**INTERVIEWER:** What do you mean by "basic"? Do you mean the easier steps, or the steps that are more fundamental to operating the system?

**MARY:** I mean the steps that are more fundamental to operating the system. Fortunately, the first tasks the user had to perform were also the simplest. Other tasks were more complex and also less important to know.

**INTERVIEWER:** So your sequence was both simple to complex and step by step.

**MARY:** Yes. Both influences were operating simultaneously.

**INTERVIEWER:** Was the organization for certain documents prescribed? That is, were you required to follow a given organization for some documents?

**MARY:** Our new standards and guidelines include templates (format models) for typical documents. The templates display the side margin captions we've been using for years, and they also show some formats adapted from Information Mapping, a writing methodology developed by a group called Information Mapping, Inc. (IMI). The methodology's main thrust is to put certain types of information in certain formats: procedural steps into procedural tables, decisions into "if-then" tables, and so on. Another fundamental precept of Information Mapping is to keep information in chunks, separating the chunks with a thin line (so readers can see that they are going to deal with only a little information at a time) and putting no more than five to nine chunks on a page. We've reduced that number to three to seven because seven seems like plenty on the pages that we're doing.

**INTERVIEWER:** How does the subject matter itself help you determine the organization of a document?

MARY: You take the whole system, even though you're still learning about it, and you think of some potential ways to explain it. I think my choice has to do with how *I* understand the system, how *I* was able to process the information. I must have broken it down in some way as I absorbed it. Maybe it's just a process of coming up with the *schema* by which I've absorbed the information. At any rate, I think of it as an intuitive process.

INTERVIEWER: How does the purpose of a document help determine organization?

MARY: If it's reference material, the driving factors are these: What can be presented concisely? What's easy to look up? As a result, you might go toward an alphabetic or numeric organization for the material. You'd try not to duplicate information.

But if it's procedural, the driving factors might be these: What concepts do readers have to know before doing tasks? What tasks do they have to do? What tasks are done most often? Then you don't put anything else in; you stick to the bare bones of the procedure.

INTERVIEWER: Do you use outlines to organize your work?

MARY: It's the only way to write efficiently. I hardly ever put a *traditional* outline onto paper, but I work with task lists in which the tasks are grouped into clusters. A table of contents is formed from that list and reviewed (and the order of clusters is refined). In the table of contents, I figure out what the main ideas and the minor ideas are.

INTERVIEWER: So you use nontraditional outlines to organize the material. How important are those outlines?

MARY: All forms of planning—outlining, researching, listing, and so on— save time. Nothing wastes time more than rewriting a gloppy ten-page mess over and over again. You can get into some real problems if you try to write that way.

INTERVIEWER: Could you define "gloppy"?

MARY: [Laughter] Unstructured and disorganized. I run into such writing more often in a system overview, for which many approaches to organization exist, than in a description of a procedure, which goes from point *A* to point *B*. There have been times when I had to write out an outline for an existing document so that I could figure out what was going wrong with it.

INTERVIEWER: How would you respond to the skeptic who says, "I really don't have time to outline"?

MARY: I guess I would say you would ultimately save time by organizing before you get to the detail level. Because it's a lot harder to move chunks of information around in a text when they are already in full paragraphs. Lots of times you've already written transitions between paragraphs. Then, when you move a paragraph to another page, you have a lot more rewriting to do. It's just a lot easier to rearrange things when you're working at the more general level of an outline.

INTERVIEWER: Can you offer any general tips on outlining?

MARY: Note cards are the best advice I can give. They are so flexible. I generally use my note cards more or less as a working outline. Then I draft the document. And then I look at the structure of the draft. In other words, I don't always button everything down before I start, but I do have a flowing organization before I start writing.

For smaller items, such as a progress report, I might enter ideas in the online document, sort them, and then flesh them out with details. That's one advantage of writing with a text processor: you can stretch entries into paragraphs without having to rewrite.

INTERVIEWER: Do you incorporate visual aids or illustrations into your outline?

MARY: It's part of organization to decide where an illustration goes in the text. Are you going to describe all the fields and then show the screen sample? Are you going to define the screen, tell how to

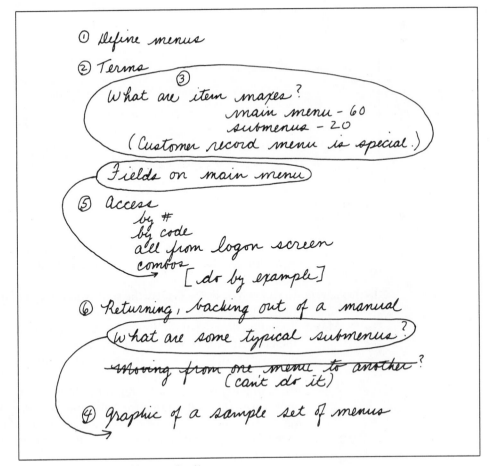

*Figure C.3*   A Rough Chapter Outline

## Using Menus_____

_____

**What's in this section?**

▸ definitions of menus and menu terms

▸ a sample menu structure

▸ examples of the four ways you can use menus

▸ how to return from a screen or submenu to a
previous menu

*Figure C.4*  Final Table of Contents for the Chapter

access it, and then show it? In the outline, I might draw a box and enter the screen name at the point where I will have a screen sample.

**INTERVIEWER:** Could you describe briefly the steps you took to plan the organization of the *Introduction* manual?

**MARY:** First, I made a list of the topics I wanted to put in the manual—all the tasks that users needed to know. Next, I put tasks into related groups. Then, within each group (which eventually became a separate chapter within the manual), I numbered the tasks, placing them in the order in which I felt they might best be explained. (In most cases, that order came from the notes I took as I was carrying out the procedure myself.) These listings of related tasks were essentially my rough outlines for the different chapters of the manual. [Figure C.3 shows Mary's rough outline or listing for the chapter "Using Menus" in the *Introduction* manual.]

INTERVIEWER: Are the skeletons of these rough outlines still visible in the document?

MARY: Yes, they're right there in the tables of contents that appear at the beginning of all the chapters. [Figure C.4 shows the final table of contents for the chapter "Using Menus" in the *Introduction* manual.]

# 7

# Integrating Visuals

What Kinds of Visuals Should I Use?

Where Will the Visuals Appear?

When Should Visuals Be Prepared?

Who Will Prepare the Visuals?

Visual aids, or graphics, often express ideas or convey information in ways that words alone cannot. They communicate by showing how things look (in photographs, drawings, and maps), by visualizing numbers and quantities (in graphs and tables), by depicting relationships (in charts, schematics, and diagrams), and by making abstract concepts and relationships concrete (in visualizations and organizational charts). They also condense information and emphasize key concepts by setting them off from their surrounding text. Information made concrete and concise—and set off for emphasis—focuses reader attention, promotes understanding, and helps readers remember its significance.

The power of graphics to enhance communication was demonstrated in a study conducted several years ago at the Wharton School of Business (Oppenheim et al.). When three test groups were presented two sides of an issue by two teams not using graphic aids, the test groups divided evenly on the issue. When each team, in turn, supported its position with graphics, it won decisively. The authors went on to make the following points about the role of graphics in communications:

- They are persuasive.
- They aid decision making.

- They promote consensus.
- They make a good impression.
- They shorten meetings.

Whether they are used in presentations or in documents, graphics can produce other on-the-job benefits as well. Carefully chosen and professionally designed visuals allow firms to produce smaller, easier-to-read documents; promote communications to an international audience, as they require little or no translation; and train a larger pool of people more effectively, in that they allow for a lower level of competency among trainees (Bennet and Samolyk 29).

The occasions for which visuals can be used and the types available for these occasions are limited only by the human imagination. Accordingly, this chapter does not attempt to exhaust the entire range of visual options available to professional writers or the graphics principles governing each of these options. Instead, it answers specific, practical questions that arise for anyone preparing documents that contain visual materials:

- What kind of visuals should I use?
- Where will they appear?
- When should they be prepared?
- Who will prepare them?
- How do I work with graphics designers?
- How can I prepare them?

Answering these questions—as well as the additional questions their answers stimulate—will ensure that the graphics chosen mesh smoothly with the surrounding text to create an effective document.

## WHAT KIND OF VISUALS SHOULD I USE?

To select the most effective visuals to communicate with your readers, consider the forms best suited to the type of information in the document, as outlined in the following table.

| *Type of Information* | *Type of Visuals* |
|---|---|
| Literal and hypothetical representations | Photographs<br>Drawings and renderings |
| Geographical information | Maps |
| Organizational information | Charts |
| Sequential information | Flowcharts |
| Quantitative information | Tables<br>Graphs<br>Charts |
| Symbolic information | Schematic diagrams |

Think about your audience when deciding which of the graphics options available will be most effective in a given context. How might the audience influence your selection? Consider the following example.

You are writing a public information brochure for the Community Education Department of the hospital where you work about the hospital's new kidney stone treatment program. The treatment, called lithotripsy, uses high-energy sound waves to shatter kidney stones and does not require surgery. Several options to illustrate the brochure come to mind. You think about including a photograph of the physician who heads the program but dismiss the thought quickly. Such a photograph would convey no information about the new treatment and would be more appropriate for an article in the hospital's employee newsletter than in a brochure going to the public. A photograph of the apparatus used to generate the sound waves likewise conveys no useful information about the treatment. Nor does a photograph of a kidney make sense. Not only would such a photo show extraneous detail, but the fact that kidney stones form *inside* the kidney means such a photo would show nothing meaningful. X-ray photographs of a kidney before and after treatment might be an acceptable alternative, but upon further thought you reject this option, too. Although many people have seen X-ray photographs of a baby's stomach showing a safety pin or coins, most medical X-rays are hard for lay audiences to interpret. Your goal is simplicity, yet accuracy, for a nonexpert audience. How about a cutaway line drawing of a kidney? The chief advantage of this option is that it can show the interior of the kidney. But how much of the interior is essential to show? Medical textbooks contain cross-sectional drawings of kidneys, but these, you reason, offer far too much detail. Instead, your audience and purpose lead you to choose spare, simple line drawings of a kidney before, during, and after treatment (see Figure 7.1).

Although many different types of graphics exist, each identified by a different name, most graphics are variations on one of a number of basic forms. These forms are presented on pages 128–132, along with a sample illustration and a listing of the primary features of each.

The visuals chosen for your documents should conform to the standard conventions for depicting information graphically. The coverage in this chapter assumes that you are familiar with basic graphics conventions and will already understand, for example, that flowcharts are read from top to bottom and left to right, that the segments of a pie chart should total 100 percent, and that the zero point for a line graph appears at the intersection of the $x$ (horizontal) and $y$ (vertical) coordinates. The following guidelines apply to most graphics materials, whether they are used to supplement text in a document or for presentations at meetings.

- Keep the information concise and simple.
- Present only one type of information or key relationship in each visual.
- Label or caption each visual informatively but concisely.
- Include a key that identifies all symbols used.
- Specify the units of measurement used or include a scale of relative distances.

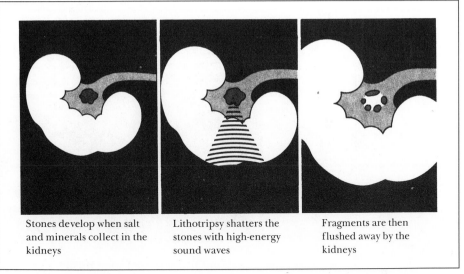

Stones develop when salt and minerals collect in the kidneys

Lithotripsy shatters the stones with high-energy sound waves

Fragments are then flushed away by the kidneys

***Figure 7.1*** Cutaway Line Drawing of a Kidney for a Nonexpert Audience, with Explanatory Captions

- Keep terminology consistent. (For example, do not refer to something as a "proportion" in text and a "percentage" in the graphic.)
- Omit visually distracting details.
- Allow adequate white space around and within the graphic.

Additional information on specific graphic forms and their conventions can be found in Brusaw, Alred, and Oliu and in other valuable sourcebooks listed in the "Further Reading" section at the end of this chapter.

## WHERE WILL THE VISUALS APPEAR?

There are two important considerations when thinking about where visual aids will be used: where they appear within a document and the setting in which the document itself will be used.

### Placement of Visuals

The location of visual aids in most documents follows a set of well-established guidelines.

- Position visuals to appear as close as possible to their first mention in the text. They should not precede their first text mention, however, because their appearance without an introduction will confuse or distract readers.

*(text continues on page 133)*

## *Photographs*

○ Show the actual physical image of something
○ Record the development of phenomena over time
○ Record an event as it occurs
○ Record the immediate aftermath or as-found condition of a situation for an investigation

## *Line Drawings*

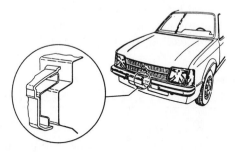

○ Depict literal or imagined images
○ Depict objects or situations difficult or impossible to photograph
○ Emphasize and highlight certain features of a complex topic by focusing on only the parts viewers need while eliminating unnecessary details
○ Show internal parts of equipment in a way that makes their relationship to the overall equipment clear using cutaway drawings
○ Show the proper sequence in which the parts of complex equipment fit together using exploded views

## Maps

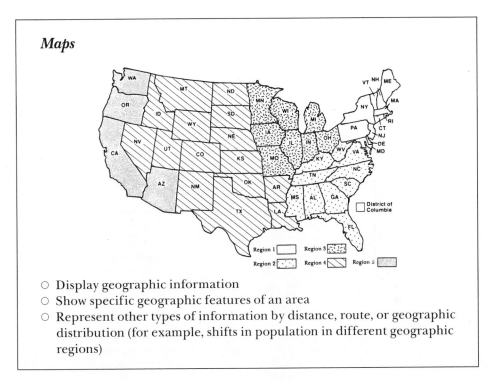

○ Display geographic information
○ Show specific geographic features of an area
○ Represent other types of information by distance, route, or geographic distribution (for example, shifts in population in different geographic regions)

## Organization Charts

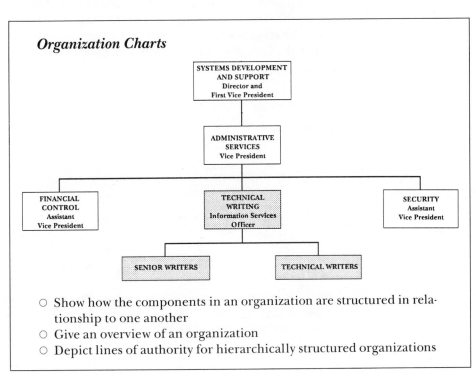

○ Show how the components in an organization are structured in relationship to one another
○ Give an overview of an organization
○ Depict lines of authority for hierarchically structured organizations

## *Flowcharts*

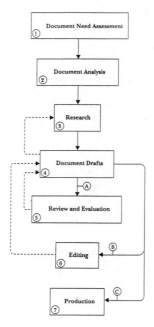

○ Show how the parts or steps in a process or system interact
○ Show the steps or stages of a process in the correct direction of flow, with the sequence of steps shown from beginning to end, including when the flow must double back to repeat one or more steps

## *Tables*

| Storage facilities | Gross volume planned | Gross volume completed |
|---|---|---|
| **Phase I sites:** | | |
| Bayou Choctaw | 48.3 | 48.2 |
| Bryan Mound | 74.5 | 72.8 |
| Sulphur Mines | 27.4 | 27.3 |
| Weeks Island | 73.1 | 73.1 |
| West Hackberry | 51.1 | 50.6 |
| **Total** | **274.4** | **272.0** |

○ Display quantitative information in a way that facilitates comparisons
○ Present large quantities of specific, related data in a form more concise than if presented in text
○ Flesh out trend and graphical information with precise data

## *Schematics*

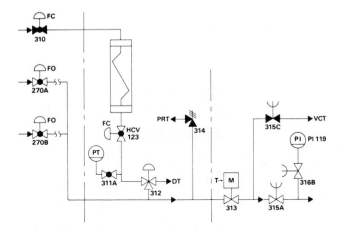

○ Show how components in electronic, chemical, electrical, and mechanical engineering systems interact and are interrelated
○ Use standardized symbols rather than realistic depictions of system components

## *Line Graphs*

○ Show trends over time in amounts, sizes, rates, and other quantifiable measurements
○ Allow comparisons among one or more kinds of data over the same period of time
○ Show trends, forecasts, extrapolations of data
○ Give readers an immediate visual impression of the numbers and their relationship
○ Complement data shown in tables

## *Bar and Column Graphs*

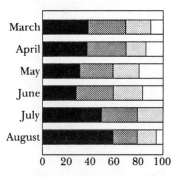 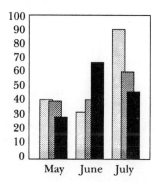

Vertical and horizontal bars (or columns) show numerical trends or relationships in several different configurations.

○ The same type of data at different times (for example, a county's annual school budget over a decade)

○ Quantities of different types of data for the same period of time (for example, energy consumption for residences, industries, and transportation for one year)

○ Varying quantities of the same kind of data during a fixed period of time (for example, the population of several cities over the same decade)

○ Quantities of different parts that make up something (for example, a bar divided into the appropriate number of cents to show how the typical municipal tax dollar is spent)

## *Pie Charts*

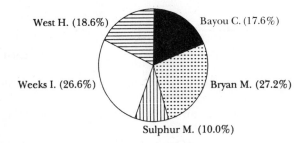

○ Show parts that make up a whole
○ Give readers an immediate visual impression of the parts and their significance
○ Complement data in a table or list

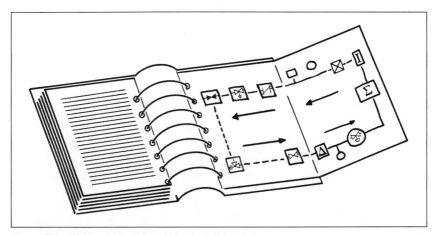

***Figure 7.2***    Foldout Page at the End of a Manual

- Place visuals that readers must refer to throughout a document in an easy-to-reach location—for example, on pages at the back of the document that fold out and lie flat for convenient reference (see Figure 7.2).
- Position visuals in "portrait" (vertical) orientation so that readers can view them without having to turn the page sideways (see Figure 7.3). If the information will not adequately fit in this orientation, position it in "landscape" (horizontal) orientation (see Figure 7.4).

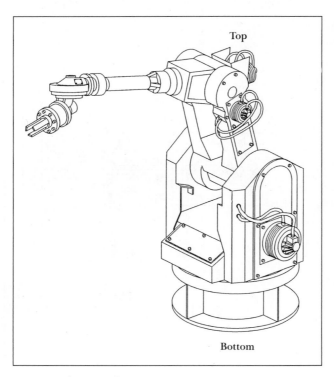

***Figure 7.3***
Portrait Orientation

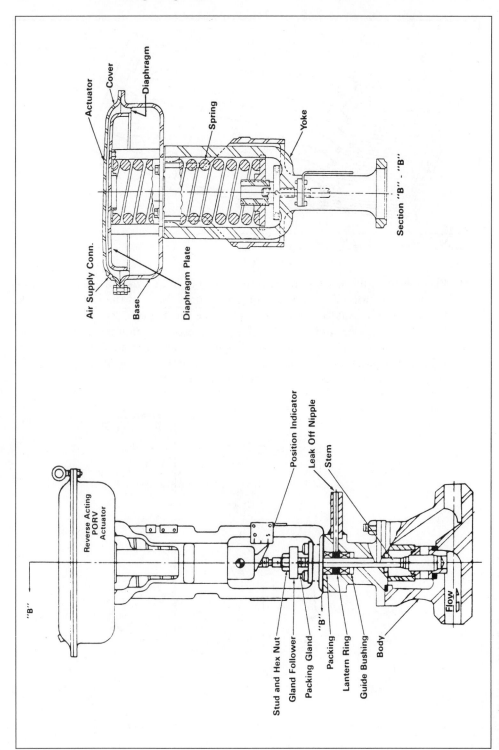

***Figure 7.4*** Landscape Orientation

*Figure 7.5* Image Properly Straddling Gutter Seam

- Large visuals can also appear straddling facing pages (see Figure 7.5), although this practice should be avoided because imperfect alignment between left- and right-hand pages may occur. Unmatched alignment of facing pages not only looks unprofessional but also makes information that straddles the gutter seam (the vertical fold dividing the pages) difficult to follow.
- Visuals may be set off from the surrounding text by boxing them in, by allowing white space around them, or by placing them in a screened (or shaded) area.
- Provide visuals with concise descriptive titles. (Detailed explanations of what the visual depicts or of its significance should appear in the text.)
- Give each visual a figure number if the document in which it appears contains more than a few figures.
- Refer to the visual in text by the figure number or title, as appropriate.
- Place visuals that are voluminous (such as multipage computer printouts of data) or peripheral to the focus of the discussion in an appendix.

## Circumstances of Document Use

The setting in which readers use the document can affect how the visuals are prepared, too. Desk or benchtop use under normal lighting conditions may require few special graphics preparations beyond the necessity that foldout pages lie flat. Careful planning is necessary, however, when readers view graphics in a confined area (airplane cockpit, crane cabin), under low or unusual lighting (red night-vision lights), or under other conditions where normal reading may be difficult.

> [For] an electrical engineer who is troubleshooting some computer circuitry or . . . a mechanic working in the wheel well of an aircraft's landing gear, . . . legibility and the ability to locate relevant information in a graphic at a glance without turning pages is extremely important. (Winn 37)

On the job, seek out the help of experienced graphics designers or human factors engineers to meet such special requirements and collaborate closely with them. Provide them with explicit information about the setting since it will affect many and perhaps all of the following features:

- Type of visual (photograph, line drawing, etc.)
- Size and proportion of the graphic
- Size and style of the typeface for captions and callouts
- Weight (thickness) of lines
- Color of ink or inks
- Color of paper
- Size of paper
- Use of alternatives to color (crosshatching, dots of varying density)
- Type of paper stock (dull or glossy)
- Use of recognizable words, abbreviations, symbols, icons
- Reader expectation of graphic conventions (right-to-left and top-to-bottom orientation)

## WHEN SHOULD VISUALS BE PREPARED?

Begin thinking about your visuals requirements as you gather and organize your written information. Treat visuals as an integral part of your outline, noting approximately where each should appear throughout the outline. At each such place, either make a rough pencil sketch of the visual, if you can, or write "illustration of . . ." and enclose each suggestion in a box. Like other information in an outline, these boxes can be moved, amended, or deleted as required. Planning visuals from the beginning stages of your outline will ensure their harmonious integration through each successive version of the draft to the finished document.

Another important advantage of thinking about visual requirements at the outline stage of a project is that it allows time for visuals to be prepared for you by designers or obtained from other sources. Although preparing a cover and a few bar graphs for a report might not take longer than an hour or two for a graphic designer, the assignment could take much longer if the visuals staff is busy. Discuss your requirements with the art staff early and plan accordingly. If you intend to use an illustration or data from a copyrighted publication, allow time in your schedule for obtaining permission to use it. (See page 137 for more information about obtaining permissions.)

# WHO WILL PREPARE THE VISUALS?

You have a variety of sources for visual materials. You can use existing visuals or stock art. You can have art prepared by a visuals specialist. Or you can prepare visual materials yourself.

## Using Existing Visuals

Consider using visual materials from existing publications. Sources for such materials include in-house publications, works published elsewhere, and stock art and photographs from commercial sources.

### In-House Publications

As you review in-house documents in conducting research for your project, you may discover visual aids in them that fulfill your needs. If your organization has a file system for previously used in-house artwork, review that too. Both sources may contain work that can be used as is or with a little modification; either way, using such art will save you time and money. If your organization does not keep artwork in a central file, develop your own file. "Whether used or not, art represents money spent that may not need to be spent again if you can find a use for it (or parts of it) later" (Pickens 216).

### Visuals Published Elsewhere

If you find an illustration, photograph, map, or graph in an outside publication that perfectly suits your needs, ask the publisher for permission to use it. If the work is copyrighted, you must secure a copyright release, for which you may have to pay a fee. The sample permission letter in Figure 5.1 in Chapter 5 can be adapted for this purpose. Generally, permission is granted for one-time use only. Subsequent use requires another release and may entail an additional fee. (Acknowledge such borrowings in a credit line below the visual in your finished publication.) You will also have to obtain permission far enough in advance to meet your document deadline.

Visual materials published by the federal government—the Bureau of the Census, the U.S. Geological Survey, and other agencies—are not copyrighted. You need not obtain written permission to reproduce them, although you should acknowledge their source in a credit line.

When writing for permission to use photographs or visuals with color, you may need to ask for the camera-ready (or prepress) artwork as well. This is the original art version prepared for the printer, and it alone will suffice for continuous-tone visuals, a category that includes black-and-white photographs and color illustrations. (For photographs, ask for the halftone negatives, and for color illustrations, ask for the color separation.) Line art, a category that includes pen-and-ink drawings, charts, graphs, schematics, and other visuals made up of solid black lines and images on a white background, will reproduce faithfully from a photostat (a copy made on photographic paper) of the visual in the document from which it is being copied. You must receive prior permission in writing to use such material if it is copyrighted, of course.

Finally, be aware that visuals borrowed from another document may look different in style from the other visuals in your document in many respects. You may have to modify at least the type style and the size of the callouts and captions to get as close a match as possible. Do not take liberties with the actual words or data of the original, however.

## Stock Visual Materials

Stock art refers to the thousands of existing "generic" designs, illustrations, photographs, and other materials that are available from art supply dealers, bookstores, and mail order suppliers. The variety available is virtually limitless. It includes lettering and numbers in various styles and sizes, symbols, icons, designs, emblems, pictures, arrows, ornaments, borders, decorative devices, rules, and brackets. A small sample of stock design available is shown in Figure 7.6. Some of these materials, called press-on or transfer art, appear on transparent paper sheets from which they are rubbed (pressed) or lifted off (transferred) to another surface. These designs are not copyrighted and offer both student and professional writer significant advantages.

> Such transfer art will be especially valuable if you have limited needs, skills, and budget. . . .
>
> For symbols to complete a map, arrows to clarify a diagram, or printers' marks to break up columns of type, an investment in transfer art will be minimal compared with the bill you would otherwise have from an artist. (Pickens 217)

## Working with Graphic Designers

Graphic designers are trained to communicate information visually by using images, symbols, and a variety of typefaces. Your close collaboration with them will ensure that the visuals in your documents reinforce the text. The guidelines in this section apply equally to your collaboration with in-house artists, with those who work in commercial design studios, or with freelancers.

Graphic designers need detailed information about your graphics requirements. Plan to meet with them and, if possible, submit thumbnail sketches, engineering drawings, photographs, or sufficiently detailed written descriptions. If no such starting points are available, provide the artists with access to the equipment, location, or facility that is the subject of the assignment, or arrange to have them spend time with a technical expert who can provide sufficient information.

Graphic designers must understand the dimensions and relative proportions of objects and the angle of view necessary for the viewer. When adapting engineering drawings or other densely detailed drawings, for example, designers need to know which details to emphasize and which to eliminate. The intended audience for the document will help the designer and author determine the

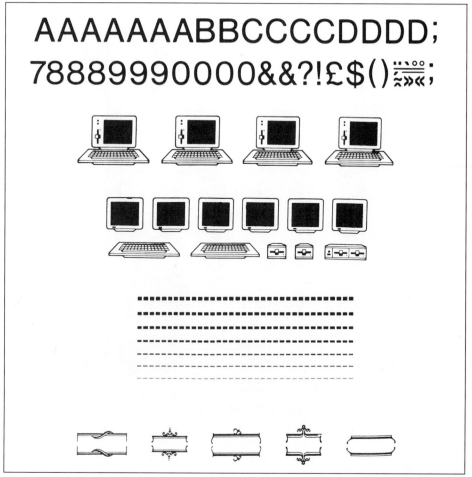

***Figure 7.6***   Sample Stock Art

answers to most such questions. As a rule, less technically knowledgeable audiences require simpler, less detailed visuals than knowledgeable readers, although exceptions to this rule exist. Generally, beginners learning complex technical subjects require detailed information, but they need it in manageable steps in the proper sequence.

The level of detail required also affects the size of the visual. Electrical schematics for large systems or risk assessment decision trees (also called logic diagrams) require adequate space if they are to be legible. Oversize or foldout pages (as shown in Figure 7.2) may be required for these kinds of visuals. The decision tree diagram shown in Figure 7.7, for example, clearly requires an oversize page for legibility.

The environment in which the visuals will be used must also be taken into account. It will influence the artist's use of the features listed on page 136. When

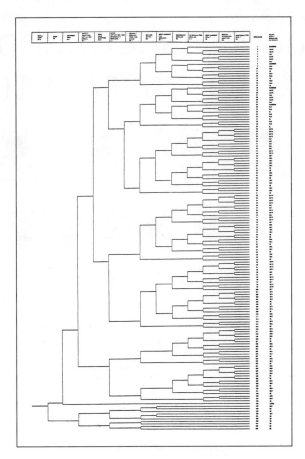

**Figure 7.7**

Example of a Graphic Reduced Too Small for Legibility

you submit more than one assignment at a time, explain whether each figure will appear in one document or in a related series so that all can be designed consistently. Otherwise, type size and style, shading, line widths, and other details may vary from one figure to the next in the finished document, making for an inconsistent, amateurish look. Work with the designer at the beginning of each project to establish design standards and visual consistency throughout. These design standards will usually be affected by the overall layout and design specifications chosen for the document. The visuals then become one of the elements of the document, like text columns, headings, and type size and style, that are arranged on the page in accordance with the prescribed design specifications. Chapter 8 provides a detailed discussion of document layout and design principles.

Professional etiquette and mutual respect should govern working relations between writers and artists. Adhering to the following practical guidelines will fulfill your part of the relationship and win the respect and cooperation of your graphics colleagues.

1. Submit your work in person so that you can explain its purpose and answer questions.
2. Submit all work on only one side of each sheet, and type or write the title of your publication at the top.
3. Write your name and phone number on the back of each sheet for the artist's convenience.
4. Specify the final size of the document so that the artist knows the image area within which the finished work will be viewed. An 8½-by-11-inch page for a newsletter or a report allows for more flexibility than the image area for a book (with 6-by-9-inch pages) or a typical three-panel brochure (with 3¼-by-8½-inch panels).
5. Carefully proof everything, including the spelling on callouts and captions, before submitting it. Wording and numbers will appear exactly as you submitted them.
6. Draw a circle around instructions meant for the artist to distinguish them from wording to appear on the graphic.
7. When proofing artwork received from the graphics staff, photocopy the original and make corrections on it. *Never* write on the original.

Enlist artists' unique abilities in your cause. Artists conceptualize visually and can bring new insights to a project to enhance your communication goals. That's why it's important to provide them with as much information as possible about your requirements, particularly if they see only the visual components and not the text for the project. The better able you are to set the context of these visual requirements for them, the more likely it is that they will produce the most appropriate visuals for your project.

## Preparing Your Own Visuals

Assignments in settings where you do not have access to professional help with visuals, such as the classroom or a small business office, may require that you produce your own credible visuals for your documents. In those cases, you can either prepare them manually using a host of aids available from art supply houses, or you can prepare them by using one of a growing number of computer graphics software programs.

### Manual Visuals

Many types of charts, graphs, and drawings can be prepared using nothing more elaborate than a T-square, a ruler, a compass, and a few curves and templates. Figures 7.8 and 7.9 show examples of hand-drawn illustrations that could be submitted to a graphics designer for professional rendering or are adequate as is for many informal reports. The basic designs created with these tools can be embellished with press-on or rub-on letters, arrows, symbols, borders, and other stock visual materials.

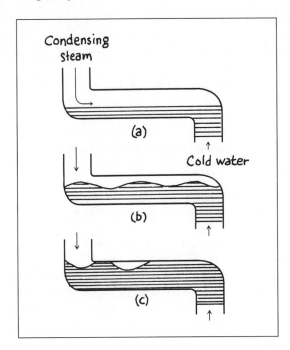

*Figure 7.8*
Hand-drawn Piping Sections

## Computer Graphics

The widespread availability of graphics capabilities on personal computers makes their use by non–graphics professionals increasingly common. With access to the proper hardware, software, and printer, you can create tables, graphs, charts, drawings, and even maps of near-professional quality. Once an image has been created on the computer screen, you can save it in a file, print it out again and again, or recall it for subsequent modification.

Computer graphics for personal computers fall into two broad categories: business graphics and free-form graphics. We shall provide a concise introduction to graphics programs for personal computers and show examples typical of each kind.

*Business Graphics*   Business graphics programs, also called charting programs, produce preformatted charts, graphs, and tables based on numerical data taken either from a computer spreadsheet program, as in Figure 7.10, or from text and data entered from the computer keyboard. Figure 7.11 shows a range of typical bar and pie charts created by business graphics software.

*Free-form Graphics*   In free-form graphics, as the name implies, images are not automatically preformatted to become charts, graphs, and tables that reflect numerical data. Instead, the user controls the image by manipulating a variety of strokes, lines, shapes, and symbols, much as an artist wields pencil, brush, charcoal, templates, and other tools to create images on paper or canvas. Many such

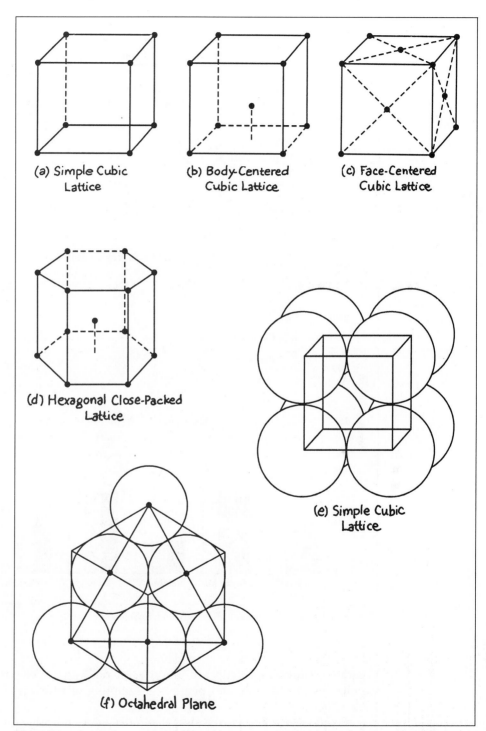

*Figure 7.9* Hand-drawn Crystal Lattices

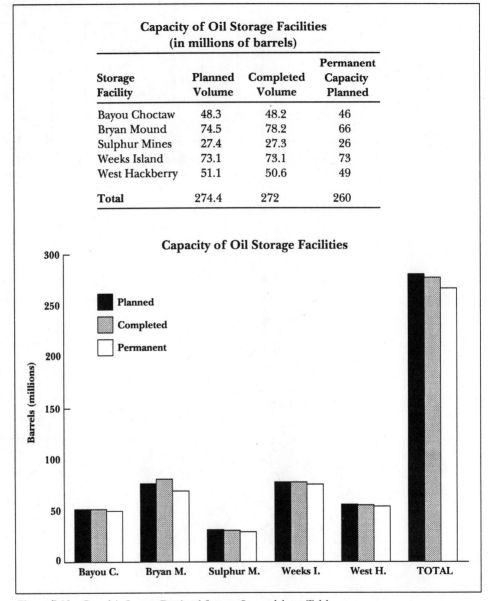

**Capacity of Oil Storage Facilities**
**(in millions of barrels)**

| Storage Facility | Planned Volume | Completed Volume | Permanent Capacity Planned |
|---|---|---|---|
| Bayou Choctaw | 48.3 | 48.2 | 46 |
| Bryan Mound | 74.5 | 78.2 | 66 |
| Sulphur Mines | 27.4 | 27.3 | 26 |
| Weeks Island | 73.1 | 73.1 | 73 |
| West Hackberry | 51.1 | 50.6 | 49 |
| Total | 274.4 | 272 | 260 |

*Figure 7.10*    Graphic Image Derived from a Spreadsheet Table

programs also contain "clip art" libraries of ready-to-use electronic images of various symbols, shapes, and pictures, a sampling of which is shown in Figure 7.12.

Free-form programs are subdivided into paint and draw programs, categories defined by the software technology each uses.

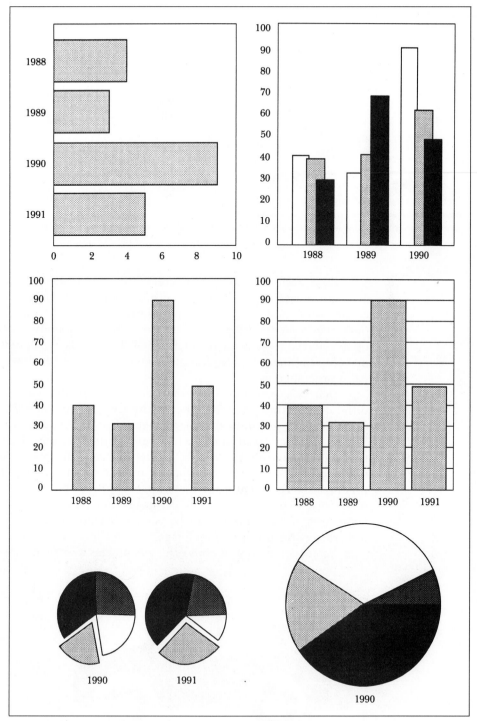

***Figure 7.11*** Typical Computer-generated Bar and Pie Charts

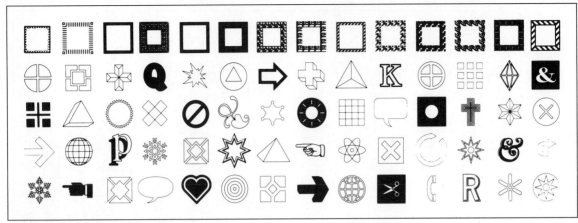

***Figure 7.12***    Sample Computer Clip Art (Courtesy Adobe System, Collector's Edition 1)

*Paint programs* allow users to create freehand images on screen, usually with an input device called a mouse (defined on page 148). You can control images created with paint programs down to the individual dot, "bit," or picture element (known as a *pixel*) on the screen. Because of this capability, these programs are often referred to as pixel- or bit-mapped. The pixels are similar to the dots that make up the images on a television screen. Figure 7.13 shows an example of a paint program image.

*Draw programs* allow users the freedom to create images on the screen but differ from paint programs in that the user manipulates predefined shapes, called *graphics primitives* (boxes, triangles, arcs, lines, etc.), rather than pixels. That is, the user creates images by manipulating existing "building block" configurations, like those shown in Figure 7.14. The images are manipulated and saved to memory by the coordinates, or vectors, of the beginning and ending points of the lines that make them up. For this reason they are called vector- or object-oriented programs. These images can be combined to form graphs, flowcharts, and organizational charts, as well as line drawings. Figure 7.15 shows a characteristic draw program image.

Vector-oriented programs offer advantages over pixel-mapped programs in

***Figure 7.13***
Paint Program Image

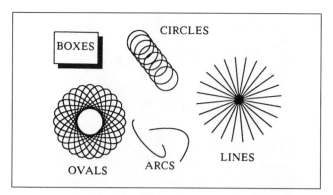

*Figure 7.14*
Graphics Primitives

how images are manipulated and in the clarity of the image when enlarged. Consider a user who needs to move a circle that's superimposed over a square so that each figure stands alone. Using a pixel-mapped program, the user "must make a copy of the two shapes, and then use an 'eraser' to remove the circle. . . . This is necessary because the computer cannot tell the difference between the dots that make up the circle and the dots that make up the square" (James 18). With a vector program, by contrast, the user "selects the circle and moves it away from the square. This is possible because the computer can differentiate between the two objects—they are not just dots on the screen" (James 18). The clarity of pixel images also suffers when they are enlarged. An image consisting of dots can be enlarged only by increasing the size of the dots. The resulting image looks "jagged" when printed, as in Figure 7.16.

*Specialized Graphics Programs*   Other graphics programs, generally beyond the capacity of personal computers, are available for creating complex maps, statistical analyses, animation, and computer-aided design (CAD). They offer specialized applications for the design, engineering, architectural, or scientific professional. Their capabilities are beyond the scope of this text.

*Graphics Manipulation*   Once the basic components of an image have been created, most graphics software allows the user to manipulate them on screen in a variety of ways. Some or all of the following options may be available, depending on the software used.

*Figure 7.15*
Draw Program Image

**Figure 7.16**
Enlarged Pixel Image

- Change chart features, such as line thickness, bar width, or colors
- Rotate, flip, or duplicate images
- Size and align visuals using grids and rules
- Fill in or erase areas
- Specify a variety of type styles and sizes
- Add, move, and delete terms in the callouts and captions
- Highlight text with boldfacing and italics
- Zoom in on details and pan back and forth across an image (see Figure 7.17)
- Overlay one or more images

*Graphics Input*   Users create and modify images on the screen using such devices as *light pens*, *mice*, and *graphics pads* that mimic the artist's use of brushes, pens, and pencils. *Light pens*, not much used today, permit users to create and modify freehand images by moving the point of the pen directly across the screen. A *mouse* is a hand-held device that the user pushes across a tabletop pad to move the cursor on the screen and to draw freehand images. A *graphics pad* is a traylike drawing surface laid out in a grid that senses the tip of a penlike stylus. Lines and images drawn directly on the pad with the stylus appear on the screen.

Rather than creating images using these devices, users can bring existing paper-copy visuals on screen by means of an electronic scanner. The steps involved in this process are shown in Figure 7.18. Many computer systems allow scanned images to be enlarged, reduced, duplicated, and manipulated in other ways after they are on screen. Chapter 13 describes how scanners are used in conjunction with electronic publishing systems.

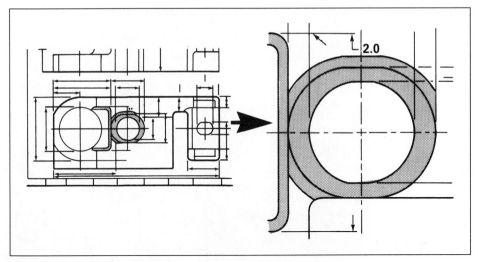

***Figure 7.17***   Example of Zoom Capability

*Graphics Output*   Getting the image from the screen to another medium, usually paper, requires a printer, a plotter, or some other output device. Such devices vary in their technology and in the quality of image they produce. The visual quality and legibility of a printed image is referred to as its *resolution* and is

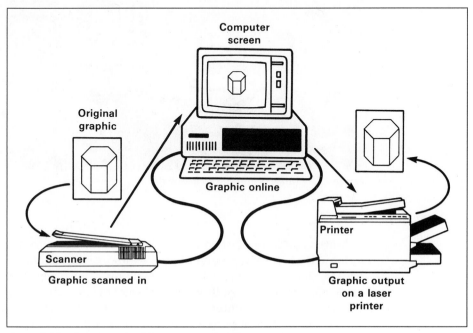

***Figure 7.18***   Computerized Image-scanning Process

***Figure 7.19***    Image Reproduced at 150 dpi

measured in dots per inch (dpi) (that is, by how densely the dots cluster when they form each image). Graphics prepared on low-resolution dot-matrix printers (usually less than 150 dpi) may be acceptable for draft review copies but are not of sufficient quality or legibility for final camera-ready manuscripts. The next best output option is a medium-resolution laser printer. Such printers produce crisp-imaged text and visuals (except for halftones and color), typically at a resolution

***Figure 7.20***    Image Reproduced at 300 dpi

of 300 dpi, that may be used for camera-ready copy. Note the difference in clarity between the image reproduced at 150 dpi in Figure 7.19 and the same image shown at 300 dpi in Figure 7.20.

Top-of-the-line, high-resolution, output devices (1,200 dpi and above) include film recorders, phototypesetters, image setters, and plotters. These devices are expensive and are designed for and operated by professionals in the visual arts, typesetting, and printing fields. Advances in computer technology, however, are closing the resolution gap between laser printers and phototypesetters. Some laser printers are now capable of 1,000 dpi resolution. Plotters use a variety of colored pens, computer-controlled to produce multicolored visuals that look hand-drawn. They are used primarily for architectural and engineering applications. The quality needed for your camera-ready version will determine the type of output device required.

## Checklist 7.1

## Placement of Visuals

☐ Position visuals *following* first mention in the text, as close to first text mention as possible

☐ Place visuals that are referred to frequently in an easy-to-reach location

☐ Choose the appropriate orientation for visuals
- Portrait (vertical) orientation whenever possible
- Landscape (horizontal) orientation when information will not fit adequately in portrait orientation
- Straddling facing pages only when absolutely necessary (for especially large visuals)

☐ Set visuals off from surrounding text
- Surround with a box
- Allow extra white space
- Place in a screened area

☐ Provide brief descriptive titles for visuals

☐ If the document contains more than a few visuals, assign each visual a figure number

☐ Place exceptionally large or peripheral visuals in an appendix

## Checklist 7.2

## Options for Preparation of Visuals

☐ Use visuals from previous publications
- In-house documents
- Documents published elsewhere

☐ Use stock visual materials

☐ Work with graphic designers

☐ Prepare your own visuals
- Manual (hand-drawn) visuals
- Computer graphics
  ○ Business graphics
  ○ Free-form graphics

---

## WORKS CITED

Bennet, Garner R., and Chris Samolyk. "To Select an Illustration System: Know Which Functions Are Needed." *Electronic Publishing and Printing* May 1988: 29–32.

Brusaw, Charles T., Gerald J. Alred, and Walter E. Oliu. *Handbook of Technical Writing*. 3rd ed. New York: St. Martin's, 1987.

James, Geoffrey. "Artificial Intelligence and Automated Publishing Systems." *Text, Context, and Hypertext*. Ed. Edward Barrett. Cambridge: MIT Press, 1988. 15–24.

Oppenheim, Lynn, et al. *A Study of the Effects of the Use of Overhead Transparencies on Business Meetings*. Philadelphia: Wharton Applied Research Center, U of Pennsylvania, 1981.

Pickens, Judy E. *The Copy-to-Press Handbook*. New York: Wiley, 1985.

Winn, William. "The Role of Graphics in Training Documents: Towards an Explanatory Theory of How They Communicate." *Creating Usable Manuals and Forms: A Document Design Symposium*. Communications Design Center Report No. 42. Pittsburgh: Carnegie Mellon U, 1988.

## FURTHER READING

Bailey, Robert W. *Human Performances: A Guide for System Designers*. Englewood Cliffs, NJ: Prentice, 1982.

Bann, David. *The Print Production Handbook*. Cincinnati: North Light, 1985.

Barton, Ben F., and Marthalee S. Barton. "Toward a Rhetoric of Visuals for the Computer Era." *Technical Writing Teacher* 12 (1985): 126–145.

Briles, Susan M., and Henning P. Jacobshagen. "Improving Communication between Artists and Writers." *Technical Communication* 26.4 (1979): 12–14.

Caernarven-Smith, Patricia. "It May Be Computerized Art, But . . . ." *Technical Communication* 34 (1987): 117–118.

Carliner, Saul. "Help on a Shelf: Developing a Customer Service Manual." *Technical Communication* 32.2 (1984): 8–11.

Craig, James. *Production for the Graphic Designer*. New York: Watson, 1974.

Curry, Robert. "Visual/Graphic Aids for the Technical Report." *Journal of Technical Writing and Communication* 9 (1979): 287–291.

Enrick, Norbert Lloyd. *Handbook of Effective Graphic and Tabular Communication*. Huntington, NY: Krieger, 1980.

Field, Janet N., et al., eds. *Graphic Arts Manual*. New York: Arno-Musarts, 1980.

Filley, Richard. "Opening the Door to Communication through Graphics." *IEEE Transactions on Professional Communication* 25 (1982): 91–94.

Glen, John. "Can Flow Charts Be Overdone?" *Technical Communication* 34 (1987): 172.

Grange, Charles, and Amy Lipton. "Word-free Setup Instructions: Stepping into the World of Complex Products." *Technical Communication* 31.3 (1984): 17–19.

Gross, Alan G. "A Primer on Tables and Graphs." *Journal of Technical Writing and Communication* 13 (1983): 33–35.

Hanna, J. S. "Six Starts toward Better Charts." *Technical Communication* 29.3 (1982): 4–7.

Kleper, Michael L. *The Illustrated Handbook of Desktop Publishing and Typesetting*. Blue Ridge Summit, PA: Tab, 1987.

Magnan, George. "Technical Drawings and Illustrations." *IEEE Transactions on Professional Communication* 20 (1977): 239–245.

Manning, Alan D. "The Semantics of Technical Graphics." *Journal of Technical Writing and Communication* 19 (1989): 31–51.

Marra, James L. "For Writers: Understanding the Art of Layout." *Technical Communication* 28.3 (1981): 11–13.

Mracek, Jan. *Technical Illustration and Graphics*. Englewood Cliffs, NJ: Prentice, 1983.

Newcomb, John. *The Book of Graphic Problem Solving*. New York: Bowker, 1984.

Parker, Roger C. *Looking Good in Print*. Chapel Hill, NC: Ventura, 1988.

Redish, Janice C. "Integrating Art and Text." *Proceedings of the 34th International Technical Communication Conference* (1987): VC-4–VC-7.

Richardson, Graham T. *Illustrations*. Clifton, NJ: Humana, 1985.

Robertson, Bruce F. *How to Draw Charts and Diagrams*. Cincinnati: North Light, 1988.

Schmid, Calvin F., and Stanton E. Schmid. *Handbook of Graphic Presentation*. 2nd ed. New York: Ronald, 1979.

Seybold, John, and Fritz Dressler. *Publishing from the Desktop*. New York: Bantam, 1987.

Standera, Oldrich. *The Electronic Era of Publishing: An Overview of Concepts, Technologies and Methods*. New York: Elsevier, 1987.

Tufte, Edward. *The Visual Display of Quantitative Information*. Cheshire, CT: Graphic Press, 1983.

Vogt, Herbert. "You Don't Have to Be an Artist to Produce an Illustrated Document." *IEEE Transactions on Professional Communication* 26 (1983): 108–112.

Zimmerman, Donald E., and David G. Clark. *The Random House Guide to Technical and Scientific Communication*. New York: Random, 1987.

### Chapter 7: Integrating Visuals

**INTERVIEWER:** When do you begin to think about what visuals, such as drawings and tables, should appear in a document?

**MARY:** There are two types of graphics that are always included in our system documentation as a matter of convention: screen samples and report samples. I also use some of our department's Information Mapping guidelines to determine when to include a graphic [see Chapter 6 Case History, page 119]. For example, these guidelines recommend using "if-then" tables for decisions, so I put tables in at the point in a document where the reader has to make a decision.

For other graphics, I start thinking about them when, during an interview on some technical aspect of a project, the technical expert starts drawing something to explain some point. Chances are that a drawing would be helpful to the reader as well as to me. When somebody has to explain something visually, that tells me, "Hey, maybe this illustration should go into the document."

**INTERVIEWER:** So during your information gathering, you or the expert may draw a rough sketch that helps you understand a concept, and that rough sketch could form the basis of an illustration in the document.

**MARY:** Yes. Of course, sometimes the visual goes in and sometimes it doesn't.

**INTERVIEWER:** How do you determine when you will use a visual and when you will not?

**MARY:** I find out whether or not the readers understand the concept already. If they do, if it's common knowledge for the readers of the manual, then I don't put the graphic in because that would be explaining the obvious. I need to check my knowledge against that of the readers, because I often document systems for which I don't have a background understanding—systems in which *I* need a little help on the basics that the actual readers would not need.

**INTERVIEWER:** How do you generate graphics in your department? Do you use an artist, or are graphics part of your text processing system?

**MARY:** It's part of the text processing system we use.

**INTERVIEWER:** Describe the graphics capability of your system.

**MARY:** Our system has vector-oriented graphics features and a very powerful table-generating feature. The graphics are shapes that can be stretched or magnified and filled in with various patterns. The objects can be layered and can have lined edges, broken-line edges, or invisible edges. Multiple objects can be joined into a single object, and any shape or group of shapes can be moved or copied into other parts of the illustration or to other documents. You can also do precision drawing (setting line length using inches or other measurements). Most important, you can type whatever you want inside the graphic, and you can place it anywhere in a document.

INTERVIEWER: Since you don't have a professional graphic artist on your staff, is there someone in your office who has become especially good at using the graphics features, someone who is something of a graphics expert?

MARY: Yes. We have a few people who are really good with graphics. One of our writers, Eric, has become very skilled at producing graphics on the system. And of course, as writers come up with something, they routinely share it with other writers. We now have a collection of shared graphics that includes some really great key graphics and keyboards, as well as a variety of charts that we use for scheduling. Sharing graphics saves a lot of time; it does take time to develop a graphic, so if you can just edit someone else's graphic, you're way ahead.

INTERVIEWER: Then you yourself, as a writer, are also being asked to do artwork?

MARY: Yes. That's something relatively new in the industry. Suddenly, with electronic publishing, writers are being asked to draw effective graphics.

INTERVIEWER: What's your concern about that as a writer?

MARY: I'm apprehensive that since I don't have any training in drawing, I may draw something that's not effective or not the right kind of design. So far, I've just done things like time lines and diagrams. I'm still new to drawing, so I draw something simple and hope that it will be effective. Simple graphics *are* generally effective, but our system's graphics capabilities are far more extensive than my use of them right now would indicate.

INTERVIEWER: What are you doing to counter your apprehension?

MARY: I've asked Eric to teach me some of the basics. I also plan to take a class on mechanical drawing.

INTERVIEWER: What graphics or visuals did you use in the *Introduction to Using the International Banking System* manual?

MARY: I used drawings of the screen [see examples in the chapter titled "Using a Password" in the *Introduction* manual, reproduced on pages 398–402], a diagram explaining how menus work [see "Using Menus," page 411], and a diagram explaining the "behind the scenes" connections that take place as you log on the system from a personal computer [see "Logging On and Off," page 403].

INTERVIEWER: How did you generate them?

MARY: Right on our text processing system. In the past, I might have simply used underlining and vertical lines to draw crude boxes. But with the new system, I can call up perfect shapes drawn by the system and then easily manipulate them to create appropriate graphics.

INTERVIEWER: In what ways are the graphics in the *Introduction* manual valuable or useful?

MARY: The menus diagram enabled me to show something that I could not adequately describe in prose form. I might have tried, but it would have taken longer, would have required more than the one page needed for the diagram, and would have been a lot harder for the reader to understand. A prose description just would not have been as effective.

Graphics can present information that might not be presentable in another format, especially when you consider that while some people learn by hearing or by reading, other people tend to learn *visually*. With graphics, you can reach the portion of the audience that learns visually. Readers can also study the diagram and the prose description together, which enhances their understanding by giving them the same information in two different ways.

It's also pleasing to see a picture as you read. It gives your eyes a break.

INTERVIEWER: Especially with technical material.

MARY: Right.

INTERVIEWER: Let's talk about some of the other decisions you've made concerning graphics. How do you determine the placement of a visual aid?

MARY: I usually put the graphic immediately after the explanation for it. But for screens in the International Banking System manuals, each screen description begins with a screen sample, then shows the purpose and access for the screen, and then segments of the screen are repeated immediately before the field descriptions for each segment. Those decisions came from writing standards that were in place before I arrived on the project.

We used to have a guideline that said screen samples should always be placed at the top of a page, but we've since dropped that. We found that the guideline worked when the screen itself served as a heading—in other words, when the screen was the main thing being described—but not in other cases where a task was being described and the screen played only a small part in the task.

INTERVIEWER: Did you make rough versions of the illustrations in the *Introduction* manual before producing the final ones?

MARY: Yes. In most cases, I first did a rough paper version by hand, then a rough online draft using the text processing system, and then a final online form. [Figures C.5, C.6, and C.7 show, respectively, the rough paper, rough online, and final online versions of the menu diagram in the chapter titled "Using Menus" in the *Introduction* manual, page 411.]

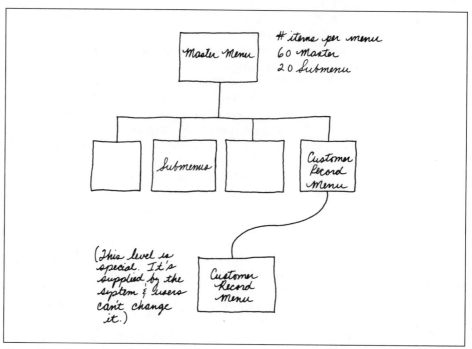

*Figure C.5*   Rough Sketch of a Visual

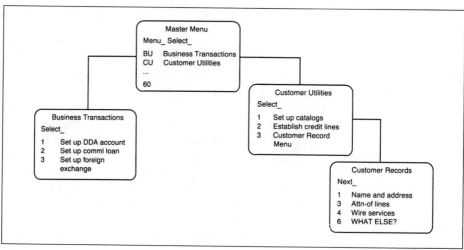

*Figure C.6*   Online Draft of the Visual

158

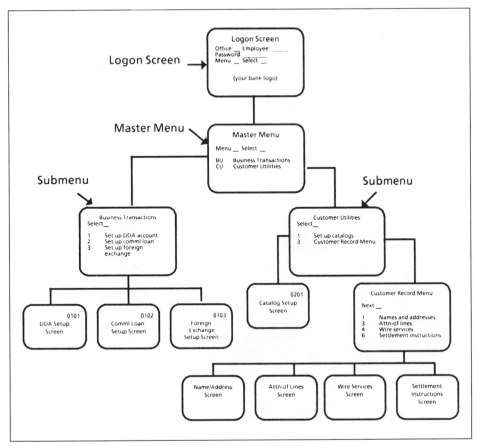

***Figure C.7*** Final Online Version of the Visual

# 8

# Layout and Design

Theory of Design

Using Typography

Using Highlighting and Finding Devices

Designing the Page

Packaging the Document

Good visual design is crucial to the success of a document. Both research and the practical experience of writers and graphic designers show that documents are most effective when information in them is accessible "both visually and syntactically" (Benson 35). In fact, a well-designed document should make even complex information look accessible and give readers confidence that they can master the information. Notice how inviting the well-designed version of the document at the right appears in Figure 8.1. Because it is more readable, a document with good design can also save money by improving productivity and reducing errors.

No book—or college course, for that matter—can substitute for the professional skills and knowledge of graphic designers, illustrators, and other experts.[1]

---

[1]For information on printing and production, see Chapter 13; for pointers on creating visuals, see Chapter 7; and for advice on designing computer screen layouts, see Chapter 15.

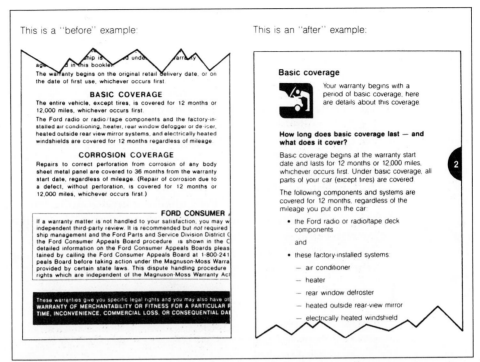

*Figure 8.1* The Importance of Graphic Design  (From *Simply Stated in Business*, March 1987; reprinted with permission from the Document Design Center, American Institutes for Research.)

If you work in an organization that employs graphic designers, make use of their expertise, or at least seek their advice. If an organization lacks such expertise, it is wise to seek outside graphic design services (McGhee 127–132).

To work productively with graphic designers, you must learn some of the language of graphic design. Doing so will enable you to talk with them and remain credible. Furthermore, in some circumstances, you will not have access to a graphic designer. Limited publishing budgets, tight schedules, or limited audiences can make employing a graphic designer a luxury. In such cases, you may be expected to make design decisions yourself. Moreover, increasing access to electronic publishing systems enables writers to control page layout to a far greater degree than in the past—but it has also put an almost bewildering array of design possibilities before professional writers. As a result, writers who do not know the basic principles of layout and design are professionally handicapped.

This chapter—together with the references listed at its end—can enable you to participate fully in graphic design decisions and to make full use of the graphics capabilities of powerful electronic publishing systems.

## THEORY OF DESIGN

Initial reader response to the visual design of a document is pivotal to its acceptance and its utility. Readers tend to look at a document in much the same way that a viewer looks at a painting; that is, a reader's initial response is subjective, either positive or negative at the moment (Rubens 77; Flower 136). To make a reader want to read the document, its "look" must communicate a sense of order and economy.

Like most writing decisions, creating the design of a document requires that you consider its audience and its purpose. In fact, the general design of a document—modular, online, one of a series of documents—must be determined at the outset when you assess audience, purpose, and scope. (Review Chapter 4 for other features of audience and purpose that help determine these initial design features.)

As you learn about your audience and consider design, anticipate which of the following reading styles the reader is likely to use, as discussed in Chapter 4:

- Skimming
- Scanning
- Search reading
- Receptive reading
- Critical reading

To determine the audience's reading style, consider how readers will use the document as well as their expectations based on what similar readers have seen in the past. You also need to consider the audience's reading ability. The lower the perceived reading level, the more non-prose elements you will need (pictures, graphics, white space, etc.).[2]

Good design must blend aesthetics with utility. In most functional documents, although aesthetics is important to the overall impression, utility must take precedence: the reader must find, understand, and use information. For a design to be useful, it must accomplish four goals:

- It should be visually simple and uncluttered.
- It should highlight structure, hierarchy, and order.
- It should help readers find information they need.
- It should establish the organization's image.

First, good design strives for visual simplicity on the page. Visual simplicity results from forming compatible or harmonious relationships, such as using the same family of type in a document or the same highlighting device for similar

---

[2]Because many military enlistees read at an eighth- or ninth-grade level, technical writing departments that prepare military manuals are being required to create intensively visual documents called "new look" publications. The basic unit for the "new look" text with companion illustration is called a *module*. This module is similar to a storyboard panel for a film or a videotape, containing the narrative elements for a single scene or topic (Meyer 17).

items. By simplifying and then repeating design features, you also achieve "rhythmic repetition." Notice the repetition of elements in the bank document in the Appendix. The appearance of topic headings above the thick line at the top of each page, for example, assures readers that they will always be able to confirm the topics as they turn pages. Such repetition of elements creates a visual expectation for readers and provides them with visual maps through the document (Barton 12; White 40).

Second, useful design reduces the complexity of prose by highlighting structure and helping readers see how the document is organized. Useful design, for example, should create "chunks" of information that can be quickly noticed and absorbed (White 61). Visual "chunking" is effective because it makes use of the mind's natural inclination to seek patterns in information, as discussed in Chapter 4. The layout shown in Figure 8.2 shows an effective use of this technique. The captions on the left lead the eye to skim for the appropriate information, which can be located quickly on the right. The explanatory text to the right of each caption is both concise and visually discrete ("chunked") on the page.

Design should also reveal hierarchy. Page arrangement should signal what is more important and what is less so, what are topics and what are subtopics, what is primary and what is secondary, what are general points and what are examples. Notice the arrowheads with the inset "examples" in Figure 8.2. If readers know the overall structure of a procedure from visual cues, for example, they can follow the specific instructions more easily. Visual cues should also lead readers from one part to the next, as illustrated by the boxed arrows in the lower right corner of pages of the bank document in the Appendix.

Third, visual cues should make information easy to find the first and every later time it is sought, as Figure 8.2 also illustrates. Visual cues make information accessible to different segments of an audience by allowing them to scan and find the information they need. Research also suggests that when readers are able to select what they want, they gain control over their learning, and comprehension improves (Rude 65).

Finally, the design of a document projects an image of an organization. Just as an effective design attracts readers, an ineffective design alienates readers. Schoff and Robinson point out that "a poorly designed, confusing, or unreadable manual may cast doubt on the quality of the product itself or convince a user not to buy from a certain company in the future, even if the products of that company are good" (x). Good design, by contrast, can build goodwill for your organization and respect for its products or services.

Effective design should also reflect the image that the reader expects. If the document is intended for customers who are paying a high price for a professional service, the audience may expect a sophisticated design that befits a successful, high-quality operation. If a document is intended for personnel inside an organization, however, the audience may accept—even expect—a standard company binder suggesting frugality. And if the organization is tax-supported, a modest though well-thought-out design will project an appropriate image.

Some organizations, of course, have established design and format guidelines for their publications that you will need to follow. If your organization

**8**                                                                    **System security**

---

**System security**     The International Banking System includes the following
                        security features:

*Passwords*             Each employee must establish a personal password and
                        use it every time he or she logs on. The employee is
                        responsible for all transactions entered while he or she is
                        logged on.

                        Employees must change their passwords at least every 30
                        days.

*Consecutive error*     If an employee makes three consecutive errors while
*lockout for passwords* logging on, the employee's password becomes inoperative
                        until a supervisor makes it operative again. The system
                        'remembers' consecutive errors made on previous days.

*Automatic logoff*      To minimize the chance of an unauthorized person using
*after inactivity*      the system, the system automatically logs off terminals
                        that have been inactive for more than a few minutes.
                        (Your office determines the exact amount of time.) A
                        warning appears on the screen before the terminal is
                        logged off. Any incomplete transactions are not saved.

*Authority requirements* Your office's registration of an employee defines what
                        activities the employee can perform.

                        ▷ **example:** In your office, a letter of credit clerk may be
                        authorized to create letters of credit but not foreign
                        exchange contracts. (Such an employee might still be
                        able to view foreign exchange contracts created by other
                        employees.)

*Authority indexes*     As each employee is registered on the system, he or she is
                        given *authority index numbers* that limit the amount of
                        credit the employee can authorize.

                        ▷ **example:** In your office, a letter of credit clerk may
                        authorize letters of credit for up to $1,000.00, while a
                        credit officer may authorize letters of credit for up to
                        $1,000,000.00.

---

Introduction to Using the International Banking System   2/88

***Figure 8.2***   Example of Visual "Chunking"    (Courtesy of First Wisconsin National Bank
of Milwaukee)

does not, management should be encouraged to create design standards for routine publications. (We discuss this in Chapter 14.) If you are preparing a document for an agency or a unit of government, you will no doubt need to follow rigidly prescribed format guidelines (ANSI 27–31; U.S. GPO 7–22).

## USING TYPOGRAPHY

Graphic designers often refer to type as the "vocabulary of design." In other words, the elements of typography (typefaces, type size, spacing, etc.) are the raw materials used to lay out the page. You must understand typography before you can arrange the text and other elements on the page.

Well over 10,000 different typefaces are available, each with a "personality" (formal, informal, bookish, legal-looking, exotic, etc.), as shown in Figure 8.3. Some typefaces have many variations within the same "family," as shown in Figure 8.4. Typefaces are often named for their designers (like John Baskerville) or their historic use. Bank Script, for example, was a style used in bank transactions before the invention of the typewriter.

| | |
|---|---|
| Albertus Light | Berner |
| **Albertus** | *Berner Italic* |
| **Albertus Bold** | **Beton Bold Condensed** |
| American Uncial | Bodoni |
| Aster | *Bodoni Italic* |
| *Aster Italic* | **Bodoni Bold** |
| **Aster Bold** | ***Bodoni Bold Italic*** |
| BANK GOTHIC LIGHT | Euro Bodoni |
| BANK GOTHIC MEDIUM | **Euro Bodoni Demi** |
| **BANK GOTHIC BOLD** | ***Euro Bodoni Demi It.*** |
| *Bank Script* | **Broadway** |
| Baskerville | **BUFFALO BILL** |
| *Baskerville Italic* | Bulletin Typewriter |
| **Baskerville Bold** | Bulmer |
| ***Baskerville Bold Italic*** | Burgondy Right |

***Figure 8.3*** Sample Typeface Styles

Helvetica Light
*Helvetica Light Italic*
Helvetica Regular ·
*Helvetica Regular Italic*
**Helvetica Medium**
***Helvetica Medium Italic***
**Helvetica Bold**
***Helvetica Bold Italic***
Helvetica Regular Condensed
*Helvetica Regular Condensed Italic*

***Figure 8.4***
Family of the Helvetica
Typeface

A complete set of all the characters available in one typeface and size of type is called a *font*. Keep in mind that not all fonts have the same assortment of characters. A complete font in a single typeface style appears in Figure 8.5.

Type is measured in points, and line length is usually measured in picas. Figure 8.6 shows the relations in size of points, inches, and picas. Many typefaces are available in 6- to 72-point sizes, with a complete font in each size. It is a good idea to own a book or dictionary of type fonts and styles; some are listed at the end of this chapter.

The letters of type have a number of components, some of which are shown in Figure 8.7. Although there are many classifications of typefaces (none of which are precise), Michael Bruno makes the following helpful classifications (36–37) based on style and use.

abcdefghijklmnopqrstuvwxyz
[.,¶§:;!?¿¡/-–—*†‡]1234567890($¢£%#@)
ABCDEFGHIJKLMNOPQRSTUVWXYZ&
ABCDEFGHIJKLMNOPQRSTUVWXYZ&:;!?$¢£1234567890
′″○©®™«» ⅛ ⅜ ⅝ ⅞ ⅓ ⅔ ¼ ½ ¾ + − × ÷ = ±
áàçêîñôŏüÁÀÇÑÔÜáàñôü

***Figure 8.5***   Complete Font in a Typeface (Garamond ITC)

$$1 \text{ point} = \frac{1}{72} \text{ inch} = \frac{1}{12} \text{ pica}$$
$$12 \text{ points} = \frac{1}{6} \text{ inch} = 1 \text{ pica}$$
$$72 \text{ points} = 1 \text{ inch} = 6 \text{ picas}^3$$

*Figure 8.6*
Points and Equivalents

**Oldstyle:** Open, wide, with printed serifs
**Modern:** Thick/thin stroke, squared serifs
**Square serif:** Even serifs—use for headlines only
**Sans serif:** Simple look, even strokes
**Script:** Simulates handwriting—use in invitations
**Text letters:** Old script—use on certificates
**Decorative:** Contemporary—use for novelty

## Selecting a Typeface

For technical writing, you should select a typeface for legibility first and aesthetics second.

### Legibility

Examine several typefaces for their "legibility," or the speed with which each letter and word can be recognized. Look particularly at the upper portion of a line, since it should be more legible than the lower portion, as illustrated in Figure 8.8. Serifs (discussed later) and the distinctive tops of letters, particularly the ascenders of lowercase characters, contribute to the legibility of the top half of a line.

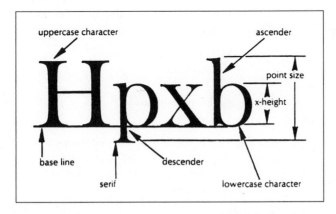

*Figure 8.7*
Primary Components of
Letter Characters

---

[3]One point does not equal *exactly* 1/72 inch; it is actually 0.1383 inch. Thus 12 points actually equal 0.16596 inch, and 72 points equal 0.99576 inch.

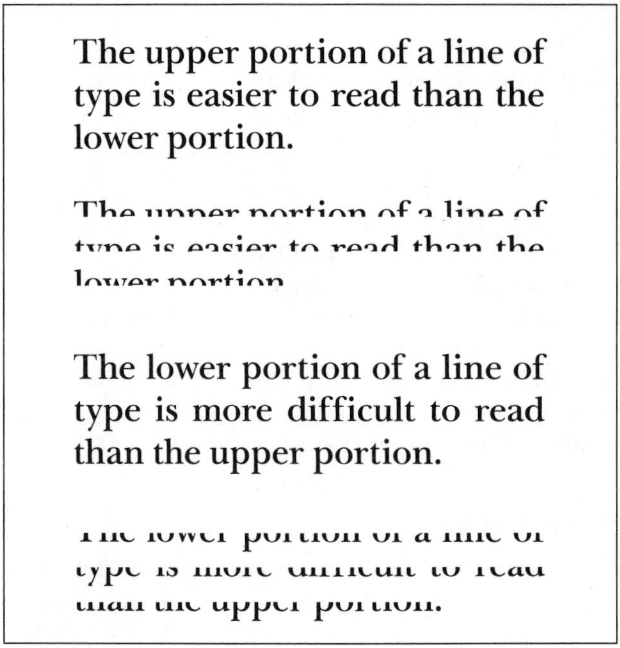

The upper portion of a line of type is easier to read than the lower portion.

The upper portion of a line of type is easier to read than the lower portion.

The lower portion of a line of type is more difficult to read than the upper portion.

The lower portion of a line of type is more difficult to read than the upper portion.

**Figure 8.8**
Upper and Lower Portions of Type (Courtesy of *The Typencyclopedia* by Frank J. Romano, copyright © 1984 by the R. R. Bowker Company)

Research suggests that legibility is impaired when a letter has extreme contrasts in the thickness of various components (called *weight*), contains heavy and overly long serifs, or is overly narrow (Ettenberg 118). Avoid typefaces that may distract readers with contrasts in weight or with odd features on certain characters or the whole set, as is often the case with script, or cursive, typefaces. In addition, avoid typefaces that fade when printed or copied.

A reader's familiarity with a typeface also helps make it legible. Some very common typefaces have been in use for centuries (Dugdale 3; Gottschall 120):

- Baskerville
- Bodoni
- Caslon
- Century
- Futura
- Garamond
- Gill Sans
- Helvetica
- Times Roman
- Univers

In a survey (Stermer 72), ten leading graphic designers were asked, "If you were limited to five families of type for the rest of your professional career, what would they be?" These designers selected the following typefaces (the number of times each was named follows in parentheses):

- Times New Roman (6)
- Helvetica (5)
- Garamond (5)
- Caslon 540 (4)

- Goudy Oldstyle (4)
- Bodoni (3)

### Practical Considerations

In choosing a typeface, you also need to determine the following practical facts:

- Know which character sets contain all the necessary and special characters you need.
- Know which character sets are available for use with your electronic publishing system or are held by the commercial printing company you select.
- Make sure that the sizes and weights you might need are available.

As you are considering typefaces, keep in mind that you should not use more than two families of typeface in a document, *even though you may have access to many more.* One leading graphic designer, in fact, suggests that you use the same typeface "family" for headings and other displays as for text: "Doing so results in a smoother, more unified product that has a more 'designed' appearance" (White 82). However, if you wish to contrast headlines dramatically with the text, as in a newsletter, use a typeface that is distinctively different from the text. You may also wish to use a noticeably different typeface *inside* a graphic element (Rubens 84). In any case, you should experiment with different typefaces before making final decisions.

### Serif versus Sans Serif

One of the most basic distinctions of typefaces is between *serif* and *sans serif* type, as shown in Figure 8.9. Serifs are the small projections at the end of each stroke in a letter. Serif type styles have these lines; sans (French for "without") serif do not. For years graphic designers have debated whether serif or sans serif type is easier to read. Some believe that serifs guide the eye; others say that they clutter. Some readability studies suggest that serif typefaces are easier to read because "the serifs tend to 'seat' the letters on the line and pull the eye along to the next word" (Schoff and Robinson 66; Ettenberg 118). Philippa Benson further observes that "letters in sans-serif typefaces are less distinct from each other not only because they lack the extra, ornamental strokes . . . but [also] because the widths of the lines that constitute each sans-serif letter are often uniform" (36–37).

The conclusion of the research so far is that although sans serif type has a modern look, serif type is easier to read, especially in the smaller sizes. Sans serif, however, does appear to work well for headings. If you do choose sans serif type-

| Times Roman | Helvetica |
|:---:|:---:|
| (serif) | (sans serif) |

**Figure 8.9**
Serif and Sans Serif
Typefaces

**Figure 8.10**
Samples of Counters

faces for text, pick one with large "counters," the fully or partially enclosed white spaces in letters like *c* and *b* illustrated in Figure 8.10.

## Choosing Type Size

Type size (or point size) is based on the height of the block that was used in traditional typesetting (as illustrated in Figure 8.11), not on the height of individual letters, which often vary slightly among typefaces. In fact, the same letter with the same type size may vary in height and width from one type style (or font) to the next. The resulting "image size" of typefaces in the same point size can vary

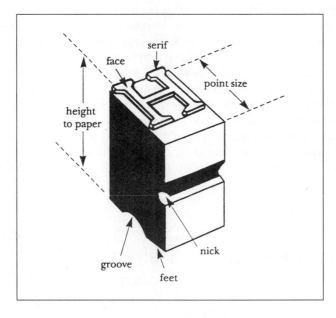

**Figure 8.11**
Block of Type

Once upon a time

Once upon a time

Once upon a time

*Figure 8.12*
Identical Point Size with
Different "Image"
Sizes   (Courtesy of
*Designing Instructional Text*
by James Hartley;
copyright © 1985 by
Kogan Page)

dramatically, as shown in Figure 8.12 (Hartley 20). The *x-height* is usually the key factor in the image size. The x-height is the height of the letter *x*, as shown in Figure 8.13 (Romano 166).

In general, don't use a type that is too small because (1) it may be skipped over by readers, (2) it will cause eye strain, (3) it will make text look crammed and uninviting. Six-point type is the smallest that can be read without a magnifying glass (see Figure 8.14). Type that is too large can also cause problems: (1) it may use more space than necessary, (2) it can be costly, (3) it can make reading difficult and inefficient, and (4) it can make readers perceive words in parts rather than wholes.

General recommendations for ideal point sizes for text range from 8 to 12 points (see Figure 8.14); 10 points is the most popular. Considering the impact of

*Figure 8.13*   Illustration of x-Height   (Courtesy of *The Typencyclopedia* by Frank J. Romano, copyright © 1984 by the R. R. Bowker Company)

| | |
|---|---|
| 6 pt. | Type size can determine legibility. |
| 8 pt. | Type size can determine legibility. |
| 10 pt. | Type size can determine legibility. |
| 12 pt. | Type size can determine legibility. |
| 14 pt. | Type size can determine legibility. |

*Figure 8.14*     Samples of 6- to 14-Point Type

x-heights, however, look at examples of text in various type sizes and typefaces— and trust your reaction to samples. Remember, too, that the longer the lines, the larger the type size should be; hence shorter lines can accommodate smaller type.

The distance from which the document will be read should help determine type size. For example, a manual on a table with the user standing requires a larger typeface. Consider also the audience's age, physical condition, and reading experience. One study found, for example, that older adults need 12-point or larger typefaces (Nagy 32).

Be aware that a reduction of type size for a block of text combined with a reduction of column width can signal that the material is less important. You might consider using a smaller typeface in sections on the page aimed at secondary audiences. A smaller type size, for example, is often used for footnotes.

## Adjusting Spaces between Letters and between Words

Letter spacing and word spacing used to be the province of professional typesetters and printers. Today, powerful electronic publishing systems enable writers to adjust spacing between words and letters. Word spacing in professional typesetting has followed the guideline to "use as little space as needed between words while still providing a clear optical break between them" (Bruno 38; Doebler 113). Traditionally, the standard word space is the width of space allowed for the lowercase letter *i*, as illustrated in Figure 8.15. Too much word space can create

Once|upon|a|time

Onceiuponiaitime

Once upon a time

*Figure 8.15*
Standard Word Spacing Width   (Courtesy of *Designing Instructional Text* by James Hartley; copyright © 1985 by Kogan Page)

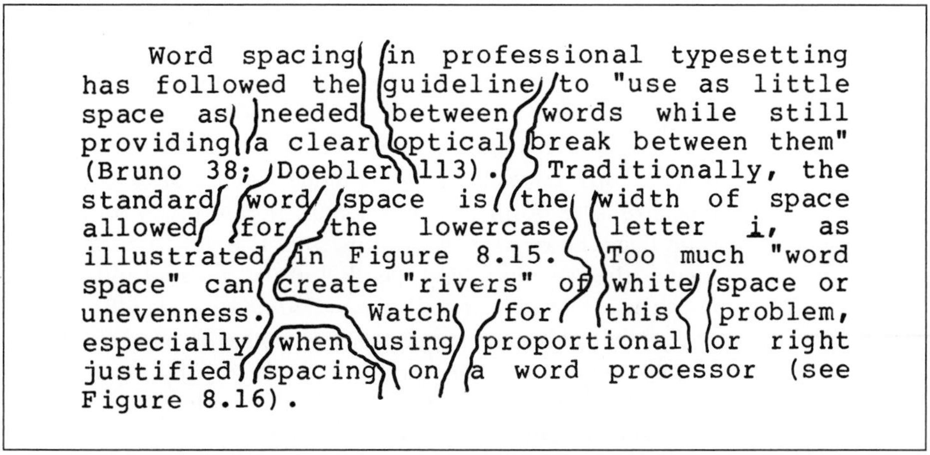

```
     Word spacing in professional typesetting
has followed the guideline to "use as little
space as needed between words while still
providing a clear optical break between them"
(Bruno 38; Doebler 113).  Traditionally, the
standard word space is the width of space
allowed for the lowercase letter i, as
illustrated in Figure 8.15. Too much "word
space" can create "rivers" of white space or
unevenness.       Watch for this problem,
especially when using proportional or right
justified spacing on a word processor (see
Figure 8.16).
```

***Figure 8.16***   White-Space "Rivers"

"rivers" of white space or unevenness. Watch for this problem, especially when using proportional or right-justified spacing on a word processor (see Figure 8.16).

Letter space, the amount of space between letters, is usually fixed on small word processors or changed equally between all letters on larger ones. Some electronic publishing systems are capable not only of changing letter spacing uniformly but also of adjusting individual spaces and even "kerning." Kerning is removing space between letters until they overlap, as in Figure 8.17. Certain pairs of letters are typically kerned (Romano 74). Kerning, like letter spacing, can be used to fit text (often headings or headlines) into tight spaces to make it more readable. However, you should use kerning and letter spacing *very* selectively.

## Adjusting Line Length and Leading

When determining line length, consider your audience's reading style: short lines are better for reading in spurts; long lines are better for prolonged reading. Many researchers recommend a line length of 1½ alphabets (about 40 characters) in the

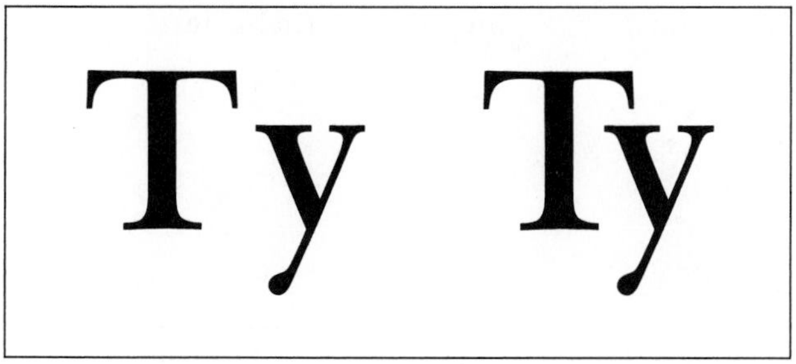

**Ty Ty**

***Figure 8.17***
Unkerned and Kerned
Letter Pair

typeface. Generally, a line that is much shorter forces the eye to jump around, causing fatigue and decreasing comprehension. If the line is much longer, the eye has too far to travel back to begin the next line and may settle on the wrong line of type.

Designers sometimes use the following rule of thumb for line length in typeset copy: *Double the point size of the typeface for line length (in picas)*. Thus if the type size is 12 points, the line length should be 24 picas (or 4 inches). A normal *typed* page, however, violates this rule. It has a line length at least half again as long as the recommended length, but we are so used to it that readability seems not to be affected, especially when double-spaced (Schoff and Robinson 64).

Other designers suggest that a maximum of 10 to 12 words is acceptable in many typefaces (Carter, Day, and Meggs 87). Further, line length decisions are affected by the amount of white space on the page, as well as other elements discussed later in this chapter.

*Leading* refers to the space between lines. (The term *leading* goes back to metal typesetting days, when thin strips of lead were inserted by hand between lines.) Leading should be in proportion to line length and point size—about 20 percent of the point size (Romano 88). The table in Figure 8.18 shows standard type sizes with recommended proportional line lengths and leading. Ranges are given (in parentheses) because some typefaces can affect the leading you should use. For example, you can follow the standard leading for a typeface with normal-length ascenders and descenders; however, if they are short in the typeface you select, you may need to add an extra point or two of leading for legibility. Boldfaced type also requires more leading—add one point across the board.

You may also use extra leading between paragraphs as well as indention for paragraphs to achieve maximum readability.

## Choosing Justified or Ragged Right Margins

Research has revealed little difference in reading *speed* or *comprehension* between ragged and justified text, but one study found that poor readers had more difficulty with justified text (Felker et al. 86). Ragged right may be *easier* to read

| Type Size (in points) | Line Length (in picas) | (in inches) | Leading (in points) |
|---|---|---|---|
| 6 | 12 | 2 | 1(0–1) |
| 8 | 16 | 2⅔ | 1½(0–2) |
| 9 | 18 | 3 | 2(0–3) |
| 10 | 20 | 3⅓ | 2(0–3) |
| 11 | 22 | 3⅔ | 2(1–3) |
| 12 | 24 | 4 | 3(2–4) |

*Figure 8.18*   Optimal Type Sizes, Line Lengths, and Leading

than right-justified because the uneven contour of the right margin provides our eyes with more landmarks to identify. Certainly, unjustified text is more helpful for less able readers, such as younger children and older adults (Hartley 31). Ragged right also has the advantage of cutting printing costs, since the irregular margins allow room for corrections without the need to reset whole blocks of text. Ragged right may also be preferable if the justification or proportional spacing software in your word processor inserts irregular-sized spaces between words, producing the white-space rivers discussed earlier.

Because ragged right margins look slightly informal, justified text is appropriate for publications aimed at a broad readership that expects a more formal, polished appearance. Further, justification is often useful with multiple-column formats because the spaces between the columns (called *alleys*) need the definition that justification provides. However, vertical rules (described later) between columns can also help define and give a finished look to ragged right margins.

Regardless of your choice, do not justify short lines. When lines are short, justification leaves huge gaps of space between words.[4]

## USING HIGHLIGHTING AND FINDING DEVICES

A number of means are available to emphasize important words, passages, and sections within documents:

- Typographical emphasis
- Headings and captions (called *display type*)
- Headers and footers
- Rules and boxes
- Icons and pictograms
- Color and screening

When used thoughtfully, such highlighting and finding devices give a document a visible sense of logic and organization. For example, they can set off steps or examples from surrounding explanations.

Keep in mind also that typographical devices and special graphic effects should be used in moderation. Just because a feature is possible does not mean that you should use it. In fact, too many design devices will clutter and actually interfere with comprehension. A good way to check for clutter is simply to count the *number* of visual elements in a text and the frequency of the ones you do use.

Remember that consistency is important. When you choose a highlighting technique to designate a particular feature, be sure that you use the same technique for that feature throughout your document.

---

[4]For this book, the regular text type is 10 point New Baskerville, with a leading of 2 points. The line length, at 30 picas (5 inches), is somewhat longer than that recommended for most reference and technical documents. But the optical properties of the typeface, the use of lists, and other textbook requirements and characteristics make the ten- to twelve-word average line length discussed earlier appropriate for this book.

## Typographical Emphasis

One easily available method of typographical emphasis is the use of uppercase letters.[5] BUT ALL UPPERCASE LETTERS ARE DIFFICULT TO READ BECAUSE THEIR UNIFORMITY OF SIZE AND SHAPE DEPRIVES READERS OF IMPORTANT VISUAL CLUES AND THUS SLOWS READING. Letters in lowercase have ascenders and descenders that make the letters easier to identify; therefore, a mixture of uppercase and lowercase is most readable. You should use all uppercase letters only in short spans—three or four words, as in headings (Brusaw, Alred, and Oliu 396). Remember, too, that the more capital letters you use in a heading, the more important you make the words seem.

As with all uppercase letters, use italics sparingly. *Continuous italic reduces legibility and thus slows readers because they must expend extra effort.* Because italics slow readers, they may be useful where you wish readers to be slow, as in cautions and warnings. Jan White, a leading graphic designer, argues against using italics for *any* emphasis because "they are a development of an informal, handwritten style, and so are lighter and paler and weaker—which are precisely the characteristics you don't want for emphasis" (90). He admits, however, that conventional uses of italics (signaled by underlining when typewriting) will remain for some time. Research has found **boldface** to be a better cuing device than any other because it is visually different yet "retains the flavor of the text" (Rubens 79).

## Headings and Captions

Headings are signposts that give the reader a sense of what is covered in a section of a document. Headings also reveal the organization of the document and indicate hierarchy. Many readers will often scan a document's headings before reading sections carefully. Headings should therefore help readers decide which sections they should read. Using too few headings in a document forces readers to work to find their way; conversely, using too many headings can confuse readers and make a document look like an outline. Captions, shown in Figure 8.19, are key words or *glosses* that highlight or describe illustrations or blocks of text. Captions can be easily overused and do not emphasize hierarchy as well as headings.

Numerous heading formats are possible, and organizations often set standards to ensure a consistent look for headings in documents. Headings may appear in many typeface variations (boldface being most common) and often use sans serif styles. Although others are possible, the most common positions for headings and subheadings are centered, flush left, indented, and floating. Major section or chapter headings should normally appear at the top of a page. Generally, avoid headings in the lower third of a page, and never leave a heading on the final line of a page. Instead, carry the heading over to the start of the next page. Insert one additional line of space or extra leading above a heading to emphasize that it marks a logical division.

---

[5]Capital letters were called "uppercase" because they were stored in the upper case of the printer's cabinet; small letters were stored in the lower case.

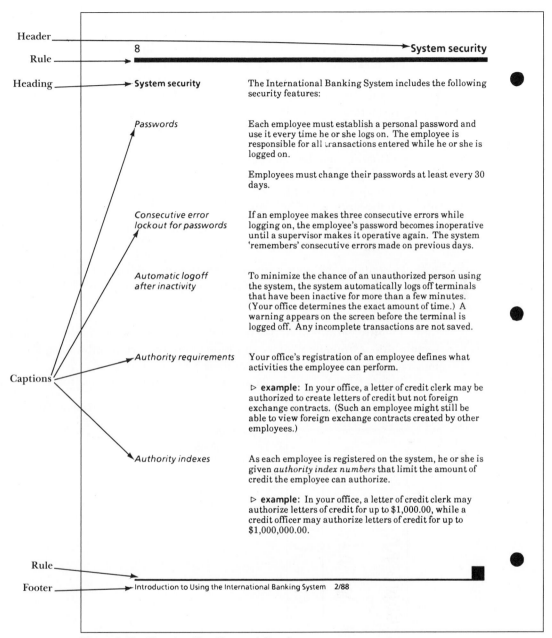

Header

Rule

Heading

8                                                              **System security**

**System security**          The International Banking System includes the following
                             security features:

Passwords                    Each employee must establish a personal password and
                             use it every time he or she logs on. The employee is
                             responsible for all transactions entered while he or she is
                             logged on.

                             Employees must change their passwords at least every 30
                             days.

Consecutive error            If an employee makes three consecutive errors while
lockout for passwords        logging on, the employee's password becomes inoperative
                             until a supervisor makes it operative again. The system
                             'remembers' consecutive errors made on previous days.

Automatic logoff             To minimize the chance of an unauthorized person using
after inactivity             the system, the system automatically logs off terminals
                             that have been inactive for more than a few minutes.
                             (Your office determines the exact amount of time.) A
                             warning appears on the screen before the terminal is
                             logged off. Any incomplete transactions are not saved.

Authority requirements       Your office's registration of an employee defines what
                             activities the employee can perform.

                             ▷ **example**: In your office, a letter of credit clerk may be
                             authorized to create letters of credit but not foreign
                             exchange contracts. (Such an employee might still be
                             able to view foreign exchange contracts created by other
                             employees.)

Authority indexes            As each employee is registered on the system, he or she is
                             given *authority index numbers* that limit the amount of
                             credit the employee can authorize.

                             ▷ **example**: In your office, a letter of credit clerk may
                             authorize letters of credit for up to $1,000.00, while a
                             credit officer may authorize letters of credit for up to
                             $1,000,000.00.

Captions

Rule

Footer                       Introduction to Using the International Banking System    2/88

*Figure 8.19*    Floating Heading and Captions

Research suggests that one of the most effective positions for headings and captions is the left-hand outside column because readers notice this area of a page first (Hartley 50; White 16). As illustrated in Figure 8.19, such "floating headings" make excellent use of white space, limit lines to a readable length, and eliminate the need to use a larger type or space after the heading. Floating headings and captions not only make the words more visible but also make them useful as reference points if there are many pages of text for the reader to cross-reference. This placement is especially useful when readers must look back and forth from a document to a device to perform a procedure.

Set floating headings or captions flush left or flush right in the column, and consider using vertical hairline rules to split the text from captions, as shown in the thumbnail sketch in Figure 8.20. (We will return to thumbnails shortly.) For advice on the language of headings, see Chapter 9.

## Headers and Footers

A *header* contains the identifying information carried at the top of each page; a *footer* contains similar information at the bottom of each page.[6] Pages may feature only a header or a footer, or they may have both (see Figure 8.19). Although practices vary, headers and footers carry such information as the topic or sub-topic, identifying numbers, document date, page number, document name, and section title. Although headers and footers are important reference devices, too much information in a header or footer can create clutter.

Allow at least 2 picas (24 points, or 1/3 inch) between the text and the

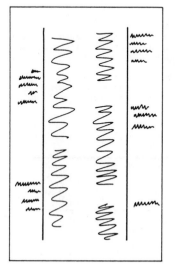

*Figure 8.20*
Thumbnail Sketch with
Rules and Floating Captions

---

[6]In traditional printing terminology, headers are referred to as *running heads* and footers as *running feet*.

header or footer. As also illustrated in Figure 8.19, *rules* (horizontal lines) can effectively set off headers and footers from the rest of the page.

## Rules, Icons, and Color

*Rules* are vertical or horizontal lines that are used to divide one area of the page from another (see Figures 8.19 and 8.20). Rules can also be combined to box off elements on the page. When using rules, make sure that their *weight* (thickness) does not visually overpower the type or figure inside them. In general, a ½ -point rule is the most useful size (Romano 129). Do not overuse rules. One study found that pages with heavy rules added to headers and footers and with boxes around all illustrations had a negative impact on readers. Readers considered the boxed material an "afterthought" and hence "unimportant" (Rubens 77).

An *icon* is a pictorial representation of an idea that you can use as a symbol to identify specific actions, objects, or sections of a document.[7] Today some of the most widespread icons are the symbols used to designate men's and women's restrooms and the symbol indicating wheelchair accessibility or parking for the handicapped.

In documents, icons are useful to identify sections and to warn readers of danger, especially non–English speakers and semiliterate users. To be effective, icons must be simple and intuitively recognizable—or at least easy to define, like those illustrated in Figure 8.21. For icons to meet these criteria, they must fit the purpose of the document as well as the conventions (cultural and professional) familiar to your audience. Icons can be placed in headers, in footers, next to

**Figure 8.21**
Typical Icons

---

[7]The word *icon* comes from the Greek word for "likeness, image, or figure." The use of icons predates the use of written language.

headings, or in the open left column of a page. Icons are especially useful in creating warnings (discussed in Chapter 14).

*Color and screening* can distinguish one part of a document from another or unify a series or group of documents. (Screening refers to gray shaded areas that are actually composed of patterns of tiny black dots. Screened panels appear on the sides of pages in the Case History sections throughout this book.) Color and screening can set off sections within a document, highlight examples, or call attention to warnings. Studies indicate that colors are often easier to identify and remember than size, placement, or shape (Waller, Lefreri, and MacDonald-Rosse 80). Be aware that colors are affected by lighting conditions and background and that 6 percent of the population is color-blind. For these reasons, and because color printing is expensive, screening can be a good alternative. (The processes of producing screened and color images are described in Chapter 13, pages 328–330.)

## DESIGNING THE PAGE

Page design is a recursive process; each step affects the others. For example, the size of type you use affects the line length, and the line length affects the page

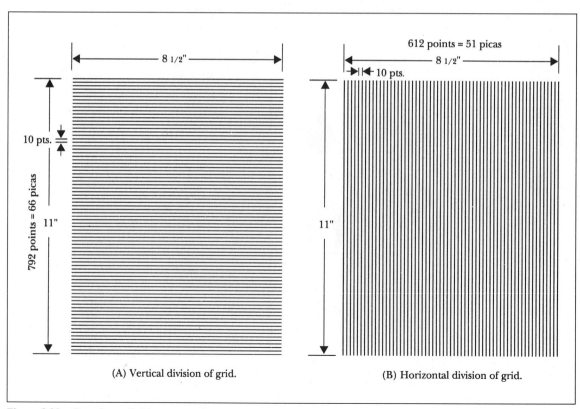

**(A)** Vertical division of grid.                    **(B)** Horizontal division of grid.

*Figure 8.22*   Creating a Grid

size. However, page size may be dictated by the purpose of the document, which would in turn affect line length and thus type size. Each time you plan a design, you need to remain flexible as you make decisions about specific elements. When you determine how the document will be reproduced and the budget available, you will know just how flexible and lavish your design can be.

## Using a Grid

As an aid to arranging pages, consider creating a grid to serve as a visual framework to position blocks within the image area (Romano 105). To create a basic reference grid, divide the page into points.

To construct a master grid that you can duplicate for many projects, divide an 8½-by-11-inch page into 10-point units, as shown in parts A and B of Figure 8.22. You may not wish to draw every line, of course. Subtract one point from each of the four sides to make the height 790 points and the width 610 points. Next, round off points so that 610 points equals 61 units and 790 points equals 79 units on the grid (Kung 40). These units can then serve as measuring devices, as shown in part C of Figure 8.22. After establishing the margins, such as those for the 46-by-70-unit page shown in part C of Figure 8.22, you can then arrange

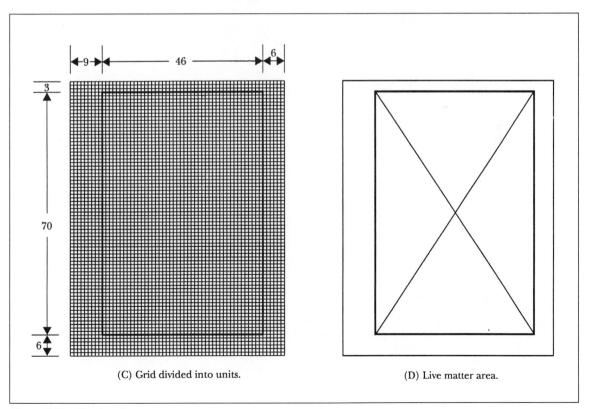

(C) Grid divided into units.                    (D) Live matter area.

*Figure 8.22*    Continued

blocks of material inside the "live matter area" marked with the × in part D of Figure 8.22. As also shown in that part of the figure, the left ("inside") margin often needs extra space to allow for hole punching or binding. Further, you may need to accommodate prescribed dimensions for the live matter area (ANSI 27; *Chicago* 567).

Within the live matter area, you can draw vertical lines to create columns and horizontal lines for blocks of material, as illustrated in part A of Figure 8.23. These divisions could form the resulting page shown in part B of the figure. Before you put actual text and illustrations on a page, you may wish to create a thumbnail sketch with blocks indicating the placement of elements (text, illustrations, headings, etc.; see Figure 8.24). You can go further by laying out a rough assembly of all the thumbnail pages showing size, shape, form, and general style of a publication. This mock-up, called a *dummy*, allows you to see how a publication will look. If you are having a professional graphic designer lay out a publication, the designer should prepare a dummy for you to examine and

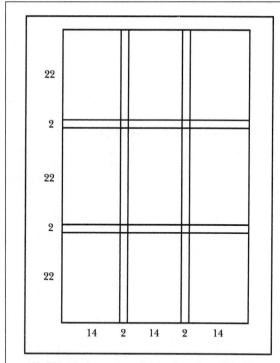

(A) Three-column page division based on 46-by-70 grid. Page is divided into three columns of 14 units each and three vertical divisions of 22 units each, with 2-unit spacing.

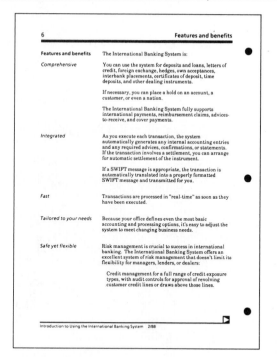

(B) Page resulting from three-column grid.

*Figure 8.23*   Three-Column Grid and Resulting Page

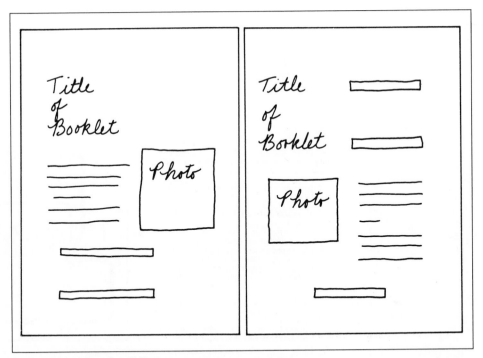

***Figure 8.24*** Thumbnail Pages

approve (White 15–38). Figures 8.25 and 8.26 show professionally designed pages.

As you work with elements on the page, experiment with different designs. Often what may seem useful as a concept in principle turns out not to work in practice.

## Defining Columns

The size and number of columns should be early considerations as you design the pages. Although more variations are possible, Figure 8.27 shows eight ways of placing text on a page (Craig 150). Pattern A provides maximum text; patterns B and C provide more white space; patterns D and E combine maximum text and readability; and patterns F, G, and H provide ample illustrative space. For typewritten material, such as reports, the maximum block (pattern A) is acceptable if double-spaced. For fairly solid prose set in type, the traditional two-column structure (pattern D) is readable. For much technical writing, the more open formats (B, C, F, and G and variations) may be more appropriate.

For a visual effect, it is best not to have a new paragraph start on the last line of a column. A word at the end of a column is called a *widow*. This term is also used for carried-over letters of hyphenated words. If a widow is carried over to

**ELEC**    SOLENOID PULL-IN TEST/SOLENOID RETURN TEST

## SOLENOID PULL-IN TEST

**WARNING**

**Perform this test outside the boat to avoid bodily injury or boat damage due to possible explosion.**

1. Disconnect field coil connector (4) and insulate it carefully.

2. Connect solenoid to a 12 volt DC power source as shown.

3. With 12 volts applied, momentarily close switch (5). The solenoid should push the pinion out to the pinion stop. Due to the high amperage draw, open the switch as soon as the pinion is at the stop. The pinion should remain against the stop until the battery is disconnected.

   a. If solenoid does not move the pinion out, remove the solenoid and manually move the shift lever back and forth. If lever moves smoothly, the pull-in windings have failed and solenoid must be replaced.

   b. If solenoid moves pinion out, but then chatters instead of staying firmly engaged, the hold-in windings have failed and solenoid must be replaced.

## SOLENOID RETURN TEST

**WARNING**

**Perform this test outside the boat to avoid bodily injury or boat damage due to possible explosion.**

1. Disconnect field coil connector (1) and insulate it carefully.

2. Connect solenoid to a power source as shown.

3. Close switch (2).

4. Pull the pinion drive out until it contacts the stop. Release the pinion drive and it should return without any hesitation. Replace solenoid if any delay is noticed.

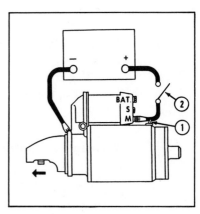

1 - **Voltmeter**
2 - **Carbon Pile**
3 - **Large Gauge Wire**
4 - **Field Coil Connector**
5 - **Switch**

2-16

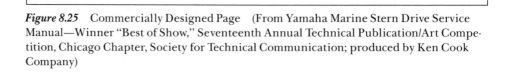

***Figure 8.25*** Commercially Designed Page (From Yamaha Marine Stern Drive Service Manual—Winner "Best of Show," Seventeenth Annual Technical Publication/Art Competition, Chicago Chapter, Society for Technical Communication; produced by Ken Cook Company)

**ANNUAL REPORT**

············ D E S I G N E R   C A S E   S T U D I E S ··············

# Illustration sets pace for Emulex annual's visuals

RIK BESSER
RIK BESSER/
DOUGLAS JOSEPH A.D.S
STEVEN GUARNACCIA, ILLUS.
BESSER JOSEPH PARTNERS, INC.
SANTA MONICA CA
CLIENT: EMULEX CORPORATION

Emulex Corporation designs, manufactures and markets high performance computer enhancement products in the areas of storage and system/user connectivity. Most people don't have the slightest idea what that means. A typical annual report approach of photographing their products wouldn't make it much clearer. In this report we chose to illustrate examples of how the company's products are used. This was an instance where illustration offered a creative and conceptual freedom not readily available with a camera.

The images in this book are based on hypothetical scenarios depicting how different Emulex products are used. We wanted to keep them friendly and easy to understand. It was also important to work with an illustrator who could take the information we had and complete it with his own ideas. Steven Guarnaccia was a good choice. As we began to see roughs of the images, their style started to influence the attitude of other elements in the book. Therefore, in keeping with the relaxed, nontechnical style of the illustrations, the bar charts and diagrams were drawn freehand.

When this project began, we realized that the subject matter for the visuals would be one that had been used frequently: pictures of people sitting in front of computers! It's been done by most of the high technology companies and even companies in unrelated industries. If they've got them, most likely you've seen them. If I had a dollar for every photograph I've seen in an annual report of a person standing or sitting in the near proximity of a computer, I would be retired now and writing this from the beach on Bora Bora. We had to look for a way to communicate this idea from a fresh point of view. Illustration gave us that opportunity. We thank the people at Emulex for supporting that view.

*Graphic Design: USA 44*

*Figure 8.26*  Commercially Designed Page  (From *Graphic Design: USA*, September 1990, showing a sample layout from the 1989 Texaco, Inc. Annual Report)

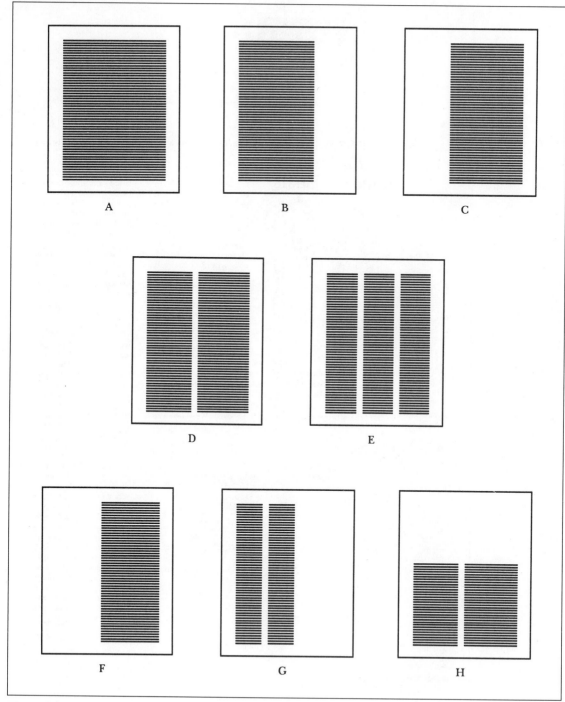

***Figure 8.27*** Eight Ways of Using Columns   (Courtesy of *Designing with Type* by James Craig; copyright © 1971 by Watson-Guptill Publications)

the top of the next column or page, it is called an *orphan*. Avoid both widows and orphans.

## Using White Space

White space should be used functionally as an active design element, not simply as a blank negative background. Think of white space as a unit on the page, just as you think of text and illustrations as units.

Research indicates that clarity can be enhanced "by a rational and consistent use of white space" and that ample white space communicates "user-friendliness" to readers (Felker et al. 83; Hartley 27). Conversely, a cluttered, tightly packed page can intimidate any reader. Adequate white space simply makes reading the page physically easier by allowing the eyes to rest.

More important, white space makes difficult subjects seem easier to comprehend by visually framing information and breaking it into manageable chunks. Even the simple use of white space between paragraphs helps the mind to see the information in that paragraph as a unit. Use extra white space between sections to signal to the reader that one section is ending and another is beginning—a visual way of indicating the organization of a document. You can also emphasize text—a word, a phrase, or a paragraph—by framing it with white space. You need not have access to sophisticated equipment to make use of white space. You can easily indent and skip lines for paragraphs, lists, and other chunks of material.

White space should be balanced with text. Some research even suggests that white space (including margins) should occupy 50 to 60 percent of a page (Sadowski 30; Brockman 75). Although white space may add to the length of a document, it is economical. For example, Schoff and Robinson observe:

> Careful use of white space may add a tiny bit to the cost of a manual because not every page is crammed corner to corner, but it will help ensure that the manual is used. If the manual sits untouched on the shelf, the whole cost of producing it is wasted. (69)

## Using Lists

You may not think of them as elements of document design, but lists provide an effective way to use white space. Lists combine both linguistic and visual elements to highlight words, phrases, and short sentences. They are particularly useful for certain types of information:

- Steps in sequence
- Materials or parts needed
- Items to remember
- Criteria for evaluation
- Concluding points
- Recommendations

However, watch out for both too many lists and too many items in lists. A list may not really be necessary—say, for only three items. Such a simple list should normally appear within a sentence. You can tick off the items with numbers or lowercase letters in parentheses, as in (1) or (a), if you so desire.

When you do use a list, recall from Chapter 4 that most typical readers are able to retain no more than seven items in short-term memory at a time. If your list must contain many items, cluster them in groups of five or fewer.

Follow the standard advice on lists, such as listing only comparable items, following parallel structure, and using only words, phrases, and *short* sentences (Brusaw, Alred, and Oliu 385–386). The surrounding text should provide a context for the list. Don't use *for example* by itself followed by a list of phrases. The sentence or phrase that introduces a list should end with a period or a colon. Here are other practical considerations to simplify the look of lists (Plunka 37):

- Avoid punctuation at the end of listed items, unless they are complete sentences.
- If any item in a list is punctuated, punctuate all.
- Follow a lead-in term or phrase by a period and two spaces, then write the text, as shown here. (Do not use a dash.)

    1. **Move Required.**    Move text by pressing . . .

- Establish a standard for capitalizing the first letter of listed items, and consistently follow that standard.

Precede listed items with numbers or lowercase letters only when sequence is important. Prefer bullets or other graphic devices like the "greater than" symbol (>) when sequence is not the issue.

Use bullets or lowercase letters for the sublist of a numbered item. Consider using hyphens or numbers for the sublist of a bulleted item. Of course, you should remain open to particular needs of specific documents. It may be useful to establish a standard for graphic devices that begin lists and other format issues.

## Using Illustrations

Readers notice illustrations before text and large ones before small ones. Thus the size of an illustration is the reader's gauge to its importance. But a picture is large only in contrast to its frame and the surrounding text, as shown in Figure 8.28 (Marra 13). Consider the proportion of the illustration to the text. Proportion often means employing the three-fifths rule: page layout is "more dramatic, more appealing when the major element (photo, copy block, etc.) occupies 3/5ths rather than half of the available space." In fact, any odd fraction proportion is more suitable by "avoiding the pitfall of humdrum which is a result of page layout halves" (Marra 11). Finally, remember that some illustrations are not readable if they are too small.

Although illustrations can be gathered in one place (as at the end of some reports), placing examples and illustrations within the text provides visual relief

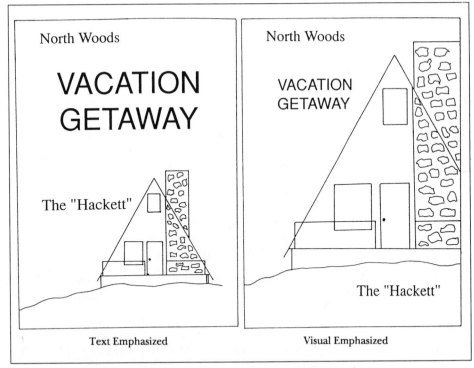

***Figure 8.28***    Variation in Size of Visual

and also makes them more effective by putting them closer to the accompanying explanations. Placing illustrations to the left of text is one of the best positions because we process information from left to right, so that layout provides readers with an overview and a "quick concept from the picture before they read the details in the text" (Rude 71). When you cannot place illustrations to the left of the text, consider aligning the edges of visuals with the margins of text to bring a sense of order to the page (Sadowski 29–30). Use runarounds, in which the text wraps around an illustration, very sparingly; a number of graphic designers recommend against their use (White 106).

## PACKAGING THE DOCUMENT

The selection of document size (height and width) and resulting page size should come early in the document process, since that decision helps determine such elements as type sizes, typefaces, and line lengths. The page size of the document should depend in part on its purpose. Will the document be read thoroughly? Used for quick reference? For assembling a frequently modified device or system? Or as a comprehensive reference document? The answers to such questions will help determine if you need a 4¼-by-5⅜-inch booklet-size page that will fit in a shirt pocket, an 8½-by-11-inch hole-punched page that can be easily inserted into a binder, or a 5½-by-8½-inch trade book page size that can be shelved with other reference books.

The size of the document also depends on where the reader will use it: on a desk near a VDT, on a repair bench, or in the cabin of an aircraft or a construction vehicle. A secretary may need a 6-by-9-inch spiral binder, a card for a desk tray, or even a wall chart for quick reference. Size may also depend on where the document will be stored. An odd size, such as 3 by 6½ inches, may be a good size for an auto owner's manual that will be stored in a glove compartment. If a document must be packed with a product, the document should not add to the size of the packaging. Other factors in choosing page or document size include the following:

- What the reader expects or prefers
- How large the illustrations must be
- How production costs are affected

Some page sizes are more easily available and economical to produce since they make use of the ways standard sheets can be cut (Bruno 169):

| | |
|---|---|
| 4 by 9 inches | 9 by 12 inches |
| 6 by 9 inches | 8½ by 11 inches |
| 5½ by 8½ inches | 4¼ by 5⅜ inches |
| 4½ by 6 inches | |

Some industries gravitate to certain sizes. Some suggest, for example, that the 5½-by-9-inch page size is becoming a new standard page for external computer documentation, for a number of reasons (Brockman 79):

- It looks less intimidating than 8½ by 11 inches.
- It is physically easier to handle when the user is working at a computer terminal on a crowded desk.
- It automatically decreases column width, thus making the pages more readable.
- It is more difficult to photocopy and thus pirate, because this size is difficult for automatic loaders.

## Binding the Document

The binding, like page size, should be based on the purpose of the document and the environment in which it will be used. If the document will be used by a reader who must have both hands free, for example, the document must lie flat. If the document must be updated often, the pages must be easy to remove and replace. A repair manual for industrial equipment might require a heavy-duty binding that will hold up under rough conditions.

Essentially, there are four types of binding: wire-stitched, glued, thread-sewn, and mechanical. Examples of these bindings are shown in Figure 8.29. Wire

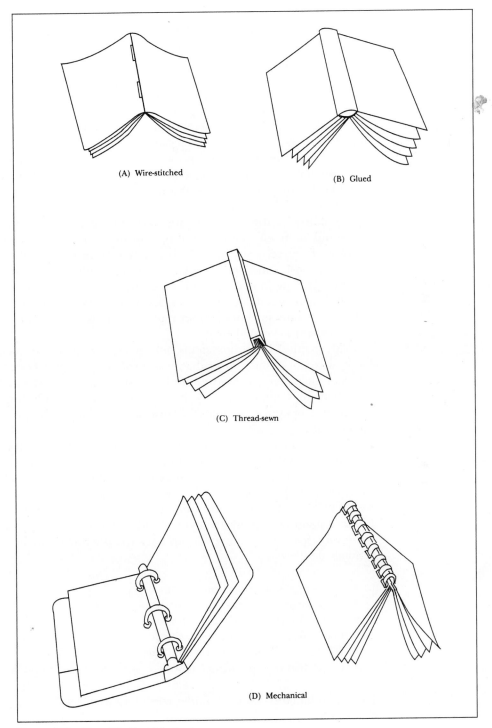

(A) Wire-stitched

(B) Glued

(C) Thread-sewn

(D) Mechanical

*Figure 8.29* Four Binding Methods

stitching, a popular binding for informational booklets and industrial operator manuals, includes saddle-wire and side-wire stitching. In saddle-wire stitching, a machine punches wire through the backbone fold of the publication and crimps the ends (similar to stapling). Saddle-wire bindings can be bent to lie flat like three-ring and spiral-bound bindings. Side-wire stitching punches the wires from the front cover through the back edge of the booklet. Side-wire-stitched publications will not lie flat.

Gluing is used for a method called *perfect binding*: the back edges of the folded sections of pages (called *signatures*) are ground (similar to sanding) until they are rough enough to hold a coat of glue. The cover is then attached with glue that also holds the pages together. Perfect bindings look more finished and booklike, but they are not durable and may come apart if the spine is "broken" to lie flat.

Saddle-wire stitching and gluing are appropriate for publications to clients or the public in which a polished or "finished" image is important. Such bindings are, of course, costly since they often require heavy machinery.

The most expensive permanent binding is thread-sewn. Thread sewing actually sews the folded signatures together with a fabric thread. A hard or soft cover is then glued to the spine. Because of its expense, a sewn binding is appropriate only for publications that must "look like a book," such as a corporate or organizational history.

Mechanical binding systems hold pages together with metal rings, plastic loops, wires, or other devices that pass through holes punched along the binding edges of the pages. Mechanical binding systems are of two types: permanent and loose-leaf (or semipermanent). Permanent mechanical bindings are useful for repair and operator manuals because it is important to prevent pages from coming loose.

Semipermanent, loose-leaf systems are the most common types of bindings for both internal and frequently updated publications. If a document is likely to be updated often, a loose-leaf, three-ring binder is the best choice because it updates easily, protects the paper, remains durable, and stands up and shows up well on a shelf. Choose well-made binders rather than off-the-shelf or "school" binders; good ones hold up better. Binders come in a variety of shapes with a variety of features. Among the options you should consider when you select binders are these:

- Binder capacity (number of pages)
- Ring fittings ("slant D" are excellent)
- Tab dividers (should be reinforced)
- Page lifters (should be hard plastic)
- Display easel and flip chart options
- Cover features: printing, artwork, color, material, and pockets

For a thorough discussion of binder considerations, see d'Agenais and Carruthers (208–231).

## Publication Cover

Through its clarity, authority, and suggestion of accessibility, a cover should invite the audience to read the document. Covers contain, as appropriate, any of the following:

- The name or logo of the organization
- The title of the publication
- The name of the device, system, procedure, etc.
- Publication data: date, numbers, etc.
- A picture or drawing of the device
- The precise model number

Large organizations often use a standard design and color for publication covers. If a publication is to be used in a harsh environment, the cover should protect the inside pages by being made of heavy paper stock or vinyl.

The cover for a document not only provides information but also projects an image of the organization. The image the cover projects ought to match the audience's expectations. For a publication produced by a charitable organization, readers may expect a modest cover. In contrast, a professionally designed, high-quality cover may suggest authority or quality to clients or customers. Of course, expensive padding (such as that used for wedding albums) for three-ring binders would seem extravagant in most settings.

## Selecting Paper

When selecting the appropriate paper, you must depend on the advice of printing experts. However, a knowledge of paper grades and sizes will enable you to communicate what you want and help your organization get the most for its money. You can visit printers and paper distributors, who may provide samples you can keep on file. You may also wish to learn something about how paper is made in order to understand its characteristics (see Bruno 156). For example, due to the way in which it is manufactured, paper has a grain, much like wood. Knowing that the paper is stronger in the direction of the grain can be important in choosing the direction of a fold, as in a three-fold brochure, for example.

### Size and Weight
The size of paper a printer uses depends on four factors:

- The size available in the grade selected
- The sizes in stock
- The size of the press
- The size of the publication

You should also consider the most economical cuts, listed on page 194. Paper mills weigh paper according to the weight in pounds of 500 sheets (a *ream*) of the paper in its basic size. For example, some 8½-by-11-inch paper is called 20-pound paper because in its uncut, basic size (17-by-22-inch), 500 sheets weigh 20 pounds. However, printers and suppliers usually list the price according to 1,000 sheets. For an overview of papers, see Bruno (169) and Field et al. (447–514).

### Porosity and Durability

The two major qualities of paper are porosity and durability. Porous paper like that used for newspapers accepts ink easily, so photos and thin letter strokes tend to blur. Porous paper also easily absorbs substances such as oil and dirt—a liability if a document must be used for industrial applications. Less porous paper, by contrast, holds ink on the surface cleanly and thus prevents "bleedthrough" (that is, it does not allow one side to be seen from the other).

If a document will be referred to often, consider a durable paper with cotton fibers or other ingredients designed to help the paper hold up. Although they are more expensive, coated papers are excellent for both density and durability. The advantage of coating is that it allows for a sharper, denser image on the page. When images are printed on coated paper, the ink does not soak into the paper as much as it does with noncoated paper. However, coated paper can be so glossy that it reflects overhead light and makes reading difficult.

### Grades of Paper

Grades of paper are defined by their weights, coatings, and similarities in uses. Although experts differ about grades of paper, nine basic categories are common. Each is described here, with the standard dimensions shown in parentheses.

- **Bond** (17 by 22 inches).   Used for business letters and forms—hard surface good for typewriters, impact printers, and pen and ink; usually cut to 8½ by 11 inches.
- **Text paper** (25 by 38 inches).   Used for higher-quality jobs than book paper—announcements, menus, folders, booklets, brochures, and newsletters.
- **Book paper** (25 by 38 inches).   Used for books, magazines, photocopies, and general printing; a wide variety of weights; less expensive than text paper.
- **Offset paper** (25 by 38 inches).   Similar to book paper but manufactured to resist shrinking after exposure to the moisture present in offset printing.
- **Bristol** (22½ by 28½ inches).   Sometimes called *index paper*; stiff and durable for use as index cards, tickets, greeting cards, and the like.
- **Tag or cardboard** (24 by 36 inches).   Heavy (100–250 pounds); often used for file folders and tags.
- **Newsprint** (24 by 36 inches).   Cheapest paper available; lacks strength and durability and yellows after a short time; used for newspapers.
- **Coated paper** (25 by 38 inches).   Paper with a smooth coating, ranging from dull to glossy and coated on one side or both; used for high-quality printing because of its smooth surface and uniform ink receptivity.
- **Cover paper** (20 by 26 inches).   Coated and text papers made of heavier weights and matching colors; used as covers on booklets, etc.

## *Checklist 8.1*

# Layout and Design

The order of elements in this checklist corresponds roughly to the order in which you need to make decisions rather than the order in which layout and design are typically learned.

☐ Determine packaging
  • Publication and page size
  • Type of binding
    ○ Stitching
    ○ Gluing
    ○ Sewing
    ○ Mechanical binding
  • Cover design
  • Paper grade

☐ Arrange elements on the page
  • Grid, thumbnail, or dummy
  • Number and width of columns
  • Adequate and useful white space
  • Listed elements
  • Illustration size and placement

☐ Determine typographic features
  • Typeface          • Justified or ragged margins
  • Type size         • Leading
  • Line length       • Word and letter spacing

☐ Determine highlighting and finding devices
  • Typographic emphasis (uppercase, italics, boldface)
  • Headings and captions
  • Headers and footers
  • Rules and boxes
  • Icons and decorative features
  • Color or screening

---

### WORKS CITED

American National Standards Institute. *American National Standards for Information Sciences Scientific and Technical Reports: Organization, Preparation, and Production.* ANSI Z39.18-1987. New York: American National Standards Institute, 1987.

Barton, Ben F., and Marthalee S. Barton. "Simplicity in Visual Representation: A Semiotic Approach." *Journal of Business and Technical Communication* 1 (1987): 9–26.

Benson, Philippa. "Writing Visually: Design Considerations in Technical Publications." *Technical Communication* 32.4 (1985): 35–39.

Brockman, R. John. *Writing Better Computer User Documentation: From Paper to Online*. New York: Wiley-Interscience, 1986.

Bruno, Michael H., ed. *Pocket Pal: A Graphic Arts Production Handbook*. 13th ed. Memphis: International Paper, 1988.

Brusaw, Charles T., Gerald J. Alred, and Walter E. Oliu. *Handbook of Technical Writing*. 3rd ed. New York: St. Martin's, 1987.

Carter, Rob, Ben Day, and Philip Meggs. *Typographic Design: Form and Communication*. New York: Van Nostrand Reinhold, 1985.

*The Chicago Manual of Style*. 13th ed. Chicago: U of Chicago P, 1982.

Craig, James. *Designing with Type*. New York: Watson, 1971.

d'Agenais, Jean, and John Carruthers. *Creating Effective Manuals*. Cincinnati: South-Western, 1985.

Doebler, Paul D. "The Fundamentals of Type." *Graphic Arts Manual*. Ed. Janet N. Field et al. New York: Arno-Musarts, 1980. 110–115.

Dugdale, Juanita. "Picking the Typeface: A Designer's View." *Simply Stated* Aug. 1985: 3.

Ettenberg, Eugene M. "Typographic Legibility." *Graphic Arts Manual*. Ed. Janet N. Field et al. New York: Arno-Musarts, 1980. 116–118.

Felker, Daniel B., et al. *Guidelines for Document Designers*. Washington, DC: American Institutes for Research, 1981.

Field, Janet N., et al., eds. *Graphic Arts Manual*. New York: Arno-Musarts, 1980.

Flower, Linda. *Problem-solving Strategies for Writing*. Orlando, FL: Harcourt, 1981.

Gottschall, Edward M. "Type 'Whys.' " *Graphic Arts Manual*. Ed. Janet N. Field et al. New York: Arno-Musarts, 1980. 120–122.

Hartley, James. *Designing Instructional Text*. 2nd ed. London: Kogan Page, 1985.

Kung, Hans. "The Grid System." *Graphic Arts Manual*. Ed. Janet N. Field et al. New York: Arno-Musarts, 1980. 39–42.

Marra, James L. "For Writers: Understanding the Art of Layout." *Technical Communication* 28.3 (1981): 11–13, 40.

McGhee, Brad. *The Complete Guide to Writing Software User Manuals*. Cincinnati: Writer's Digest, 1984.

Meyer, B. D. "The ABCs of New-Look Publications." *Technical Communication* 33 (1986): 16–20.

Nagy, Patricia M. "Writing Educational Material for Patients." M. A. thesis. U of Wisconsin-Milwaukee, 1985.

Plunka, Gene A. "The Editor's Nightmare: Formatting Lists within the Text." *Technical Communication* 35 (1988): 37–44.

Romano, Frank J. *The Typencyclopedia: A User's Guide to Better Typography*. New York: Bowker, 1984.

Rubens, Philip M. "A Reader's View of Text and Graphics: Implications for Transactional Text." *Journal of Technical Writing and Communication* 16 (1986): 73–86.

Rude, Carolyn D. "Format in Instruction Manuals: Applications of Existing Research." *Journal of Business and Technical Communication* 2 (1988): 63–77.

Sadowski, Mary A. "Elements of Composition." *Technical Communication* 34 (1987): 29–30.

Schoff, Gretchen H., and Patricia A. Robinson. *Writing and Designing Operator Manuals*. Belmont, CA: Lifetime Learning, 1984.

Stermer, Dugald. "A Desert Island Question Concerning Type." *Communication Arts* 25.4 (1983): 72–77.

U.S. Government Printing Office. *Style Manual*. Washington, DC: GPO, 1984.

Waller, R., P. Lefreri, and M. MacDonald-Rosse. "Do You Need That Second Color?" *IEEE Transactions on Professional Communication* 25 (1982): 80–86.

White, Jan V. *Editing by Design*. 2nd ed. New York: Bowker, 1982.

## FURTHER READING

Barton, Ben F., and Marthalee S. Barton, eds. Spec. issue of *The Technical Writing Teacher* 17 (1990): 189–270.

Bibus, Connie Martin. "The New Visual Techniques: Learning and the Page." *Proceedings of the 34th International Technical Communication Conference* (1987): VC-25–VC-28.

Caernarven-Smith, Patricia. "Using a Word Processor for Page Design." *Word Processing for Technical Writers.* Ed. Robert Krull. Amityville, NY: Baywood, 1988.

"Document Design Moves into the Next Decade." *Technical Communication* 36 (1989): 313–381.

Duffy, Thomas M., and Robert Waller, eds. *Designing Usable Texts.* Orlando, FL: Academic, 1985.

Felker, Daniel B., et al. *Document Design: A Review of Relevant Research.* Washington, DC: American Institutes for Research, 1980.

Goswami, Dixie, et al. *Writing in the Professions.* Washington, DC: American Institutes for Research, 1981.

Kelly, Nichols. "Improving Your Page Layout Skills." *Proceedings of the 35th International Technical Communication Conference* (1988): VC-53–VC-55.

Keyes, Elizabeth. "Information Design: Maximizing the Power and Potential of Electronic Publishing Equipment." *IEEE Transactions on Professional Communication* 30 (1987): 32–37.

Kostelnick, Charles. "A Systematic Approach to Visual Language in Business Communication." *Journal of Business Communication* 25.3 (1988): 29–48.

Pakin, Sandra, and Associates. *Documentation Development Methodology.* Englewood Cliffs, NJ: Prentice, 1984.

Southard, Sherry G. "Practical Considerations in Formatting Manuals." *Technical Communication* 35 (1988): 173–178.

Spyridakis, Jan H. "Signaling Effects: A Review of the Research, Part 1." *Journal of Technical Writing and Communication* 19 (1989): 227–240.

"Using Icons as Communication." *Simply Stated* Sept.–Oct. 1987: 1 + .

### Chapter 8: Layout and Design

INTERVIEWER: What is the ultimate purpose of design, as you see it?

MARY: To make it easier for people to understand the information on the page.

INTERVIEWER: How have you learned about layout and design?

MARY: I've read articles on it. I've looked at manuals that I've liked and tried to emulate their design. Then, of course, in conversations with other writers and in presentations by the page layout and design committee, I've learned why some formats work better than others.

INTERVIEWER: At what point in the publication process do you begin to consider layout and design issues?

MARY: Very early—in the planning stages.

INTERVIEWER: What general form (size and binding) was used for the *Introduction to Using the International Banking System* manual?

MARY: A three-ring binder with 8½-by-11-inch paper.

INTERVIEWER: Any specific reason why you used the binder?

MARY: We use a three-ring binder and 8½-by-11 paper for almost all our manuals. The manuals we do are generally updated too frequently to warrant any type of binding other than a three-ring binder.

INTERVIEWER: So the workstation was big enough to accommodate the full-size manual? Some manuals come in the 6-by-9-inch format because it fits better in the workspace—it's less awkward.

MARY: Well, in the International Banking Department, employees have a generous amount of desk space (they work at big old desks). Since these were the people we were preparing the manual for, we decided that the normal 8½-by-11 binder was acceptable.

INTERVIEWER: What general aspects of audience and purpose determined the overall design of this document?

MARY: Because I wanted the manual to be accessible to its users, who were nonnative English speakers who were skeptical about the system, I kept chapter size to 10 to 15 pages, included plenty of white space and illustrations, and generally made the manual pleasant to look at. I didn't want to put any stumbling blocks in the users' way.

INTERVIEWER: What are the general design capabilities of your computer system?

MARY: It allows us to do a lot. We can have paragraphs set up in more than one column; we can left-justify, right-justify, or center text; we can change the size of type anywhere from 6-point to 72-point; and we can choose from about 15 typefaces (although not every face exists in every point size). We can change line height, or leading, at will. We have running headers and footers, which we can alternate (one for the left page, another for the right page); we can have page numbers appear with or without prefixes. We can draw, resize, and place graphics right within the document. We can create rules within text, and we have a very powerful table feature that lets us enter information into a basic table format and then have the table stretch to fit the text we enter; we can also size tables.

INTERVIEWER: How much of a manual's design is standardized (that is, dictated by company standards)?

MARY: Quite a few design decisions are set by our standard format [the International Banking System Manuals Style Sheet, which lists such standards, appears in the Case History for Chapter 14]. I do have the freedom not to follow every single standard to the letter, but unless there is a convincing reason not to, it's to my advantage to use the standard format.

INTERVIEWER: Why?

MARY: Because documents are easier to read *and* easier to revise if they are in a standard format.

INTERVIEWER: In preparing the *Introduction* manual, how much control did you have of the design and layout?

MARY: Although I used the standard page format with only a few variations, I made plenty of design decisions that weren't covered by the standards. [Mary devised the "design decision table" in Figure C.8 to present some of the design choices that were made for the *Introduction* manual, the reasons for these choices, and additional comments.]

| Design Element | Decision and Rationale |
|---|---|
| Typefaces | We have about 15 typefaces available. We decided on a face with serifs called Classic for body text because it's the most readable and it comes in the widest range of type sizes. |
| | For side-margin captions, we use a boldfaced sans serif typeface called Modern. |
| | For key names, we use a 10-point type size in the same face as the text surrounding the key name (e.g., POST key). |
| | For screen and report samples, we use a typeface called Terminal that resembles the type that appears on actual computer screens or in reports. |
| | Several books recommend this type of combination for text and headings. But because too many different type styles can quickly overwhelm the page, making it look too busy, we made a conscious effort to limit the number of typefaces on the page. (We don't even touch about seven available faces on our system.) |
| Leading (space between lines, which we generally refer to as "line height") | We can set any leading we want. For this 12-point type, we first tried a 14-point line height but later switched to a 13-point line height because the 14-point seemed too "floaty," too open. |
| Margin type (ragged right or justified right) | Ragged right was chosen because studies have shown that it's more readable. |
| Heading style | Headings are in 18-point boldface Modern (sans serif) on the first page of a topic. Headings are repeated as 14-point boldface Modern headers on subsequent pages. |
| | Major side-margin captions are 12-point boldface Modern. Only the first word in the caption is capitalized. |
| | Minor side-margin captions are in 12-point italic Modern. Only the first word in the caption is capitalized. |

*Figure C.8*   Design Decision Table

| Design Element | Decision and Rationale |
|---|---|
| Headers and footers | We chose to use them and selected their format for a number of reasons. |

- It is important to distinguish the first page of a topic so that readers can recognize when they reach a new topic. Thus the header on the first page looks different from headers on other pages.
- The rules that appear as part of the headers and footers give them definition. They also separate the headers and footers from the text area of the page without requiring the type in the headers and footers to be "giant-sized" (in order for them to stand apart). That way, we can fit meaningful headers into the manual—we don't have to shorten them because of a large type size.
- The arrow icon at the end of a footer rule indicates a continuation from a previous page. The square icon at the end of a footer rule indicates the end of a topic.

| Design Element | Decision and Rationale |
|---|---|
| Graphic devices for emphasis | A number of choices were made in this area. |

- We chose the triangular arrowhead bullet for bulleted listings because it looked sharper than the round bullet. We also had a choice between a 10-point bullet and a 12-point bullet. We decided to go with a 12-point bullet because it fits the body text better.
- We use italics to introduce new terms and for emphasis.
- We use boldfaced Classic for the exact text a reader is supposed to enter on a screen. We don't use quotation marks because they make the text look busy and because it's awkward to have to explain to readers that they don't actually enter the quotation marks, only text within the quotation marks.

**Figure C.8** *Continued*

| Design Element | Decision and Rationale |
|---|---|
| Margin and column size | Industry recommendations call for a 1-inch border around the edge of the page. The columns on the left of every page, where side-margin captions appear, are indented 25 spaces from the left margin. This width was chosen to align the body margin with an underlying grid. The design concepts we followed in the International Banking manuals say that objects on a page are most balanced when they are aligned with an underlying grid that creates a functional center based on dividing the page roughly into thirds. |
| Size of illustrations | We have a consistent, predetermined set of sizes for reports and screens: one for all reports, one for detailed screens, and a few smaller ones for use in diagrams and screen flowcharts showing screen flow. |
| | We have conventions for the appropriate use of the various screen sizes: the large one is used as the main sample and the small ones may be used in diagrams. We're not allowed to size each screen to fit the text that appears on it; that would create too much of a jumble. |

**Figure C.8**   *Continued*

# 9

# Drafting a
# Document

Developing Confidence
Using Time Management Tactics
Communicating with the Reader
Opening the Document
Special Previewing and Finding Devices
Concluding the Document

This chapter is intended to bolster your confidence in meeting professional deadlines, suggest tactics to enable you to be productive, and offer strategies to help you communicate effectively with the reader. The chapter also deals with what many professionals find to be the toughest chores in writing the draft: the opening and the closing. The final section provides advice on using previewing and finding devices.

## DEVELOPING CONFIDENCE

Professional writers must deal with the pressure of constant deadlines. A grant proposal, for example, *must* be submitted by the date prescribed by the governmental or private funding agency; otherwise, the agency will not consider the proposal. Other documents *must* be finished in time to be distributed with the products they are to accompany, and the products must be distributed by specific target dates in order for the company to remain competitive. Understandably, such deadlines can create anxiety as a writer begins the first draft. Since as a professional writer, your livelihood will depend on your ability to get going, you need to use the four tactics that experienced professional writers follow to approach their drafts with confidence:

- Prepare adequately for the task.
- Focus on writing instead of revision.
- Avoid procrastination.
- Develop a positive attitude.

Nothing builds a writer's confidence more than adequate *preparation*. As the Case History at the end of this chapter reveals, if you haven't done enough research to feel comfortable with the material, you will no doubt face great anxiety as you begin the draft—perhaps even "writer's block." Furthermore, if you start without an outline adequate to the task, as described in Chapter 6, not only will you face writing anxiety, but you will also be frustrated because you will spend far too much time producing a first draft (Weiss 140). Adequate preparation also means developing good work habits, as described later in this chapter.

Second, to avoid undermining your confidence, keep in mind that writing and revising are two very different tasks. When you write the draft, consider yourself a *writer* communicating with your reader; when you revise, then—and only then—become your own toughest *critic* as described in Chapter 10. As Peter Elbow points out, one of several dangers of "trying to write things right the first time" is that it puts pressure on you—pressure that can become debilitating (42). In fact, *any* attempt to correct or polish your writing only stimulates the "internal critic," which can undermine your ability to complete the draft (Mack and Skjei 31–38).

One way to avoid the temptation to revise is understanding that the first draft is *necessarily* rough and unpolished. As Edmund Weiss puts it: "Far from scolding yourself for not being able to write a smooth, readable sentence the first time out, you should comfort yourself that it is in the natural order of first drafts to be clumsy and long winded" (138). So when you write the draft, don't worry about such issues as precise word choices, usage, syntax, grammar, or spelling. Instead, concentrate entirely on getting the message to your reader, as we shall discuss in this chapter. In other words, concentrate on *what* you are writing not *how* you are writing it.

Third, once you are prepared and understand that the goal of the first draft is simply to produce a draft, *avoid procrastination at all costs*. For example, be wary of such diversions as rearranging files, watching the clock, checking the mail, or calling to check an appointment. They may be simply ways of avoiding work. Furthermore, you cannot afford to wait for inspiration (the "marination" method, as it is sometimes called); marination may work if you have no deadline to meet, but it is often an excuse for stalling (Weiss 141).

Finally, to help you avoid procrastination and at the same time build a positive attitude toward your writing, keep in mind the time-tested advice of successful writers:

- Use past success to bolster you: you *have* completed writing tasks before; understand that you *will* this time.
- Avoid the trap of thinking that everything you write should be either effortless or impossibly difficult.
- Remember that the sooner you get to work, the sooner you'll be finished.

---

## USING TIME MANAGEMENT TACTICS

Because professional writers must deal not only with constant deadlines but also with several assignments at once, managing time is an essential part of the writing process.

### Allocate Your Time

One effective time management practice is keeping a calendar that indicates various deadlines for projects, appointments for interviews, and time periods for gathering information. Your daily calendar should also include "writing appointments," times set aside for writing that you must keep (without interruptions) as if you had an appointment with a person.

Within the deadlines set by managers and others, set your own short-term, manageable deadlines for completing sections of a draft and other tasks. Not only can concentrating on such subgoals help you meet the overall deadline, but it can also relieve some of the pressure of writing the draft (Flower 45). Some professional writers think of the completion of subgoals as building a draft "one brick at a time."

List these goals together with your other job tasks; then rank the items in your list. Time management experts advise working on the most difficult or unpleasant tasks during the time of day when your mind is keenest (Ammons and Newell 149).

### Prepare Your Work Environment

Another useful management strategy for writing the draft is to prepare your writing environment and assemble the materials before you begin. Find a place or a method of isolating yourself for writing the draft; then hang out the "Do Not Disturb" sign. Especially when a deadline is in jeopardy, managers often send writers to quiet areas, away from phones and meetings.

Put order into your writing environment by arranging your materials and supplies. Use whatever writing technology (pens and pads, word processing, tape recording, etc.) is most comfortable for you. You may even discover that certain "props" will help you get started. For example, sitting in a favorite chair or placing a reference book on your desk may symbolize your commitment to yourself and your work (Tichy 23).

## Use Boilerplate

Often in professional writing, parts of a first draft can be constituted of text from existing internal documents (manuals, proposals, brochures, reports, etc.). As long as the rights to such boilerplate are the property of your employer, you can use it freely without fear of violating copyright laws.

Using boilerplate is much like quoting from your own writing, and it can save you many hours of work on a first draft. However, seldom is boilerplate entirely appropriate to your document, and you can easily forget the needs of the current audience when you use boilerplate. So even though boilerplate can save time during the drafting stage, be *very* careful to adapt it to the style, tone, audience, and purpose of your current document.

## Remain Flexible

Start with the outline as a guide, but remember that it is not cast in concrete, and you should feel free to improve your organization as you work. Consider starting with the easiest part ("frosting first") just to get moving. You may find that writing out a statement of your purpose will also help you to get started.

Once you are rolling, keep going. You may even wish to write comments to yourself (as shown in the Case History) while you are writing the rough draft if that tactic keeps you moving.

When you reach landmarks (such as the subgoals described earlier) or feel powerfully tempted to start revising, you may need to take a break. When you do, leave a signpost, such as a note in the outline with the date and time you stopped so that you will not waste time searching for your place when you resume work. When you finish a section, reward yourself with a cup of coffee, a short walk, or another small diversion. In fact, a physical activity serves as an excellent break for writers. If possible, avoid immersing yourself in another mental activity while you are on your break. If you are not under mental pressure, you may even discover a solution to a nagging writing problem.

When you resume, you should reread what you have written up to the point where you stopped so that you can recall your frame of mind. Some writers also like to change their writing tools or environment when they resume writing the draft. In any case, remember, when you work, work; when you take a break, do that.

# COMMUNICATING WITH THE READER

As stated earlier in this chapter, when you are writing the draft, you should focus on communicating with the reader. To communicate with the reader, you must avoid the trap of writing about a subject for yourself, which is *very* easy even for experienced writers. So they can focus on their readers, some writers imagine themselves talking with their readers throughout the document, while keeping the purpose of the document firmly in mind.

## Empathize with Your Reader

As suggested by the STC Code for Communicators in Chapter 1, you should value your readers' effort and appreciate that their task may be difficult. If you are writing a grant proposal, for example, appreciate the task of a busy decision maker who may have dozens of grants to review in very little time. If you are writing a manual, learn to empathize with your audience on small things, such as the difficulty of finding an on-off switch hidden behind a printer and under the lip of its plastic frame. Remember, what was difficult for you to understand at first will also be difficult for your reader.

## Visualize the Reader and the Setting

One way to empathize with your audience is to visualize a reader in the act of reading or using your document. Together with what you learn from your audience analysis, this picture will help you predict your readers' needs and reactions. (The audience analysis checklist from Chapter 4 can help create this picture.)

You might imagine a typical reader sitting across the desk from you as you write directly to that person; taking such an approach will promote communication because your writing will be more direct and conversational. Or if you are writing to an audience of readers with widely different personalities, work environments, backgrounds, or lifestyles, picture a small group of people (who represent these differences) gathered around a conference table. However, if you find you must write to widely different audiences with different purposes, you will need to consider a modular design or multiple documents, as discussed in Chapter 4.

You should also try to visualize the setting in which readers will use your document—at a desk in an office, at a repair bench, in the cockpit of an airplane, on a table next to a computer terminal, and so on. To picture his readers in their setting, one writer of an instruction manual telling "owners of very small businesses how to use a computer program designed to help them with their bookkeeping . . . placed over his desk a photograph of an elderly couple standing behind the counter of the neighborhood grocery that they owned." He used that photo to picture his readers in their environment as he wrote his draft (Anderson 96). Even if you do not go that far, you might post a note with a short description of the reader and the setting near your desk.

## Understand Your Readers' Roles

Walter J. Ong observes that every writer "must construct in his imagination, clearly or vaguely, an audience cast in some sort of role" (12). Although Ong speaks primarily of fiction writers, this principle is as true for technical writers (perhaps even more so, as suggested by the foregoing example). For example, if you are writing instructions or procedures, you might think of your reader as an actor in a role performing certain actions. This notion is based on a concept called the *scenario principle*, which states "that functional prose should be structured around a *human agent* performing *actions* in a particularized *situation*" (Flower, Hayes, and Swarts 42). Thinking of readers as actors in a scenario (where that is appropriate) can help you see the steps that they must perform—their movements "on the stage." Thus you can give more complete information appropriate to their needs.

Early in your career, you will need to work carefully at audience analysis to determine the readers' roles. Later, as you are employed at a specific organization or in a field, you will become accustomed to the roles your readers typically play.

## Establish Your Role and Your "Voice"

Writers must also assume roles. If you are a technical writer, you most often assume the role of a teacher who guides the reader's learning process. Like a teacher, you must do more than explain—you must anticipate your readers' reactions, and their growing understanding of the subject, moment by moment. You must be alert to questions that your readers might ask while they read: Why do I need to read this document? Is this subject easy to learn? How much time must I spend? How flexible is this system? Where can I find a quick answer to a problem? By anticipating that the reader will ask such questions, you will be more likely to answer them as you write the draft.

Much like a teacher, you will discover that readers' interests (like students') do not always coincide with their *needs*. Some readers, for example, would prefer not to read your document at all; however, they are interested in completing a task as quickly as possible. You must demonstrate how your document links the readers' interests in completing a task with their needs to use the document.

As you write the draft, consider what voice your readers should hear in the document. Should it be authoritarian or friendly, formal or accessible, provocative or reassuring, or somewhere in between? In some cases, you must determine the voice you adopt by considering what is appropriate to your specific purpose. To reassure while encouraging readers to use a manual like the one shown in the Appendix, the voice is slightly formal (or "businesslike") and confident, yet helpful and nonjudgmental:

> The front pocket of this manual contains a card showing how to use the *function keys* that appear on the keyboard you will use for your work. You may find it useful when reading the chapters on Using Menus or Logging On and Off to refer to the quick reference card.

In other cases, your voice should be based on what readers expect. Readers of academic essays, for example, often assume that writers will use a cautious, somewhat formal voice, yet one that treats the reader as a colleague, as in the following:

> In short, readability has become the criterion for judging style in technical writing. That much does not concern me because reading comprehensibility and reader efficiency seem to me particular defensible goals for technical writing, especially in

> light of E. D. Hirsch's advocacy of "relative readability" as a criterion for assessing the quality of *all* writing. What does concern me is that, in giving advice to writers about style, we have often misused some readability studies and overlooked others. (Selzer 72)

Readers of some company newsletter articles, by contrast, may expect a voice that is fast-paced and reportorial:

> Author! Author!
>
> It was a number of centuries back that the Roman satirist Persius said, "Your knowing is nothing, unless others know you know."
>
> Today at Allen-Bradley some of our engineers are subscribing to Persius's philosophy. They contend it takes more than the selling of a product for a corporation to enjoy continuing success.
>
> "We must sell our knowledge as well," stressed Don Fitzpatrick, Commercial Chief Engineer. "We have to get our customers to think of us as knowledge experts." (Allen-Bradley 6)

## OPENING THE DOCUMENT

One of the most difficult tasks in writing a draft is opening the document. Yet opening segments are often crucial to the success of a document. An introduction, for example, often sets the stage for the entire document. Inexperienced writers often fail to provide any sort of opening element, and the document simply dives into the details immediately.

If the opening segment of a draft gives you trouble, consider writing it last. Many professional writers believe that an introduction *ought to* be written last. Only then does the writer have enough perspective on the material to provide an appropriate orientation for the reader. We shall review a range of opening segments appropriate to various types of documents: full introductions, openings, summaries, abstracts, prefaces and forewords, and "how to use this document" sections.

### Full Introductions

Formal reports, proposals, elaborate technical publications, and similar documents normally require full introductions. Such introductions give your readers enough general information about the subject to enable them to understand the technical details or make maximum use of the detailed information that follows. As the following example from a mainframe computer manual demonstrates, an introduction may provide a broad definition of a process that will be dealt with in specific detail in the body of the document.

> The System Constructor is a program that can be used to create operating systems for a specific range of microcomputer systems. The constructor selects requested operating software modules from an existing file of software modules and combines

> those modules with a previously compiled application program to create a functional operating system designed for a specific hardware configuration. . . .

> The constructor selects the requested software modules, generates the necessary linkage between the modules, and generates the necessary control tables for the system according to parameters specified at run time. These parameters can be input as online responses to sequentially presented display messages.

Other introductions identify the topic and its primary purpose in the first sentence or two.

> This proposal will show why INFOSYSTEM, Inc. is able to provide First Bank Network with an efficient, cost-effective reorganization of the NPC computer system. . . .

Introductions may also provide historical or background information.

> This proposal is based on our experience helping other firms manage their information systems. For example, in 1990 we reorganized the systems at Antex Corporation, working closely with their "Five-Year Committee" and senior management. The result of our effort was a 43% improvement in their productivity.

Some introductions point out the benefits of a product, what it can do for the reader, and what kind of preparations the user must make (Price 69).

> The N30 lipid-profile analyzer will provide quick and accurate results of patient samples, enabling you to provide your patients with lipid profiles during a half-hour examination. You must ensure, however, that the "sterile-guard" filter has been properly removed before you enter a new patient blood sample.

In writing introductions, you may also encounter a dilemma that is common in technical writing: you can't explain topic A until you have explained topic B, but you can't explain topic B before explaining topic A. The solution is to explain both topics in broad, general terms in the introduction. Then, when you need to write a detailed explanation of topic A, you will be able to do so because your reader will know just enough about both topics to understand your detailed explanation, as in the following example.

> The NEAT/3 programming language, which treats all peripheral units as file storage units, allows your program to perform data input or output operations depending on capabilities of the specific unit. Peripheral units from which your program can only receive data are referred to as *source units*. Units to which your program can only deliver data are referred to as *destination units* or as combination *source-destination units*.

This introduction allows the writer to explain any one of the three units in detail later because readers will have at least a general knowledge of their functions.

## Openings

Not all documents require a full introduction of one or more paragraphs; however, most documents do need an opening, which may be only the first sentence or two of a document. An opening, of course, may be used in combination with a full introduction. Like an introduction, an opening serves to frame the document, but it may do more. An opening may also serve to catch and focus the readers' attention or arouse their interest.

Be sure to integrate an opening with the information that follows; tacked-on openings are amateurish. You can use any of the following standard techniques to open a document, depending, of course, on the type of document and its purpose (Brusaw, Alred, and Oliu 452–455).

- A *statement of scope* (often used to open manuals) helps readers see the intended coverage.
- A *statement of purpose* gives readers a basis for judging if the document is appropriate to their needs.
- A *brief background* (context) provides readers with an orientation and puts the topic into perspective.
- A *statement of the problem* (often used to open reports) gives readers perspective by presenting a brief account of the problem that led to the need for the report.
- An *anecdote* (often used to open newsletter and magazine articles) tells a brief story to catch readers' interest and lead them into the main subject.
- An *interesting detail*, such as a surprising fact or statistic, is also effective in articles because it catches readers' attention and arouses their interest.
- A *definition* (often used in academic articles) offers readers some special insight into what follows by giving them the formal definition of a key technical term.
- A *relevant quotation* catches readers' attention and stimulates their interest in the topic. Often a good quotation opening to use is one that predicts a new trend or development or one that refutes conventional wisdom.
- A *forecast* (often used for general articles) arouses readers' interest by predicting a new development or trend.

## Summaries

An opening summary compresses the results, essential content, conclusions, or recommendations of your document into a brief section. The more elaborate "executive summary" condenses the entire work and usually appears separately at the beginning of a report. It tends to reflect the organization of the document and to be proportional in length to the larger work it summarizes (Brusaw, Alred, and Oliu 219–226). Executive summaries (so called because they save reading time for busy executives) cover the information in the document in enough detail to reflect its contents and development, yet concisely enough to permit the reader to digest the significance of the document without having to read it in full.

## Abstracts

An abstract indicates the essential content of a longer piece of writing, enabling readers to decide whether to read the document in full. Abstracts are used primarily with formal reports and scholarly articles or are published separately in "abstract journals" (Brusaw, Alred, and Oliu 8–11, 553). In addition to following the standard advice for writing abstracts, you must be careful to examine the type of abstract readers expect in the document.

## Prefaces and Forewords

A preface is an optional introductory statement to a formal report, a technical manual, or a book that announces the purpose, background, and scope of the work. A preface may also contain acknowledgments of help received during the course of the project or in the preparation of the document.

A foreword is also an optional introductory statement about a book, a technical manual, or a formal report. But it is usually written by someone other than the author to provide background information about the work's significance or place it in the context of other works written in the field. The writer of the foreword is usually an authority in the field whose name and affiliation, together with the date the statement was written, appear at the end of the foreword. Forewords often serve to lend authority or prestige to a document. The foreword always precedes a preface when a work has both (Brusaw, Alred, and Oliu 246–247).

## "How to Use This Document" Sections

A "how to use this document" section helps readers make the most effective use of a document, such as a manual. These sections, often set apart from other opening sections, tell readers how the document is organized, what it contains, when to use it, and how to make the best use of it. Such sections typically include a functional table of contents that indicates which parts of the document readers may want to find:

| Need to | See |
|---|---|
| Learn the basics? | Tutorial, page 21 |
| Find a command? | Quick Reference, page 100 |
| Solve a problem? | Troubleshooting Chart, page 42 |
| Log on to system? | System Logon, page 4 |

Such a section is particularly useful for readers who know little about a product, system, or technology and thus need reassurance that the document will be helpful.

# SPECIAL PREVIEWING AND FINDING DEVICES

The opening segments of a document enable readers to familiarize themselves with its content and structure. Other devices are also available to help readers by allowing them to find specific information and to preview content and structure. The most typical previewing and finding devices are tables of contents, headings, references, glossaries, and indexes. Many of the design elements discussed in

Chapter 8 (such as tabs, typographical features, and icons) also serve as previewing and finding devices. Which devices you should use depends on your purpose and the readers' needs or expectations.

## Tables of Contents

In addition to indicating how the document is organized, a table of contents permits readers to preview what is in the document and locate the information they need. Essentially, a table of contents is a sequential list of the headings in a document. It should reflect the *exact* wording of the headings as they appear in the text, with each entry followed by the page number on which the heading appears in the text.

## Headings

Headings (also called *heads*) are critically important because they serve so many useful functions. By simply scanning headings, readers can find information or preview an entire document. Without a table of contents or similar device, only the effective use of headings can help readers recognize organizational strategy by revealing at a glance the logical divisions of the subject and the relative importance of each division. Remember, if you do have a table of contents, the headings should match it.

Headings also serve to break text into "digestible" segments for readers—a powerful way to help readers learn information and remember key points, as described in Chapter 4. In addition to helping your readers, composing heads can also help you focus your own thinking.

The easiest and most effective way to ensure the good use of headings in your document is to begin by superimposing your topic outline in your draft as headings (omitting the Roman numerals and capital letters, of course). Make sure that the subordination of minor heads to major heads is logical and appropriate. Further, you may need to eliminate some heads, add others, and establish the format for your headings in the text. (Information on the design of headings was given in Chapter 8.)

You will usually need to adapt the language of the headings in your outline to suit the purposes of your draft. Headings must be informative:

*Change:* Printer Assembly
    *To:* How to Assemble the Printer

Good headings are also specific:

*Change:* Problem Areas
    *To:* Possible Errors during Software Installation

For documents that involve processes, phrases that use verbs plus nouns are usually more informative and do a better job of guiding your readers through your document than headings that use nouns only.

| *Noun Heading* | *Verb + Noun Heading* |
|---|---|
| Storage | Storing Your Document |
| Duplication | Duplicating Your Document |
| Error ID | Identifying Errors |

Check also to be sure that all major heads are parallel to each other and all minor heads are parallel to each other.

*Not Parallel:* Storing Your Document
Duplication of Your Document
Identifying Errors
*Parallel:* Storing Your Document
Duplicating Your Document
Identifying Errors

Sometimes heads in the form of questions are effective. Questions work well in documents for the general public, as demonstrated in the following heads from a brochure for patients recovering from coronary bypass surgery:

How Long Will I Be Hospitalized?

How Soon Can I Return to Work?

When Can I Resume Jogging?

Headings in the form of questions are particularly useful for readers who must cope with documents containing complicated regulations or policies, as in insurance policies or employee guidelines.

*Change:* Leave Contingencies
*To:* When Can I Take a Leave of Absence?
*Change:* Exclusions
*To:* What Is Not Covered?

The text that follows must repeat information in the heading because a heading does not *replace* text. Further, don't begin the sentence after the heading with *this*, *it*, or some other pronoun referring to the heading. Instead, start the text as if the heading did not exist.

## References

References provide full and accurate documentation to allow readers to find further information on the subject as well as to give proper credit to others whose work contributed to yours. References (used primarily in formal reports and scholarly articles) must be complete, accurate, and formatted consistently. Use references for the sources of all facts and ideas that are not common knowledge to the intended audience, as well as the sources of all direct quotations (Achtert and Gibaldi 4–5; Brusaw, Alred, and Oliu 186). For professional writers within organizations, however, "common knowledge" extends to the documents and ideas produced or used within the organization. So as with boilerplate, writers can freely use words and ideas within an organization without references, unless readers need to know where to find the detailed information.

## Glossaries

A glossary is a selected alphabetically arranged list of defined technical terms that are used in a particular document. Although a glossary can be helpful to readers as a quick reference to find the definition of a term, it does not relieve a writer of the responsibility of defining in the text any terms that readers will not know. A glossary should usually follow the body of the document, although if the terms are crucial and few in number, it could appear near the beginning.

## Indexes

An index, which falls at the very end of the document, is an alphabetical list of all major topics discussed in it. It identifies the page where each topic can be found, allowing readers to find information on particular topics quickly and easily (Brusaw, Alred, and Oliu 312). An index is particularly valuable when your document is lengthy and readers will often need to find information in it.

# CONCLUDING THE DOCUMENT

Not all documents require separate conclusions or even concluding elements. Readers of instruction or procedure manuals, for example, usually do not need a concluding element since the last step or final option lets them know that the document is complete. Instead they may find a glossary or an index more useful in tying the piece together. However, many other documents need a section or a paragraph that pulls the document together or a sentence or two that give the reader a sense of closure.

## Full Conclusions

A full conclusion for a document, such as a formal report, unifies its results or findings and interprets them or elaborates on their significance. Not only does it tie together all the main ideas, but it can also emphasize for the reader the most significant points in a document (Brusaw, Alred, and Oliu 127–130).

## Concluding Elements

Some documents (such as small reports, brochures, and letters) may require only a concluding element—a sentence or two. To conclude, you may recommend a course of action, make a prediction, offer a judgment, speculate on the implications of your ideas, or merely summarize your main points. The way you conclude depends on both the purpose of your writing and needs of your readers, as in the following examples:

- A committee report analyzing possible locations for a new manufacturing plant could end with a recommendation.
- An annual report on company sales might conclude with a judgment about why sales are up or down.
- A newsletter article about an organization's new electronic mail system could predict the future uses of communications technology.
- A letter about consumer trends could end by speculating on the implications of these trends.
- A lengthy report could end with a summary of its main points.

## Appendixes

An appendix contains material at the end of a document that supplements or clarifies. Although not a mandatory part of a document, such as a report or a proposal, an appendix can be useful for explanations that are too long for notes but that could be helpful to a reader seeking further assistance or clarification of points made in the body. Information placed in an appendix is too detailed or voluminous to appear in the text without impeding the orderly presentation of ideas. This information typically includes passages from documents and laws that reinforce or illustrate the text; long lists, charts, and tables; letters and other supporting documents; elaborate calculations; computer printouts of raw data; and case histories. An appendix, however, should not be used for miscellaneous bits and pieces of information that you were unable to work into the text.

When a document contains more than one appendix, arrange them in the order in which they are referred to in the text. Thus a reference in the text to Appendix A should precede the first text reference to Appendix B.

Generally, each appendix contains only one type of information. The contents of each appendix should be identifiable without the reader having to refer to the body of the report. An introductory paragraph describing the content of the appendix is therefore necessary for some appendixes, especially those containing computer printouts of data, long tables, or similar information.

Each appendix should begin on a new page. Identify each with a title and a head:

Appendix A
Sample Questionnaire

Appendixes are ordinarily labeled "Appendix A," "Appendix B," and so on. If your report has only one appendix, label it "Appendix," followed by the title. (To call it "Appendix A" implies that an "Appendix B" will follow.)

The titles and beginning page numbers of the appendixes are listed in the table of contents of the document in which they appear.

*Checklist 9.1*

# Confidence Building and Time Management

☐ Prepare adequately

☐ Focus on writing, not revising

☐ Avoid procrastination

☐ Develop a positive attitude

☐ Allocate your time
  • Keep a calendar
  • Set short-term deadlines
  • List tasks and subgoals

☐ Prepare your writing environment
  • Isolate yourself
  • Gather materials

☐ Use boilerplate as appropriate

☐ Remain flexible
  • Start with the easiest part
  • Once you get started, keep rolling
  • Reward yourself for reaching subgoals
  • Reread what you have written after taking breaks
  • Vary your writing tools or environment

---

*Checklist 9.2*

# Communicating with the Reader

☐ Empathize with your reader

☐ Visualize the reader
  • Picture a typical reader across a desk
  • Picture varied readers around a table
  • Reconsider the modular design (see Chapter 4)

☐ Visualize the setting
  • Visualize the workplace in which the document will be used
  • Note elements in the workplace setting

☐ Understand your readers' roles
  • Think of your reader as an actor
  • Imagine your readers' movements on stage

☐ Establish your role and your "voice"
  • Determine if your role should be that of teacher, friend, authority figure, or other
  • Determine the voice your readers should hear
    ○ Businesslike
    ○ Collegial
    ○ Reportorial
    ○ Informal
    ○ Cautious

☐ Review Checklist 4.1

---

## Checklist 9.3

# Orienting the Reader

### Opening the Document

☐ Consider a full introduction

☐ Select an opening element
  • Statement of scope
  • Benefits for reader
  • Statement of the problem
  • Interesting detail
  • Relevant quotation
  • Statement of purpose
  • Context
  • Anecdote
  • Definition
  • Forecast
  • None required

☐ Open with a summary or an "executive summary"

☐ Use an abstract

☐ Consider a preface or foreword

☐ Start with a "how to use this document" section

### Using Appropriate Previewing and Finding Devices

☐ Table of contents

☐ References

☐ Index

☐ Headings

☐ Glossary

### Concluding the Document

☐ Consider a full conclusion

- [ ] Select a concluding element
  - Recommendation
  - Speculation
  - Prediction
  - Summary
  - None required

- [ ] Determine any need for an appendix

- [ ] Reconsider the preceding list of finding devices

---

## WORKS CITED

Achtert, Walter S., and Joseph Gibaldi. *The MLA Style Manual*. New York: Modern Language Association of America, 1985.

Allen-Bradley Co. "Author! Author!" *Gossip* Nov.-Dec. 1980: 6–9.

Ammons, David N., and Charldean Newell. *City Executives: Leadership Roles, Work Characteristics, and Time Management*. Albany: State U of New York, 1989.

Anderson, Paul V. *Technical Writing: A Reader-centered Approach*. Orlando, FL: Harcourt, 1987.

Brusaw, Charles T., Gerald J. Alred, and Walter E. Oliu. *Handbook of Technical Writing*. 3rd ed. New York: St. Martin's, 1987.

Elbow, Peter. *Writing with Power: Techniques for Mastering the Writing Process*. New York: Oxford U, 1981.

Flower, Linda. *Problem-solving Strategies for Writing*. Orlando, FL: Harcourt, 1981.

———, John R. Hayes, and Heidi Swarts. "Revising Functional Documents: The Scenario Principle." *New Essays in Technical and Scientific Communication: Research, Theory, Practice*. Ed. Paul V. Anderson, John R. Brockman, and Carolyn R. Miller. Farmingdale, NY: Baywood, 1983. 41–58.

Mack, Karin, and Eric Skjei. *Overcoming Writing Blocks*. Los Angeles: Tarcher, 1979.

Ong, Walter J. "The Writer's Audience Is Always a Fiction." *PMLA* 90 (1975): 9–21.

Price, Jonathan. *How to Write a Computer Manual*. Menlo Park, CA: Benjamin/Cummings, 1984.

Selzer, Jack. "What Constitutes a 'Readable' Technical Style?" *New Essays in Technical and Scientific Communication: Research, Theory, Practice*. Ed. Paul V. Anderson, John R. Brockman, and Carolyn R. Miller. Farmingdale, NY: Baywood, 1983. 71–89.

Tichy, H. J. *Effective Writing for Engineers, Managers, Scientists*. New York: Wiley, 1966.

Weiss, Edmond H. *The Writing System for Engineers and Scientists*. Englewood Cliffs, NJ: Prentice, 1982.

## FURTHER READING

Gibson, Walker. *Persona: A Style Study for Readers and Writers*. New York: Random, 1969.

Hubbard, Francis A. *How Writing Works: Learning and Using the Processes*. New York: St. Martin's, 1988.

Keenan, John. *Feel Free to Write: A Guide for Business and Professional People*. New York: Wiley, 1982.

Lakein, Alan. *How to Get Control of Your Time and Your Life*. New York: NAL, 1974.

### Chapter 9: Drafting a Document

**INTERVIEWER:** Many people who don't earn their living as writers (as well as some who do) have trouble starting a draft. Some refer to "writer's block," for example. What techniques do you use to get started on a draft?

**MARY:** I define writer's block as a lot of anxiety, a situation in which you are disgusted with every sentence you write. When that happens to me, I figure I haven't done enough research to feel comfortable enough with the material to start drafting. But if I am past the point of not having done enough research, if I'm ready to write and just need some techniques to get going, I *start small*. For example, I'll say, "This morning, I'll just try to get the data entry rules drafted." The idea is to pick a small, unintimidating task.

**INTERVIEWER:** Are there parts of a draft that you find more difficult to write than others?

**MARY:** Opening sections. I don't write those first, you know. I tend to start with the factual parts, the simpler, smaller pieces.

**INTERVIEWER:** How do you deal with difficult sections?

**MARY:** I write chapter introductions after I've written the chapters. I use the chapter headings—the final version—as an outline for the introduction. That's a way to keep the chapter internally consistent.

With sections covering especially difficult material, I try to explain the concept to myself first, using any analogy whatsoever, even the most flamboyant, inappropriate analogy, one I know I will never show to readers. For example, when I was writing about the International Banking System's security features, my first draft began, "Did you ever see the movie *War Games*? Two soldiers in a missile silo each have keys, and they each have to turn their keys at the same time to launch the missile." Now I know that that analogy was too obscure for my readers, but I used it to "spit out" my own concept of the security system so that I could get a working draft and go on from there.

**INTERVIEWER:** Let's talk about deadlines. I think that for students, the pressure of a deadline may be either a positive or a negative force. It *might* prompt them to go ahead and get the work done. Or they might react to the stress of the deadline by putting the project off because it feels too overwhelming. Do you approach deadlines from a different point of view as a professional?

**MARY:** It's definitely different for professionals than for students. Students' study time is their own: they can do as much or as little homework as they care to. But when you are writing on your job, you're there, and you're expected to be working. You don't have the option of putting a project off until later with a plan to "pour on the steam" as the deadline nears. You don't work like that in a professional setting—the stakes are too high. Plus there's all this teamwork involved.

INTERVIEWER: People are depending on you.

MARY: Exactly. If you're smart, you break down your deadlines so that they become manageable. For example, you can make your own goal of having Chapter 3 ready by Friday, rather than just working toward the publication date for the entire document three months down the line. You can't be ruled by thinking only of the end point.

In a way, being a professional writer places you in a protective environment. For example, today I was reading some source documents. Because I was at work, I was forced to continue my research even when it became tiresome and somewhat overwhelming. In other words, I worked harder and longer at the research than I would ever have worked on a school writing project.

INTERVIEWER: What advice would you give to a student who is experiencing writer's block but doesn't have your "protective environment" for writing?

MARY: Go to a writing center or computer lab or meet with other students—anywhere where you can get support. That way you'll become part of a writing community instead of writing by yourself. Make short-term deadlines for yourself. Learn techniques for facilitating writing, such as freewriting.

INTERVIEWER: How do you manage your time with so many projects in progress at one time? What time management techniques could you share?

MARY: I have a timer. I plan my work for the day, then set the timer as I work on each project and work for the allotted amount of time.

I don't always have a lot of writing projects at one time; instead, I usually have one major project plus several smaller administrative projects. But even within one project, there are enough tasks that you still have to manage your time.

INTERVIEWER: How do you physically write a draft? Do you compose it on a word processor?

MARY: Yes. I've always had the use of a word processor. When I first began writing, if I was making a minor revision to a document (for example, adding a paragraph), I would write and rewrite the paragraph on paper before entering it online. I still do that sometimes, but 95 percent of the time I compose online, right on the screen.

INTERVIEWER: How do you avoid polishing too much as you write?

MARY: I'm sometimes hypercritical as I write. To prevent negative critical thoughts from impeding my writing, I include them in the draft. I also don't go back. Of course, sometimes I do correct errors or reword things as I write, but I try to avoid stopping to correct a

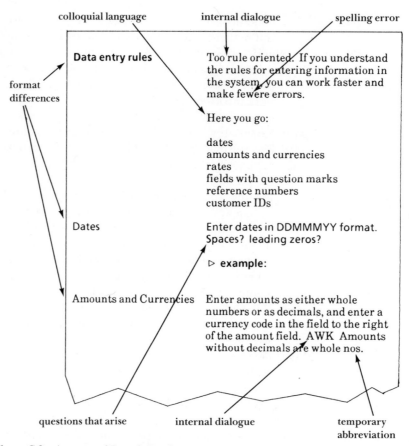

colloquial language    internal dialogue    spelling error

format differences

questions that arise    internal dialogue    temporary abbreviation

**Data entry rules**

Too rule oriented. If you understand the rules for entering information in the system, you can work faster and make fewere errors.

Here you go:

dates
amounts and currencies
rates
fields with question marks
reference numbers
customer IDs

Dates

Enter dates in DDMMMYY format. Spaces? leading zeros?

▷ **example:**

Amounts and Currencies

Enter amounts as either whole numbers or as decimals, and enter a currency code in the field to the right of the amount field. AWK  Amounts without decimals are whole nos.

***Figure C.9*** Annotated Rough Draft

typo or a minor point of format. That way, the writing has to do with clarification of the thought rather than with cleanup.

INTERVIEWER: Can you describe any particular "idiosyncrasies" that help you keep working?

MARY: The nearest thing to an idiosyncrasy on my part is that I put my internal dialogue right into the draft.

INTERVIEWER: Like this "AWK" in your draft of the Introduction chapter of the *Introduction to Using the International Banking System* manual? [See Figure C.9, which shows a portion of Mary's rough draft, with annotations.]

MARY: Yes. I put it in there and went on. Sometimes I even have to search for expletives that I want to delete before I send the draft out for review! I'll also write questions in the draft as they occur to me. I've begun putting the questions in a special typeface that I can easily recognize later (our text processing system allows me to change faces as I type).

INTERVIEWER: In this example from your rough draft [Figure C.9], could you point out some of the weaknesses you knew were there but ignored for the sake of completing the draft?

MARY: The comments to myself, of course. Also, I don't worry about format during the first draft. I write as fast as I can, then format the information later. For example, you can see that the headings on the page are inconsistent. It would slow me down to format them correctly at this point. I was typing as fast as possible and not letting myself stop to format things or spell everything out.

# 10

# Revising Your Writing

Viewing Your Draft Objectively: Distancing Techniques

Managing the Revision Process

Checking for Organization

Checking for Scope, Completeness, Accuracy, and Consistency

The Principles of an Effective Style

Good writing does not spring full-blown from the writer's mind in a flash of brilliance or inspiration; it is "worried" into existence during revision, by even the most experienced professional writers. After writing the rough draft, you must continue to improve it, through pass after pass, until it communicates your message effectively and efficiently.

In technical writing, however, it is possible to refine your draft too much. Your goal as a technical writer is simply to communicate information to your reader effectively and efficiently. When you have achieved that, any further work you do may be unnecessary because your draft will typically go through the editing, review, and evaluation processes described in Chapters 11 and 12.

One note of caution about revising your draft is in order here. On the job, the source material from which you get your information will usually contain many, perhaps even most, of the problems that will be discussed in this chapter. In fact, you will read so much bad writing on the job that you may begin to think, at least subconsciously, that you should write the same way—that *you* are the one who is out of step. It is thus very important for you to read as much good writing as you can. It can be anything from fiction to trade journals or newsmagazines.

But if you don't read *some* good writing, you could lose sight of what constitutes good writing, and your *own* writing will suffer if that happens.

## VIEWING YOUR DRAFT OBJECTIVELY: DISTANCING TECHNIQUES

Revising requires a different frame of mind from writing. You should write your rough draft quickly, but you should revise it deliberately and objectively, from the point of view of *your reader* rather than from your point of view as the writer. Here are some tricks that may help you be more objective about your writing.

- Allow a cooling period after you have written the rough draft and before you begin to revise.
- Edit your writing as if someone else had written it.
- Revise in passes.
- Watch for errors that you commonly make.
- Read the rough draft aloud.
- Ask someone else to read and criticize your draft.

### Allow a Cooling Period

Immediately after you write a rough draft, the ideas and the way you have stated them are so fresh in your mind that it may be difficult for you to imagine stating them any other way. Since you are still in the same frame of mind, a sentence that would be ambiguous to another reader will not be ambiguous to you; you remember what you *intended*, and that is how you will read the sentence. To put distance between you and your writing so that you can become more objective about it, allow some time to pass before revising the rough draft. A cooling period of a day or two is best, but if you are pressed for time, even a few hours will be of some help. As a professional technical writer, you will normally have several jobs in some stage of completion at all times; thus you will usually be able to switch from one job to another for a brief time to allow a cooling period.

### Edit as a Reader

You must revise from your reader's point of view rather than your own; in other words, you must *become a reader*. Revise your rough draft as if someone else had written it and you have been asked to edit it. It is always easier to be objective about someone else's writing than about your own.

### Revise in Passes

Don't try to find everything that is wrong with your draft in only one pass through it. Make several passes. On the first pass, look at only one aspect of your writing, such as completeness and accuracy. Then make another pass, looking at a different aspect of your writing, such as a potential problem with wordiness.

Continue to make such passes, looking for different things or different sets of things each time, until you are satisfied with your draft.

The following checklist of possible things to check for during passes over your draft could serve as a useful guide during revision.

- Organization
- Completeness and accuracy
- Clarity
- Conciseness
- Grammar and punctuation

### Be Alert for Your Most Frequent Errors

Be aware of errors that you tend to make, and watch for them as you revise. One of the benefits of taking a writing course is learning what your weak points are. Once you know what they are, of course, you should work to overcome them, and searching for them during revision will help you do so.

### Read Aloud

Some people find that reading rough drafts aloud enables them to distance themselves from their writing so that they can become more objective about them. See if it works for you. Be careful with this technique, however, because in reading aloud you can provide meaning with your vocal inflections that may in fact not be there in the words themselves.

### Get an Outside Critique

Someone who is fresh to your draft can see it much more objectively than you can and is much more likely to be able to identify problems or problem areas.

## MANAGING THE REVISION PROCESS

The more experienced you become as a writer, the more you will tend to look for global problems (such as inappropriate audience and poor organization) in your writing. The more experienced writer sees revision as a whole-text task, whereas the less experienced writer sees revision essentially as a sentence-level task.

According to John R. Hayes et al. in *Cognitive Process in Revision*, the writer of a rough draft who has detected a problem in it can choose from among the following strategies (11):

- Ignore the problem.
- Delay solving the problem for now.
- Search for more information to help solve the problem.
- Revise the text.
- Rewrite the text.

You may ignore a problem in a rough draft when you decide that it is not serious enough to justify the time and effort necessary to fix it and you know it will not confuse the reader. Experienced writers, for example, sometimes ignore certain stylistic rules, such as not ending a sentence with a preposition.

You may delay solving a problem in a rough draft because you are revising in passes and that particular problem is not one of the things you are looking for on this pass. You may also delay solving a problem because it is not an especially serious one and you have decided to solve the most serious problems first—generally a wise move. Another reason for delaying the solution of a problem is that you may not yet be quite sure *how* to solve it, so you "write around" the problem for now, trusting that a solution will occur to you before you have finished revising. This is fine, of course, as long as you do eventually find a solution.

Searching is a suitable strategy when your original research was not as thorough as it might have been. Basically, this is a matter of going back to the research stage of the writing process to get a better understanding of the topic, especially in the area where the problem occurred. It is a matter of acquiring greater knowledge of the topic you are addressing.

You would revise a draft that has good basic structure and organization but has a number of basically minor sentence-level problems, such as lack of parallel structure.

You would rewrite because you recognize that the original rough draft has too many serious problems to be salvageable. In effect, you retain the scope of the rough draft, but you reorganize it into a different structure and start over. A more experienced writer may well revise one part of a draft and rewrite another.

## CHECKING FOR ORGANIZATION

Break your rough draft back down to a topic outline. Does this topic outline conform to the outline from which you wrote the draft? Even if it does, analyze it again, one final time, for logic. If you should find problems with the logic of your organization, even at this late date, revise your topic outline—and then your draft—to reflect the better logic. Effective organization is the most critical requirement of good writing, and it is never too late to reevaluate the logic of your organization.

## CHECKING FOR SCOPE, COMPLETENESS, ACCURACY, AND CONSISTENCY

In revising your document, keep in mind the scope that was established at the beginning of the project. Your document should give your readers exactly what they need, but no more than that. Your document should not burden them with unnecessary information or sidetrack them into insignificant or only loosely related topics. Check your draft against your outline to make certain that you did, in fact, follow your plan and that you included everything you intended to include. If not, now is the time to insert any missing information into your draft. Depending on how carefully you followed your outline, this check could result in your inserting a line here and there, writing a paragraph or passage on a separate sheet of paper and attaching it to the appropriate page, or making major additions to your rough draft.

Examine the facts in your draft for accuracy. No matter how careful and painstaking you may have been in conducting your research, compiling your notes, and creating your outline, you could easily have made errors when transferring your thoughts from the outline to the rough draft. Some mistakes, such as transposing numbers, are all too easy to make when transferring information. You can quickly and easily check the facts in your draft for accuracy if you took complete notes during your research. Such a check for accuracy can be critical if it catches even a single error. Check also to be certain that contradictory facts have not crept into your draft. Such inconsistencies occur surprisingly frequently, often because of contradictory ideas that may have emerged during your research. Not knowing which idea was correct, you may have recorded both in your notes until you could determine which was correct. If you then inadvertently failed to discard the incorrect note, these conflicting ideas could have crept into your draft. Look for such inconsistencies as you revise, and eliminate them.

Be consistent with terminology. Make sure, for example, that you have not called the same item a "motor" on one page and an "engine" on another. Inconsistent use of terminology can be terribly confusing to your reader, who may not realize that the different terms are referring to the same concept or item.

Check your introduction to see that it provides a frame of reference into which your readers can place the detailed information that follows in the body of your draft. Without such an introduction, most documents would be very difficult for your readers to follow.

## THE PRINCIPLES OF AN EFFECTIVE STYLE

The revision stage is the time to work on the writing style in your document. A number of factors contribute to an effective style. Key among them are pace, tone, and sophisticated sentence construction.

### Appropriate Pace

Pace is the speed at which you present ideas to your reader. Your goal during revision should be to make certain that your pace fits both your reader and your subject. At times you may need a fast pace; at other times you may need a slow pace. The more knowledgeable a reader is, as a general rule, the faster your pace can be. The following sentences, all parallel in structure, differ in the pace at which the information is presented.

*Fast Pace:* To survive in a competitive world, industrial companies must *design*, *manufacture*, and *market* superior new products. (parallel words)

*Moderate Pace:* To survive in a competitive world, industrial companies must *design superior new products*, *manufacture these new products*, and then *market them*. (parallel phrases)

*Slow Pace:* To survive in a competitive world, *industrial companies must design superior new products, they must manufacture these new products*, and then *they must market them*. (parallel clauses)

One common practice that can create an overly fast pace is the use of a telegraphic style—by omitting articles, pronouns, conjunctions, and transitional expressions. Although conciseness is important in writing, sentences can be made too brief—and the pace uncomfortably fast—by omitting these words. Such a style forces readers to supply the missing words mentally, which means that they must work harder. Compare the following two passages, and notice how much easier the revised version reads (added words are italicized in the revised version).

*Change:* Center has had considerable experience in working with police groups and academic research organizations. Has been involved in recent years in a major study with the Center for Urban Research at Bleeker University, San Francisco. Center involved in number of projects and publications with International Association of Chiefs of Police, Southern Police Institute, and variety of state and municipal police organizations. Center has strong lines of communication with police and minority group members. Center prepared to use own regional field staff in performance of study, in collaboration with social scientists, criminologists, and writer-researchers. Will provide mature, skilled group of interviewers and community analysts who will speed process of locating and hiring staff.

*To:* *The* center has had considerable experience in working with *both* police groups and academic research organizations. *The center* has been involved in recent years in *a* major study with the Center for Urban Research at Bleeker University, San Francisco. *The* center *has been* involved in *a* number of projects and publications with *the* International Association of Chiefs of Police, *the* Southern Police Institute, and *a* variety of state and municipal organizations. *In addition, the* center has strong lines of communication with police and minority group members. *The* center *is* prepared to use *its* own regional field staff in *the* performance of *the* study, in collaboration with social scientists, criminologists, and writer-researchers. *This* will provide *us with a* mature, skilled group of interviewers and community analysts who will speed *the* process of locating and hiring staff.

Although you may save yourself work by writing telegraphically, your readers will have to work much harder to decipher your meaning.

The use of a string of modifiers preceding a noun also slows the pace of your writing.

*Change:* Your *staffing level authorization reassessment* plan should result in a major improvement.

In this sentence, the noun *plan* is preceded by a string of four modifiers that slows readers down by making them puzzle over how the modifiers individually and collectively affect the noun. Jammed modifiers occasionally occur when writers mistakenly believe that eliminating short prepositions or connectives will make their writing more concise. The problem with this idea is that these prepositions and connectives are exactly the words that make sentences clear and readable. Note how breaking up the jammed modifiers makes the example easier to read.

*To:* Your plan for the reassessment of staffing-level authorizations should result in a major improvement.

*Or:* Your plan to reassess authorizations for staffing levels should result in a major improvement.

## Appropriate Tone

In Chapter 6, we discussed why you as a technical writer might use different organizational strategies for different readers, pointing out that you would use a different strategy for a textbook on the human heart, for example, than you would use for a pamphlet designed to inform and reassure heart attack victims.

In addition to using a different organizational strategy, you would also use a different tone in the textbook than you would in the pamphlet. The tone of the textbook should be businesslike and tutorial. The tone of the pamphlet, by contrast, should be conversational and reassuring.

The following passage, for example, might be used in the textbook on the human heart:

*Formal Tone:* The human heart is a double muscular pump that lies within the thorax. It rests on the diaphragm, between the lungs. Each side of the heart has two chambers, one (the atrium) for receiving blood and the other (the ventricle) for expelling blood. These four chambers are covered by a double-walled sac called the pericardium. The innermost layer of the sac adheres tightly to the cardiac muscle, and a small amount of lubricating fluid resides between the layers of the sac. The heart of an average-size adult weighs about 300 grams.

This is a formal, impersonal approach to explaining the human heart that would be quite appropriate for a textbook. The passage uses longer, more formal words and very factual language. The following, more informal passage would be more appropriate for the pamphlet designed for heart attack victims.

*Informal Tone:* Think of your heart as a closed-circuit water pump that feeds a cascading miniature waterfall in your backyard. The pump, lying submerged in water at the base of your waterfall, takes water into one chamber and forces it, under pressure, out the other chamber through a pipe that reaches upward and empties it over the top of your waterfall. Gravity pulls the water down the waterfall to its base, where it is once more pulled into your water pump's intake chamber and again started on its journey to the top of your waterfall. Your heart is a very efficient fist-sized pump that is quite similar to this backyard water pump.

This passage uses shorter, more informal words and figurative language. The heart attack victim does not need the detailed and factual type of information needed in a textbook; thus the writer of the pamphlet can use a more informal tone that makes the information more interesting and more readily assimilated by the reader.

When revising, check your draft to make sure that it achieves a tone that is appropriate to your topic and your reader.

## Sentence Sophistication

During revision, you should also check the sentences of your draft for readability. One of the most damaging problems in technical writing is wordy, tangled, complicated sentences that are very difficult to interpret. A sophisticated knowledge of sentence construction, sentence length, and sentence variety can help you solve the problem.

### Effective Sentence Construction

An effective technical writing style is one that communicates clearly, precisely, and concisely. To create such a style, you must understand what makes an effective sentence. The normal word order of English sentences is subject-verb-object. Because your readers are always subconsciously aware of the need for these essential parts of a sentence and are looking for them, they should be made obvious in your sentences.

As Joseph Williams points out in *Style: Ten Lessons in Clarity and Grace*, an effective sentence states the *doer* of the action in the subject of the sentence and the *action* in the verb (9). That advice may seem too simple to be mentioned in this book, yet much technical writing is riddled with sentences that do not identify the doer of the action being expressed. Professional technical writers are not always quick to notice this problem because they can typically provide the missing information from their own knowledge. Their readers, however, cannot. Consider the following sentence from a training manual on systems network architecture:

This command enables sending the entire message again if an incomplete message transfer occurs.

This sentence contains no subject for the verb *sending*. The reader doesn't know who or what is doing the sending (*This command* is the subject of *enables*, not *sending*). You can improve a sentence like this one by providing the missing subject:

> This command enables *you to send* the entire message again if an incomplete message occurs.

It is also possible to bury the doer of the action within a sentence:

> Decisions on design and marketing strategy are made at the managerial level.

For your style to be effective, you should make sure that the doer of the action is positioned prominently and clearly within the sentence:

> *Managers* make the decisions on design and marketing strategy.

> OR

> *Managers* make design and marketing decisions.

Analyze the *nominalizations* in your draft to determine whether you should replace them with the verb forms of the words. Nominalizations occur when you state the action of a sentence or a clause with a noun (*conduct an investigation*) instead of using the verb form (*investigate*). Although nominalizations are not grammatically wrong, if you use them in sentence after sentence, your writing becomes sluggish with the weight of all those heavy nouns and extra verbs. Whenever a sentence seems particularly fuzzy, look for the action being expressed; if it is expressed with a noun instead of a verb, try revising the sentence to state the action in a verb.

*Change:* The Legal Department *is engaged in an assessment* of the company's liability.
   *To:* The Legal Department *is assessing* the company's liability.

Whenever you can, then, make sure that the subjects of your sentences and clauses express the doer of the action and that your verbs express the action. This, as simple and basic as it may seem, is an important feature of a good technical writing style.

Check your draft during revision for the unnecessary use of the passive voice, which can also work against a clear and vigorous style. The extra words and the indirectness that characterize the passive voice can slow your writing to a crawl. In passive sentences, the subject expresses the *goal of the action* instead of the *doer of the action*, and the doer of the action is identified in a *by* phrase—if at all. Used in sentence after sentence, this sequence defeats the easy flow of a direct style. Although the passive voice has legitimate and specific uses, a direct and straightforward writing style requires that you try to write primarily in the active voice. As a rule of thumb, write in the active voice unless you can identify precisely *why* you are using the passive voice. (For more information on use of the passive voice, see Brusaw, Alred, and Oliu 706.)

Check to make certain that your draft addresses the reader directly, in the second person and the imperative mood, whenever this is possible and appropriate. The combination of the active voice, second person, and imperative mood produces a writing style that is direct, vigorous, and easy to read. Don't worry that your readers will be offended because you are giving them "commands." They are generally reading your document specifically *because* they want you to tell them what to do.

Another problem that detracts from a clear style is wordiness. Just following the advice offered so far will help you avoid much wordiness. But there are other causes of wordiness as well. (For more information on the problem of wordiness, see Brusaw, Alred, and Oliu 122.)

Keep your references to your own writing, or efforts to direct your reader through your writing (for example, "The first point I would like to make is . . ."), to a minimum (Williams 81). Although some of this "metadiscourse," as Williams calls it, is necessary (for example, to define a technical term), some writers do it to such an extent that their message tends to get lost.

Check your draft for sexist language, especially for the indiscriminate use of masculine pronouns. Excellent books have been written on avoiding sexist language. For thorough discussions of this important issue, see Miller and Swift and Sorrels.

Hedge words—*usually, sometimes, almost, possibly, apparently, seemingly, for the most part, in some respects*—are helpful at times, but they can also be overdone. Some technical professionals tend to make too much use of hedge words, possibly to avoid stating things in absolute terms (*always, never*), which they may perceive as a risk. You should use hedges only when necessary, however, because overusing them detracts from the directness of your writing.

When you use negative words to state a thought that is not negative, your readers must translate your statement into a positive statement before they can understand the message. As the writer, you should make this adjustment when revising your draft rather than forcing your readers to do it when they try to read your sentence.

*Change:* When the error does not involve data transmission, the special function is not used.

*To:* Use the special function only when the error involves data transmission.

Check your draft for the overuse of expletives. Expletives begin with *it* or *there*, followed by a form of the verb *be*. Although expletives are sometimes necessary to avoid awkwardness, they are commonly overused, and most sentences can be better constructed without them.

*Change:* It can be shown by statistics that the higher the stress, the greater the rate of errors.

*To:* Statistics show that the higher the stress, the greater the rate of errors.

Check your draft for sentences that separate the subject from its verb with a long interrupting phrase or clause. Keeping your subject and verb close together generally improves the flow.

*Change: Chrysler,* in order to compete more effectively with the Japanese and other foreign automobile manufacturers, *began* a concerted effort to improve the quality of its automobiles.

*To:* Chrysler began a concerted effort to improve the quality of its automobiles in order to compete more effectively with the Japanese and
other foreign automobile manufacturers.

*Or:* In order to compete more effectively with the Japanese and other foreign
automobile manufacturers, Chrysler began a concerted effort to improve
the quality of its automobiles.

## Effective Sentence Length

Some people feel they should write only short sentences. Their rationale for
this is an assumption that if sentences are kept very short, they will automatically
be clear. Certainly, the longer a sentence becomes, the greater the odds are that
the writer will lose control of it. However, the best solution is not to avoid long
sentences but to construct sentences skillfully. You should *vary* the length of your
sentences. Competent writers don't avoid long sentences—they know how to
make a long sentence clear and easy to follow. As Williams puts it:

> The ability to write clear, crisp sentences that never go beyond twenty words is a
> considerable achievement. You'll never confuse your reader with sprawl, wordiness,
> or muddy abstraction. But if you never write a sentence longer than twenty words,
> you'll be like a pianist who uses only the middle octave: you can play the tune, but
> not with much richness or variation.
>
> Every competent writer has to know how to write a concise sentence and how to edit
> a long one down to a comprehensible length. But a writer also has to know how to
> manage a long sentence gracefully, how to make it as clear and vigorous as a series of
> short ones. (80)

As a writer, you must have a sense of your topic, your reader, and your
purpose and use all in determining the appropriateness of sentence length. It is
rarely even possible to deal with highly complex technical subjects in only short
sentences; the thoughts and ideas are too complex to be expressed under such
restrictive conditions. During revision, concentrate on making certain that your
sentences communicate effectively *regardless* of their length. Try to develop a feel
for when a sentence is just right. In a long sentence, present your information to
your readers in recognizable units—phrases or clauses—and make your logical
breaking points within the sentence obvious to your readers. For example, that
last sentence began with a phrase:

> In a long sentence, . . .

The phrase was then followed by an independent clause:

> . . . present your information to your readers in recognizable units . . .

The independent clause was followed by an interrupting phrase:

> . . . phrases or clauses . . .

Then the interrupting phrase was followed by another independent clause:

> . . . and make your logical breaking points within the sentence obvious to your
> readers.

As the reader, you had no difficulty following the logic of the sentence even though it is 29 words long. If this is true of your writing, your readers will have no difficulty following your sentences, regardless of their length. Consider the following much longer example.

> If you try to write the draft without having first determined your organization, you will be trying to (1) develop your topic, (2) structure your document, and (3) write the draft all at the same time—but they cannot be done at the same time; they can only be done sequentially: you must develop your topic first, structure your document second, and write your draft third.

Although this sentence is 65 words long, its logical units are obvious and its breaking points are clear; therefore, it is quite understandable. It could, of course, have been broken into four separate sentences.

> If you try to write the draft without having first determined your organization, you will be trying to (1) develop your topic, (2) structure your document, and (3) write the draft all at the same time. But they cannot be done at the same time. They can only be done sequentially. You must develop your topic first, structure your document second, and write your draft third.

Either version would be acceptable, but the writer felt that the first version was better because putting all the information into one sentence maintained the continuity of thought better than breaking the thoughts into separate sentences and that the use of the various marks of punctuation aided that continuity by sending the reader different signals than the repeated periods would have. (That sentence was 61 words long. Did you have trouble following it?)

The length of your sentences should be appropriate to your topic, your reader, and your purpose. But don't be complacent about long sentences. When a sentence begins to run on too long and contain too much information, the writer loses control and the sentence starts to ramble. The traditional way to correct a rambling sentence has been to break the offending sentence into two or more sentences. However, if the rambling sentence was not well constructed in the first place, breaking it into several sentences may only produce several poorly constructed sentences. The best solution is to extract from the rambling sentence the main point you are trying to make and let it stand as your first sentence, in a tightly constructed subject-verb-object pattern. Then follow with the remaining information in another sentence (or sentences), paying particular attention to a straightforward construction.

*Change:* The payment to which a subcontractor is entitled should be made promptly in order that in the event of a subsequent contractual dispute, we, as general contractors, may not be held in default of our contract by virtue of nonpayment.

*To:* Pay subcontractors promptly. Then if a contractual dispute should occur, we cannot be held in default of our contract because of nonpayment.

Thomas P. Johnson, in *Analytical Writing: A Handbook for Business and Technical Writers*, points out that much bad technical writing is caused by sentences that contain inordinately large numbers of prepositional phrases. This practice (which Johnson calls "catalogical" writing) is not only stylistically weak, but it implies that all the items in the prepositional phrases are of equal importance. Johnson gives the following example, with prepositional phrases set off with parentheses.

> Aside (from the need) (for adjustment) (of the heat-treating cycle) to obtain maximum properties, the execution (of all details) (of the operation) and utilization (of good facilities) will result (in optimum properties) (in most magnetic materials). Generally, the faithful execution (of a technically correct heat-treating specification) will preclude indiscriminate scheduling (of several lots) (of material) (for simultaneous treatment) (in "pool loads") and, (as a result), increase costs. Sometimes, however, these costs can be reduced (by selection) (of materials) suitable (to pooling). (38)

In this passage, 17 prepositional phrases add qualifying detail to the main ideas expressed. Not only that, but because most of the details are made grammatically equal, the most important of them do not have the impact that they should have. A good number of prepositional phrases appear in any writing. We all use them to work qualifying detail into our writing. It is when *most* of the detail appears in a series of strung-together prepositional phrases, in sentence after sentence, that the problem arises.

A judicious use of the dependent clause solves this problem. A dependent clause is a full unit of thought, even if it isn't a complete sentence, whereas prepositional phrases are not—they just contain details that by themselves are meaningless. For that reason, the use of dependent clauses leads to more natural writing. The words that introduce dependent clauses help the reader interpret: *because, when, although, unless, that, which, if, as, who,* and so on.

Johnson gives the following version of the paragraph quoted earlier, rewritten with more dependent clauses and fewer prepositional phrases:

> Heat treatment will develop maximum properties in most magnetic materials, if all the specifications are followed and if the facilities are good. But the heat-treating cycles must usually be adjusted for each different material, because the best cycle for one type will not always bring out optimum properties in another. For this reason, several different types of materials cannot generally be put through any one cycle simultaneously. When this happens, heat-treating costs increase, because each material has to be put through its own special cycle. Sometimes, however, closely related materials can undergo heat treatment at the same time. When this "pooling" of types is possible, costs can be reduced. (46)

### Effective Sentence Variety

You have probably had the experience of starting to read something and finding very quickly that your mind is wandering. So you shake your head and start again, trying a little harder to concentrate, but find again that your mind is quickly wandering. You think to yourself, "I'm having trouble concentrating today." But the problem probably isn't you at all. The problem is much more likely to be that the writer is writing sentences that are all monotonously alike.

If you know how to write an effective sentence and understand the value of sentence variety, you can write interestingly about the dullest of subjects.

However, the opposite is also true: if you don't know how to write an effective sentence and don't understand the value of sentence variety, you can bore your readers with even the most fascinating subject. Sentence variety is vital; don't underestimate its importance. If your sentences are all monotonously alike, your writing will be dull and monotonous. For example, consider the following passage.

> The product release was behind schedule, and the executive office was unhappy about it. The chief engineer tried to apply pressure to the product engineers, but they were already working a 60-hour week. The chief engineer hired contract engineers to help out, and the product was released only two weeks behind schedule.

All three of these sentences are compound sentences consisting of two independent clauses joined by coordinating conjunctions (*and* and *but*). Not only every sentence but practically every clause begins with an article and a noun (*The product release, The executive office, The chief engineer, The product*). And all three sentences are approximately the same length (14 words, 19 words, and 19 words). Although three such sentences won't have your mind wandering, three paragraphs of such sentences will. During revision, you can use sentence variety to break up this kind of monotony in your draft. You can vary sentence types, sentence construction, and sentence length.

You need to *vary sentence types*. Think back to grammar school, perhaps around the fifth grade. You were given an assignment to write something, and you might have written something like this:

> A car was driving down a road. A ditch was beside the road. A fence was on the other side of the ditch. A pasture was on the other side of the fence. A cow was standing in the pasture. . . .

These are all nice, simple little sentences, and they are appropriate for the fifth grade. But after you have grown up and finished your education and joined the adult working world, you may find that you are simply linking independent clauses together with *and* and *but*, as in the example about the product release. If you are, your sentences are very little more sophisticated than the fifth-grade example, regardless how complex and sophisticated the subject may be. You need to learn to vary your sentence types; more than that, you need to learn to use the complex sentence. The reason is easy to find: the simple sentence and the compound sentence both use only independent clauses; only the complex sentence allows you to subordinate by using the dependent clause. So if you have a series of simple and compound sentences, you still have only independent clauses, and the effect will still be repetitive and monotonous because of the lack of subordination. The occasional use of a complex sentence breaks up this type of monotony.

> Because the product release being behind schedule was making the executive office unhappy, the chief engineer tried to apply pressure to the product engineers. Upon learning that the product engineers were already working a 60-hour week, he hired contract engineers to help out, and the product was released only two weeks behind schedule.

Of course, you should not write only complex sentences either. *Vary* your use of sentence types.

You should also *vary your sentence construction*. Do not begin every sentence and every clause with an article and a noun. Vary the pattern by occasionally beginning a sentence with an introductory word or phrase or clause.

> *Delighted*, the chairman of the board authorized a bonus. (introductory word)

> *To meet the deadline*, he had to work overtime. (introductory phrase)

> *Because we have increased gross sales*, profits have also increased. (introductory dependent clause)

But don't overdo it; anything you do in sentence after sentence will become monotonous to your readers.

You should also *vary the length of your sentences*.

> Suppose that you wanted to stop—and reverse—the economic progress of this nation. What would be the surest way to do it? Find a way to cut off the nation's oil resources! Industrial plants would shut down; public utilities would stand idle; all forms of transportation would halt. The country would be paralyzed, and our economy would plummet into the abyss of national economic ruin. Our economy, in short, is energy-based.

The lengths of the sentences in this paragraph are 13 words, 11 words, 10 words, 16 words, 17 words, and 7 words. The variation in sentence length makes the paragraph more interesting than it would have been if the sentences had all been approximately the same length.

## Signaling Semantic Relationships

Check to make certain that your draft adequately highlights the relationships among ideas by the use of subordination, emphasis, parallel structure, and transition. Since these topics are dealt with in most basic writing texts, we will deal with them only briefly in this chapter.

### Subordination

Writers use subordination to show, by the way they put a sentence together, that one idea is secondary to another. For example, consider the following sentence.

> The deadline was only two months away, and it was approaching fast.

By combining two independent clauses with a coordinating conjunction, the writer indicates, intentionally or not, that the two thoughts are of equal importance. If, on the contrary, the writer wishes to indicate that one of the thoughts is less important than the other, this can be done by placing the secondary idea in a dependent clause, a phrase, or a single modifier instead of in an independent clause.

The deadline, *which was only two months away*, was approaching fast. (dependent clause)

The deadline, *only two months away*, was approaching fast. (phrase)

The *fast-approaching* deadline was only two months away. (single modifier)

(For further information, see Brusaw, Alred, and Oliu 652.)

### Emphasis

The opposite of subordination is emphasis. In fact, any time you subordinate one idea, you emphasize whatever you subordinate it to. But there are also other ways to achieve emphasis; position at the beginning or end of a sentence, paragraph, section, or document; repetition; sentence variety; and climatic sequence. (For further information, see Brusaw, Alred, and Oliu 212.)

### Parallel Structure

If the sentences in your draft do not flow smoothly, check them for parallel structure. The writer who tries to coordinate sentence parts that have different grammatical structures creates an awkward sentence; the writer who coordinates sentence parts that have the same grammatical structure creates a sentence that flows easily. (For more information, see Brusaw, Alred, and Oliu 480.)

### Transition

Check your draft for adequate transition. Transition is the writer's means of achieving a smooth flow of ideas from subject to subject, paragraph to paragraph, and sentence to sentence, even from one part of a sentence to another. Transition is a two-way indicator that links what has just been said and what is just going to be said—and sometimes establishes the relationship between them as well. (For more information, see Brusaw, Alred, and Oliu 682.)

## Monitoring Technical Jargon

Check your draft for *technical jargon*. As the professional technical writer on a project, you will become part of the development team. As an "insider," you will use technical jargon so extensively when conversing with specialists that you can easily lose sight of the fact that it is technical jargon. You must now pull back and look at your rough draft through your readers' eyes.

The use of technical jargon can be helpful if used wisely, and at times it is even necessary. When you use technical jargon, determine whether *all* your readers will understand it. If any would not, you must decide whether to avoid using it or whether to define it in your text or in a glossary (the decision should be based on the number of terms that must be defined—if there are only a few, you should define them in your text; if there are a large number, you should define them in a glossary).

## Monitoring Acronyms and Initialisms

Check your draft to make certain that you are not overusing acronyms and initials. In business and industry, acronyms and initialisms are often used by people working together on particular projects or having the same specialties, such as mechanical engineers or organic chemists. As long as such people are communicating with one another, the abbreviations are easily recognized and understood. If the same acronyms or initialisms were to be used with someone outside the group, however, they would be incomprehensible and should be explained.

Acronyms and initialisms can be convenient—for the reader and the writer alike—if they are used appropriately. Use the following guidelines to decide whether to use acronyms and initialisms:

- If you must use a multiword term as much as once each paragraph, you should instead use its acronym or initialism. For example, a phrase such as *primary software overlay area* can become tiresome if repeated again and again in one piece of writing; it would be better, therefore, to use *PSOA*.
- If something is better known by its acronym or initialism than by its formal term, you should use the abbreviated form. Examples would include A.M. and P.M., c.o.d., and MS-DOS.

If neither condition is met, however, always spell out the full term.

The first time an acronym or initialism appears in a written work, write the complete term, followed by the abbreviated form in parentheses.

The Transaction Processing Monitor (TPM) controls all operations in the processor.

Thereafter, you may use the acronym or initialism alone. In a long document, however, repeat the full term in parentheses after the abbreviation at regular intervals so that readers do not have to search back to the first time the acronym or initialism was used to find its meaning.

The TPM (Transaction Processing Monitor) controls all operations in the processor.

## *Checklist 10.1*

# Revising

☐ Employ effective distancing techniques that will enable you to revise objectively

☐ Develop a strategy for managing the revision process

☐ Check your draft for various features
- Organization
- Completeness, accuracy, and consistency
- Appropriate pace
- Appropriate tone

- Clarity and conciseness
- Sentence sophistication (construction, length, and variety)
- Appropriate semantic relationships (subordination, emphasis, parallel structure, transition)
- Correct grammar and punctuation

☐ Watch for possible problems in your draft
- Affectation and the unnecessary use of technical jargon
- Overuse or misuse of acronyms and initialisms

---

## WORKS CITED

Brusaw, Charles T., Gerald J. Alred, and Walter E. Oliu. *Handbook of Technical Writing*. 3rd ed. New York: St. Martin's, 1987.

Hayes, John R., et al. *Cognitive Process in Revision*. Communications Design Center Report No. 12. Pittsburgh: Carnegie Mellon U, 1985.

Johnson, Thomas P. *Analytical Writing: A Handbook for Business and Technical Writers*. New York: Harper, 1966.

Miller, Casey, and Kate Swift. *The Handbook of Nonsexist Writing*. New York: Harper, 1980.

Sorrels, Bobbye. *The Nonsexist Communicator*. Englewood Cliffs, NJ: Prentice, 1983.

Williams, Joseph. *Style: Ten Lessons in Clarity and Grace*. 3rd ed. Glenview, IL: Scott, 1989.

## FURTHER READING

Boston, Bruce O., ed. *Stet! Tricks of the Trade for Writers and Editors*. Alexandria, VA: Editorial Experts, 1986.

Cheney, Theodore A. Rees. *Getting the Words Right: How to Revise, Edit, and Rewrite*. Cincinnati: Writer's Digest, 1983.

Elbow, Peter. *Writing with Power: Techniques for Mastering the Writing Process*. New York: Oxford UP, 1981.

Oliu, Walter E., Charles T. Brusaw, and Gerald J. Alred. *Writing That Works: How to Write Effectively on the Job*. 3rd ed. New York: St. Martin's, 1988.

Zinser, William. *On Writing Well: An Informal Guide to Writing Nonfiction*. 3rd ed. New York: Harper, 1985.

### Chapter 10: Revising Your Writing

INTERVIEWER: What's the difference between writing a draft and revising it?

MARY: Initial drafting shouldn't be evaluative. You should be able to write without having to evaluate at the same time. But revising is a point at which you are primarily evaluating what you've done, finding the points that you don't like and making changes to them.

INTERVIEWER: Most of us have difficulty seeing the flaws and weaknesses in our own writing. How are you able to become objective about your own writing? What suggestions do you have for students who might have trouble being objective?

MARY: I don't know that I *can* be objective about my own writing. But I don't adopt an attitude when I'm writing that my personality is somehow reflected in the writing or that the writing's perfect. In fact, I'm probably a pretty strong critic of my own writing.

I know some writers go into a conference room to read their drafts aloud as they revise. Reading drafts aloud is a good idea because sometimes you can hear problems that you might not see. If I were a student, I would also ask someone else to look at the draft so I could get some feedback that wouldn't be purely subjective. I always let other people read my writings.

INTERVIEWER: Can a professional writer revise too much?

MARY: Yes.

INTERVIEWER: How? Under what circumstances?

MARY: Sometimes writers who are trying to improve clarity end up making the text *less* clear. For instance, at work, when we got our new text processor, we began putting information in tables to make it clearer and more visually appealing. But sometimes, writers found that putting information into a table made the description look more complicated than it needed to be.

There's a point of diminishing returns for revising. We work to make the writing functional, not to make it perfect. We have even heard comments from upper management like "We don't want you to write the manuals any better, just the same. We don't want 'premium,' we want 'regular.'"

INTERVIEWER: When do you know that you have revised enough?

MARY: If I have time to get the text into a shape I like, and then if I can have one day just to check that draft with which I'm basically satisfied, proofreading and making minor style changes, I generally feel that I've revised enough.

INTERVIEWER: The authors make the point early in Chapter 10 that some writers see so much bad writing on the job that they can begin to feel "out of step" if they are not writing in the same way. Have you ever found that to be true?

MARY: No. I have never felt compelled to lower my writing quality to match bad writing I may see at work. The only bad writing that *does* concern me is text in an existing manual. The manual may have been around for ten years and may be made up entirely of "wall-to-wall" writing.

INTERVIEWER: What's that?

MARY: Long, dense, jargon-ridden paragraphs without any headings. Text that includes a lot of passive voice and wordiness, topic descriptions written from a systems designer's point of view instead of task descriptions written from the user's point of view, text that's full of ambiguity. Here's an example: "Use of the Cancel key may detrimentally affect your posting files in certain situations."

INTERVIEWER: How do you keep a sense of what constitutes good writing?

MARY: I read manuals written by other writers. I read professional articles. I read vendor documentation and note what I like.

INTERVIEWER: Does that practice help you see problems in your own writing?

MARY: Yes.

INTERVIEWER: Could you provide an example?

MARY: Recently, I saw a manual that was actually a pleasure to read. It had a good, smooth flow, and it explained a lot of terms such as *amortization* that I had only vaguely understood before and had always wondered about. So from that manual I gained technical knowledge that will help me in my own writing. I also made a mental note to cull some definitions and consider the manual's organization.

INTERVIEWER: Could you describe the actual process you use to revise your own writing?

MARY: I do a lot of revising online. My personal preference is to revise on a paper copy, but since printing is expensive at my company, we are asked to edit and revise online.

When I revise, I tend to comb through the draft and change all kinds of things at once. I'm not the type of writer who can look first for sentence structure, then for format consistency, and so on. I'm more of a "jump in there" reviser. I'll also skip around if one part of the text is giving me trouble or if I remember something that I want to work on right away.

INTERVIEWER: In Figure C.10, we see your rough draft version of a portion of a page from the *Introduction to Using the International Banking System* manual. Figure C.11 shows your first revision of this material. Could you describe the changes you made in the rough draft and tell why you made them?

MARY: I was never happy with the severe, rule-based opening to the topic. When I edited the paper copy of the rough draft, I crossed out the word *rules* and replaced it with *formats*.

Then I made other revisions online to go along with the basic

```
┌─────────────────────────────────────────────────────────────┐
│ Data entry rules          Too rule oriented.  If you understand │
│                           the rules for entering information in │
│                           the system, you can work faster and   │
│                           make fewere errors.                   │
│                                                                 │
│                           Here you go:                          │
│                                                                 │
│                           dates                                 │
│                           amounts and currencies                │
│                           rates                                 │
│                           fields with question marks            │
│                           reference numbers                     │
│                           customer IDs                          │
│                                                                 │
│ Dates                     Enter dates in DDMMMYY format.        │
│                           Spaces? leading zeros?                │
│                                                                 │
│                             ▷ example:                          │
│                                                                 │
│ Amounts and Currencies    Enter amounts as either whole         │
│                           numbers or as decimals, and enter a   │
│                           currency code in the field to the right│
│                           of the amount field.  AWK  Amounts    │
│                           without decimals are whole nos.       │
│                                                                 │
```

*Figure C.10*    Rough Draft

ideas shown in the paper version's edits. The rough draft had said, "If you understand the rules for entering information in the system, you can work faster and make fewer errors." I thought that sentence was kind of patronizing. So in the revised version, I changed the sentence to give readers less reassurance and more information. The revised sentence reads, "Once you learn an item's format, you can use that format on all system screens."

I made other changes based on questions raised in the rough draft. For example, the rough draft said, "Enter dates in DDM-MMYY [date, month, year] format," but I didn't know whether or not it was necessary for users to include spaces or leading zeros in the date as well. I also knew I wanted an example at this point in the document.

After I found out the answer to my question (spaces and leading zeros could either be included or not included), I jotted down the answer on the paper copy. Then I edited the paragraph online, spelling out the basic definition (instead of using DDMMMYY), and worked out the example. Since there are four possible ways you can enter the same date, I included all formats in the example.

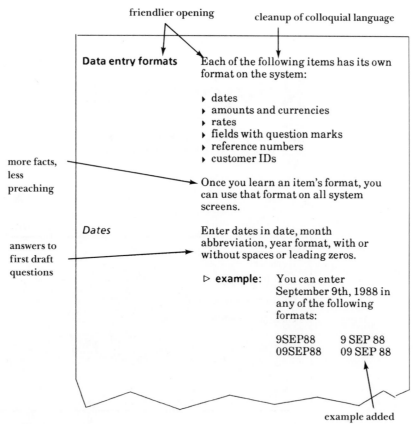

friendlier opening

cleanup of colloquial language

**Data entry formats**

Each of the following items has its own format on the system:

- ▸ dates
- ▸ amounts and currencies
- ▸ rates
- ▸ fields with question marks
- ▸ reference numbers
- ▸ customer IDs

more facts, less preaching

Once you learn an item's format, you can use that format on all system screens.

*Dates*

Enter dates in date, month abbreviation, year format, with or without spaces or leading zeros.

answers to first draft questions

▷ **example**: You can enter September 9th, 1988 in any of the following formats:

9SEP88    9 SEP 88
09SEP88   09 SEP 88

example added

*Figure C.11*   First Revision

# 11

# Review and Evaluation

Reviews

Usability Tests

Customer Validation Tests

Customer Feedback

A recent conference paper began with the following observation:

> Until recently, many companies treated [customer] manuals as an afterthought, writing them hastily and evaluating them with little more than a cursory writer-editor review. After the product and its manual were on the market, companies seldom thought of the manual again. But an increasing number of companies now realize that well-written manuals help increase customer satisfaction and so are seeking user input to help evaluate their documentation. The extra time and money a company invests in testing and revising manuals pays off with fewer customer complaints and questions, as well as a reduced need for repairing improperly handled products. (Gillihan and Herrin 116)

This realization reflects the increasing importance being placed on review and evaluation in technical writing. In this chapter, we will take *review* to mean the assessment of your document by various experts within your company and *evaluation* to mean the testing of your document by simulated customers (usability tests) or real customers (customer validation tests). Keep in mind that although reviews are most often used for manuals and other product documentation, other forms of writing, such as reports and proposals, may also be submitted to the review process.

No matter how thoroughly you have researched, written, and revised your document—and regardless of whether it is a technical manual, a grant proposal, or a sales brochure—you need to check its effectiveness from the point of view of the various people who will read and use it to make certain that you have met all their needs and concerns. The most effective way to realize this goal is to give potential users of your document an opportunity to respond to your draft *before* it is released in its final form. This chapter deals with four kinds of feedback: reviews, usability tests, customer validation tests, and direct customer feedback.

## REVIEWS

The goal of having your document reviewed is to improve its accuracy, completeness, and readability—all of which will contribute to its usefulness. Accurate, complete, and readable user manuals, for example, reduce customer phone calls and improve the company's image.

Sometimes detailed outlines can be used for the reviews when schedules are too tight to enable you to wait for the draft. If your detailed outline is as complete as it should be, it is quite adequate to be used for review and evaluation.

You may consider many kinds of reviews, and any of them may be conducted a number of times. Here is a fairly exhaustive list of possible reviews; they are listed in the order in which they are most likely to occur, except that technical, customer service, and marketing reviews would probably occur simultaneously.

- Documentation plan review
- Peer review
- Technical review
- Customer service review
- Marketing review
- Edit review
- Management review
- Legal review

Not only does using different types of reviews help you produce a complete and accurate document, but also

> setting up your documentation review cycle so that the material is reviewed for technical accuracy, usability, and management issues can make your department run smoother, improve the quality of the documentation, and give your department more time to work on the project undergoing review and other projects. (Meadows 78)

## Types of Reviews

It would not be practical for any technical writer to use all the reviews described here. You must decide which of them are applicable to your specific situation—or required by corporate, division, or department policy—and use only those.

### Documentation Plan Review

Management of the review process actually begins at the time of the documentation plan, with the definition of review checkpoints in the development process. Your documentation plan for certain user manuals, for example, should describe how you will provide information about the product to the users, the tasks to be performed by the users, the information needed to perform them, and the schedule you intend to follow.

The documentation plan review is a meeting held at the beginning of a project among all persons concerned with the project. The meeting typically includes the writer, the manager or project leader, an editor, a technical expert, and a marketing representative. The purposes of the meeting are to agree on the objectives of the document that is to be produced, to approve a tentative outline of it, and to approve a tentative schedule for producing it. These persons determine whether all needed documents have been included in the project and whether all audiences have been identified. They will also review the document after the draft has been written.

### Peer Review

A peer review is a general review of your document for logic and effectiveness by another writer whose judgment you respect or by a designated group of other writers. The reviewer evaluates your document for organization, clarity, accessibility, graphics, and typography (Sherwood and Cowan 71). Such reviews are advantageous to the whole department because not only do they strengthen each document, but they also make writers aware of other projects that are going on in the department. Peer reviews are valuable for establishing office camaraderie in addition to improving the quality of your writing. The peer reviewer should be an ideal person to evaluate the general effectiveness of your document because he or she usually has a limited technical knowledge of the subject of your document but a good knowledge of writing and document design.

When choosing a reviewer, do not select peers who will go easy on you; choose those who will challenge you. Solicit comments from peer reviewers on readability and usability, and be sure to ask for general comments in addition to specifics (Maggiore 84).

### Technical Review

This thorough critique of your document by a technical expert examines its technical accuracy and completeness. Your primary technical reviewer is usually a member of the development team (the group of engineers designing the product), but you should get as many technical reviews as you can. In addition to engineers and other technical developers, you may also be able to use technical instructors and customer service people as technical reviewers. The technical reviewers should make sure that you have correctly interpreted the material and that the information is complete, accurate, and up to date.

### Customer Service Review

This review by your Customer Service Department is especially appropriate when the document is going to be used by service people. The reviewers check your document for usability in servicing the product you are documenting. Service people are usually technically expert and therefore often catch technical inaccuracies and inconsistencies as well; in fact, they are sometimes included in the technical review for this reason.

### Marketing Review

A review by the Marketing Department can be helpful because your marketing people should know the interests, desires, and needs of your document's potential users better than anyone else. If you do not have access to a user representative, a marketing review is especially important because the representative from your Marketing Department can represent the user.

### Edit Review

The edit review is a thorough appraisal by someone who critiques documents as a full-time job. The editor evaluates the effectiveness of your organization and your writing, attempting to improve both whenever possible. The editor also checks your document's appearance and conformance to your department's documentation standards. The editor is unlikely to have a good understanding of the topic you are writing about, however, so watch carefully that he or she does not inadvertently change your meaning.

The edit review should occur after the results of all other reviews have been incorporated into your draft. If you are not conducting usability tests, the editor can also review your document from the point of view of the user. (Editorial responsibilities are discussed more fully in Chapter 12.)

### Management Review

The purpose of the management review is usually twofold: (1) to ensure consistency between your document and any related documents and (2) to ensure that your document is compatible with both the strategic and the long-range plans of the company. Your department manager may also evaluate the organization and writing of your document, although time constraints do not allow this for every document.

### Legal Review

This is a relatively rare review of the document by representatives of your Legal Department to evaluate any cautions or warnings, in addition to the wording of any claims that you might be making about a product.

## A Typical Review Procedure

The key reviews are the technical review and either the peer review or the edit review because these two reviews ensure the technical accuracy and the writing quality of your document. Since no two companies conduct reviews exactly alike, we present a review procedure that has been used successfully by one multinational company, NCR Corporation (Bartlett and Stuckey). The procedure uses basically the management review, the peer review, and the technical review, with the peer review substituting for the edit review.

### The Management Review

The manager of the department for which the writer works checks the document to ensure consistency with other related documents and compatibility with the strategic and long-range plans of the company. The manager may also evaluate the organization and writing quality of the document.

### The Peer Review

Peers are selected on the basis of their association with the project or the product being documented. If the document is a collaborative writing effort, all the writers on the collaborating team function as peer reviewers; if the document is part of a library for a specific system, the other writers involved in documenting that system serve as peer reviewers.

Copies of the document to be reviewed are provided by the document's author to all peer reviewers, and the reviewers have one week to read and study the document. The author then chairs a meeting at which the document is discussed, one page at a time, as the author solicits comments from the reviewers. The peer group reviews the document for such things as organization, clarity, ambiguity, wordiness, and grammar.

This type of peer review has several advantages:

- It develops beginning writers quickly. Beginners try especially hard to do a good job because they do not want to look bad before their more experienced peers.
- It results in consistency among documents produced by the same company and department. Without this review, documents produced by small groups headed by project leaders tend to reflect the preferences and idiosyncrasies of individual project leaders.
- It keeps writers honest. They will not attempt to give short shrift to a topic they do not adequately understand because they know that if they do, this will be pointed out in the peer review, and they will be embarrassed by their lack of knowledge.
- It helps build morale in the writing group because everyone has a broader understanding of what is happening and because the whole group produces better documents.

### The Technical Review

The technical reviewers are representatives from the Engineering Department, the Marketing Department, the Customer Service Department, and the Quality Assurance Department. The author provides all reviewers with a copy of the document to be reviewed and sets the time, date, and place of a meeting for this review. Reviewers usually have a week to read and study the draft before the meeting. The author chairs the meeting and leads everyone through the draft page by page, soliciting comments. (Of course, if the manual is especially long, the review can proceed section by section rather than page by page.) The technical review is a check for technical completeness and accuracy.

This type of technical review has several advantages:

- It enables the author to resolve most conflicting comments on the spot, whereas without such a meeting the author must go back and forth between the conflicting parties to resolve the conflict. Also, any conflict that is not resolved in the meeting can be assigned to someone *other than the writer* for resolution, leaving the writer free to update the document on the basis of the valid comments.
- It removes stress from the writer, who need not attempt to resolve conflicts without the authority to do so.
- It sometimes helps the conflicting parties understand their disagreements.

## Identifying Reviewers

If you can select your reviewers, you are indeed fortunate and should select them very carefully. Most likely, however, the technical reviewer will be assigned by the appropriate department manager. You may be able to select only additional, or secondary, technical reviewers.

Whether you select your reviewers or have them assigned to you, you should look for certain qualities in them or emphasize to them the need for such qualities. Reviewers should respect the schedules you give them because waiting for their comments can cause you to miss your deadlines. Reviewers should evaluate your document primarily from their specific area of expertise. And they must understand that you may not be able to use every one of their comments but may have to compromise between their comments and other reviewers' comments.

## Conducting the Reviews

To achieve maximum benefit from each review, you should carefully prepare the review document, brief the reviewers on their responsibilities, and make effective use of the comments you obtain.

### Preparing the Document

Submit your best effort at the current stage in the development of your document. Never ask for a review of a document you can improve without reviewer comments (that is, one that you know to contain errors, misconceptions, or avoidable omissions).

Your review document does not have to be ready for publication, but it must be complete, with all illustrations and tables in place (not necessarily finished art, but sketches that are complete enough that reviewers can evaluate text and figures together). Your review copy should be clean and free of marks. It should have a table of contents, section tabs, and as much of the glossary as you have completed (if these are to be part of the final document). If you are reviewing one section at a time, include outlines of the other sections to give reviewers a fuller picture. Each review copy should also be accompanied by a memo that gives specific instructions for each type of reviewer and specifies the responsibilities of each. Each review copy should also be accompanied by a sign-off sheet (see Figure 11.1).

### Briefing Reviewers

If all your reviewers are at a single site, hold a meeting of reviewers to discuss the general goals of the review. Discuss the scope and purpose of the document, the review criteria, specific review responsibilities, review procedures, and review deadlines, and answer any questions.

Explain to your reviewers exactly what you want from each of them (for example, the technical expert should look for completeness and technical accuracy). Many problems occur because of poor communication between writer and reviewers; writers sometimes fail to realize that reviewers do not necessarily understand writers' needs and their own role as reviewers in the writing process (Hartshorn 122). You will collect dividends if you have worked to establish a good rapport with your technical contacts and other people.

Try to impress on your reviewers, and their managers, the importance of the review *to the company* (for example, the product is not complete without effective documents; therefore, poor documentation can harm the sale and reputation of the product). Try to enlist the active cooperation of your reviewers. Explain that a thorough and timely review can help *them* because they won't have to answer so many questions later.

Although you can ask for a thorough review, you have no guarantee that you will get it (the reviewers have their own work to do). For this reason, you are wise to have more than one technical reviewer. Likewise, you can ask for a review to be completed by a certain date, but as insurance, you are wise to add the sentence, "If I have not received this document back by (date), I will assume that the information is correct and complete." Even then, it is a good idea to call your reviewer a few days before a review is due back and ask, "How is the review going?" as a reminder (Meadows 79).

If reviewers offer grammatical or stylistic comments, use them if they are good and ignore them if they are not—but don't let pride get in the way and just dismiss them out of hand. If they are valid, they can improve your document and make you look better. If accepted and acted on, constructive criticism of your document makes you appear wiser, more knowledgeable, and more talented than you really are because the final draft is the product of several minds rather than

# NCR RETAIL SYSTEMS DEVELOPMENT

## INFORMATION PRODUCT REVIEW REQUEST

DATE:

PLEASE RETURN BY:

PROJECT: ______

INFO. PRODUCT NAME: ______
NO.: ______

PRIMARY USER: ______

INFO. PRODUCT PURPOSE: ______

The attached draft information product is being distributed for technical review.

Please review the draft for clarity and usefulness. Consider the primary user and the purpose of the information product, as defined on this cover sheet.

Depending on your knowledge of the product or system described, also review the draft for the accuracy and completeness of the information presented.

You may write your comments on the draft or on the back of this cover sheet, using additional pages as necessary. If you write your comments on this cover sheet, please make appropriate references to specific pages and paragraphs in the information product.

Please fill out the information in the lower right corner of this cover sheet and return the sheet to us by the return date shown, with a copy of your comments, if any. If you are unable to meet this date and will have comments about the information product, please call us. We wil try to arrange an alternate date.

**The attached draft information product is preliminary and may contain technical errors. Do not circulate it outside your department.**

NCR CONFIDENTIAL

The content of this document is 'NCR Intellectual Property' and is the property of NCR Corporation. It is to be treated as strictly confidential and is not to be disclosed, reproduced, or used except as authorized in writing by NCR in connection with the manufacture, maintenance and use of the NCR equipment to which it pertains.

Copyright c 19 by NCR Corporation
Dayton, Ohio U.S.A.
All Rights Reserved. Printed in U.S.A.
Confidential, Unpublished Property
of NCR Corporation.

RETURN TO: Retail Systems Development
______
______

ATTN.: ______
PHONE: ______

REVIEWER: ______
ORGANIZATION: ______
PHONE: ______

REVIEWER SIGNATURE: ______

DATE SIGNED: ______

IPREVFM

*Figure 11.1* Review Request (Courtesy of NCR Corporation)

yours alone. As the writer on the project, don't let your ego prevent you from taking advantage of such a tremendous opportunity. Save your pride for the final product, and don't let it be a destructive force now. Be as objective as possible about reviewer comments.

### Using Review Comments

Prepare a master correction copy of your draft from all the comments you receive from reviewers in order to avoid conflicting and redundant corrections. If your reviews indicate that the material is too difficult, too technical, superfluous, or incomplete, don't hesitate to rewrite or change the organization as needed. Any time you do not use a review comment, note the reason on the review copy so that you can explain why.

After you have prepared a correction copy, meet with the reviewers, as a group, to discuss their comments. Explain what you have done as a result of the review and how their comments have improved your draft. Resolve any conflicting information you may have received. If such a meeting is not possible, send a postreview memo to all reviewers, detailing the significant changes you have made as a result of their participation.

## USABILITY TESTS

Usability testing is an important tool in evaluating technical user manuals and documentation in particular. It is a method for diagnosing potential problems in a document by having a test subject similar to the intended user follow the document's instructions while using the product. Although your document has gone through several reviews, a usability test can reveal overlooked gaps and ambiguities. The idea is to identify problems *before* the document is released. The emphasis is on finding and correcting usability problems by making sure that readers can find the information they need, that no important information has been left out, and that the writing is understandable. In short, the usability test verifies that the readers can find the information they need and use it to perform the necessary tasks easily and without error (Gillihan and Herrin 116).

Usability tests keep customers from discovering problems in the document in their home or office. Although usability tests can be time-consuming and expensive, the extra time and cost pay off in fewer customer complaints because the documents produced are more usable. Such tests also reduce the cost to your company of customer support services because one end result is fewer questions from customers. Testing before the document is released should save hours of frustration on the part of customers, trainers, and customer support staff.

If your budget permits, usability tests should be done repeatedly as your drafts are developed, with the data and information obtained being used to shape and organize the document to make it easier to use.

### Test Subjects

During a usability test, the document is tested by one test subject at a time. Test subjects should be a cross section of people who are as similar as possible to the actual users. The more the test subjects resemble the actual users in knowledge, education, and experience, the more useful the information gathered by the test will be. For this reason, you need to identify the relevant characteristics of your document's readers and then find test subjects who have the same characteristics. Test subjects can be recruited through college placement offices, employment agencies, or your own personnel department (if you may use employees as test subjects). If your document has multiple audiences, recruit test subjects who are typical of *each* audience because a document could be satisfactory for one audience but not for another.

The number of audiences also influences the number of subjects you should use during the test (with each tested individually, as noted). Include enough test subjects to provide a good indication of the kinds of problems the various types of users may encounter. In general, test subjects should number no fewer than 10 and no more than 50, although expense, available time, and the importance of the document are all influencing factors. An even dozen seems to be the norm. Testing only a couple of subjects makes gauging the validity of the results difficult because of the idiosyncrasies of individual test subjects. By using a larger number of test subjects, you can detect patterns. If one subject out of 15 has a problem at a given point, you can assume that the problem has more to do with the person taking the test than with the document. Conversely, if 14 out of 15 have the same problem, you can assume that the problem is in the document. In determining the number of test subjects to use, keep in mind that you must have machines and equipment for all of them. If you are documenting hardware, be sure that the hardware is available, assembled, and ready for use.

## Test Monitors

The test should be conducted by test monitors, not by the author of the document being tested. In general, test monitors gather information about the accessibility of the document, the organization of material, the clarity of the writing, the usability of the accessing aids (heads, subject index, table of contents), and the completeness and technical accuracy of the material.

Test monitors should normally be other writers because of what they can learn by monitoring the test. What writers who serve as test monitors learn about the way people read documents could by itself be worth the cost of the test. Writers who serve as monitors are very enthusiastic about the experience, and they are eager to serve again (Bartlett and Stuckey).

You could set up the test yourself, have your quality assurance department do it for you, or hire an independent testing agency. Regardless of how you do it, your test monitors should use the documentation plan developed at the beginning of the project in developing the testing strategy.

In planning a testing strategy, the test monitors should complete the following steps:

1. List the major objectives defined for the document and convert them to specific user tasks.
2. Recruit appropriate test subjects.
3. Prepare tests that require the subjects to perform the appropriate tasks.

4. Prepare test instructions (see Figure 11.2).
5. Secure copies of the document to be tested.
6. Reserve the necessary facilities and equipment.
7. Conduct the tests.
8. Observe the test subjects as they perform the specified tasks.
9. Ask the test subjects to complete the subjective posttest questionnaire.
10. Interview the test subjects after the test.
11. Analyze the test results.
12. Report the test results.

In preparing the test, the test monitors should prepare test questions or situations and answer keys. They might even include intentional errors (test bugs) to measure how carefully the test subjects are reading the document (Zirinsky 11). The writer or writing team should review the tests and answer keys to verify that they are valid and technically accurate.

In observing the test, test monitors should keep a log of all problems encountered by the test subjects. They should do all of the following:

- Record the type of problem.
- Record the number of times the subject asked for help and the places where help was needed.
- Mark passages that were confusing to the test subject.
- Determine whether explanations are in the right places.
- Determine whether any instructions were skipped (and if so, why).
- Determine whether technical terms were clearly defined.
- Determine whether the information provided was complete and correct.
- Determine whether the table of contents, index, and overviews were easy to use.

## The Test

Test monitors must establish the criteria for the test, such as "No more than 10 percent of the test subjects can have problems with any portion of the document." The strictness of the criteria should depend on the purpose of the document. If it is instructions for operating a dangerous piece of equipment, *all* test subjects must be able to understand it.

The types of tasks that test subjects should perform during a usability test depend on the end users, the document being tested, and the purpose of the test. Obviously, tests for different audiences may differ. Tasks for clerks who use a specific product might be quite different from tasks for executives using the same product because they use the product in different ways and for different purposes.

Users' guides, installation manuals, and maintenance manuals lend themselves to task-oriented activities. Tutorial manuals lend themselves to tests in which users read the manual and then attempt to perform the task. Software and programming manuals lend themselves to inspection tests and written tests. The test monitors must select the most appropriate testing method, which is often suggested by the document's objectives.

## Specific Test Instructions

### Overview

During the evaluation you will install and set up the NCR 714 communications software package and the NCR 6315 laser printer on the NCR 918 personal computer. You will also create, send, and print a message using the communications software.

### Assumptions:

- The personal computer has been properly installed and is configured for the NCR 6315 laser printer.
- If you are instructed to use it, the filename for the word processing file is memo1.
- If you are instructed to do so, use the following text for the message you will be creating.

> To:      Mr. John McCormick
>
> From:    Ms. Sally Esterhouse
>
> Subject: Conference call with Manning Sports Outlet
>
> I have scheduled the staff conference room on 3/24 from 1:00 to 2:00 to discuss the quality problems we are experiencing with Manning Sports Outlet. We will be talking to Mr. Dan Hampton, regional sales manager. Let's get together on 3/23 (in the morning) to prepare for the call and make sure we agree on what needs to be discussed. Let me know if you have any conflict with the date and time.

### Things We Want You to Do in the Test (please do in order indicated):

1. Read pages 1–1 through 1–6 of the information product.
2. Install and set up the communications software on the personal computer.
3. Install and set up the laser printer on the personal computer.
4. Create a message using the communications software.

    NOTE: Half of the participants will be instructed to read in the word-processing file memo1. The other half of the test participants will type in the memo, supplied above, on the screen. The test monitor will instruct you on which option to use.

5. Send the message to the designation name jmccormi and sesterho.
6. Print the message on the laser printer.
7. Complete an access questionnaire if instructed to do so by the test monitor.

**Figure 11.2**  Usability Test Instructions   (Courtesy of NCR Corporation)

- During an *inspection test*, the test subject reads the manual and identifies any sections that are difficult to understand. An explanation of the function of computer software would be a good candidate for an inspection test.
- *Written tests* can be used for objectives that do not involve procedural tasks. They can use matching, multiple-choice, true-or-false, or completion questions. The function of specific keys on a keyboard would be a good candidate for a written test.
- *Task-oriented tests* can be used for procedural tasks. They require test subjects to complete a procedure in a simulated environment. For this type of test, the test monitors explain all of the steps the test subjects are to perform, then record any significant incidents that occur during the test.
- An *attitudinal questionnaire*—normally used as a follow-up to the test—can be used to determine how the test subject feels about the document. It requires the subject to rate various aspects of the document by selecting one from a range of possible responses to each statement ("agree," "disagree," "neutral," "unable to judge"). It also includes a comment section for each answer.

If the test document contains a set of instructions for assembling a product, the user's task is straightforward. The test monitor sets up a plausible situation and asks the test subject to read the material and perform the task. But many tests are considerably more complex. The product and the document could be used in many ways and for many tasks. Since the test monitors cannot test them all, they must write specific test scenarios for the most representative tasks. The task for each scenario should normally fall into one of the following categories:

- Tasks that are critical to using the product
- Tasks that are critical to the goals of most users
- Tasks that are performed frequently
- Tasks that are particularly difficult
- Tasks that are performed differently for this product than for similar products (such as setting a VCR clock)

The task definitions should reflect what the audience should be able to do by reading the document. They should specify tasks that can be readily observed and therefore evaluated. A very simple set of tasks might be stated as follows:

1. Turn on the system.
2. Enter data.
3. Correct any errors.
4. Turn off the system.

The tasks stated should begin with action verbs, like the examples just given (*turn on, enter, correct, turn off*).

Keep your test as simple as possible. As Davis and Jaye put it:

> The simpler the test, the more obvious the successes and failures will be. For example, if your goal is to test how well a certain procedure is documented, focus your test on that section alone. Don't simultaneously ask subjects to test the thoroughness of the subject index or evaluate the overall organization of the document. (176)

The test monitors must be able to distinguish between problems that are caused by the equipment (or software) and those that are caused by the document.

## Conducting the Test

The process of conducting a usability test can be broken down into four phases: planning, preparing, implementing, and evaluating the test (NCR).

### The Planning Phase

During the planning phase, select the document that you want to test (if not testing the whole document, decide which parts of it to test).

Because it is neither practical nor economical to test *all* documents, you should select those that are of the greatest strategic importance. Instructional and procedural documents are prime candidates for usability testing because their accuracy is so important. Such testing is also appropriate for installation, maintenance, and operator manuals. Software and programming documents do not lend themselves to task-oriented usability testing because the test would have to use the software or program under every possible set of circumstances, which would be prohibitively costly. They can, however, be subjected to inspection tests and written tests.

### The Preparation Phase

During the preparation phase, select your test subjects and prepare any necessary test forms, such as instructions for test subjects and questionnaires (a pretest questionnaire might be used to screen the applicants for the appropriate background, and a posttest questionnaire might be used to collect subjective data about the test). Also prepare two notebooks, one for use by the author (who answers a "help" telephone line) to record the problems that were serious enough for the test subject to use the help line (see Figure 11.3), and one in which the test monitor records any other problems encountered by test subjects during the test (see Figure 11.4). Then train the test monitors who will be observing and taking notes during the test session, prepare your document for the test, and reserve the briefing room and test room.

The test monitors and the writer (or writing team) should meet ahead of time to discuss any special concerns. For example, if the writer is concerned that the language may be too technical or that a particular section may cover more information than the reader needs, he or she makes this concern known to the test monitors.

It is easy to underestimate the amount of time needed to set up the test procedure, find and correct problems encountered during the test, and retest to be sure that the finished product is as it should be. Testing may take only one or two days, but it is not unusual for it to take six weeks (depending mostly on how many test subjects are being used).

# Help-Phone Log

**Help-Phone Monitor Name:** _____Joe Holton_____

**Date:** 8/9, 8/10 _____

| Participant Test Number | Problem | Advice |
|---|---|---|
| 1 | Did not know what to do with the configuration menu. | Gave participant configuration information and told her to follow instructions given on the screen. |
| 1 | Got ''bad command or filename'' message when entered Email1. | Participant was not in the root directory. Told her to read page 4–1. |
| 2 | Did not know what to do with the configuration menu. | Gave participant configuration information and told her to follow instructions given on the screen. |
| 3 | Participant wanted to edit a typo in his message but couldn't move cursor to previous line. | Told participant to not worry about editing the message and to continue with the procedure. Need to let engineering know about this. |
| 5 | Did not know what to do with the configuration menu. | Gave participant configuration information and told him to follow instructions given on the screen. |
| 5 | Got ''message not sent'' error when he sent the message. | Participant entered the wrong name (too many characters) when he sent the message. Told him to refer to Troubleshooting in Appendix A. |
|  |  |  |

*Figure 11.3*  Usability Test Help-Phone Log   (Courtesy of NCR Corporation)

## Test Monitor's Observation Sheet

Test Participant # __5__                                       Date: ____8/10____

Test Time: Begin ___11:35___    End ___1:00___

Total Test Time ___1 hr 25 min___

What part of the information product did the participant go to when he first used the information product?

Table of contents

If applicable, what part(s) of the product did the participant first look at, touch, or examine?

Participant looked at installation diskettes and the connections on the back of the PC.

| Page Number or Task | Observations and Comments |
|---|---|
| Table of Contents | Participant said he was looking for the entry "Installing the Diskettes," said he found an entry "Installing the Communications Diskettes." |
| 2–1 | Participant turned on PC and monitor. Participant is reading "Before Starting" section while PC is booting up. |
|  | Participant glances at the connection on the printer and then returns to page 2–1. |
| 2–2 thru 2–4 | Participant looks at Step 1 and then places installation diskette #1 in the PC and enters Install a:*.* c: Participant took a little while to find the right diskette. |
|  | Participant Comment: "The wording in the manual and the labeling on the diskettes don't match. The manual uses decimal numbers and the diskettes are labeled with roman numbers." |
|  | Participant waits for the prompt and then successfully installs diskettes #2 & #3. |
|  |  |
|  |  |

**Figure 11.4**  Usability Test Monitor's Log   (Courtesy of NCR Corporation)

### The Implementation Phase

During the implementation phase, the test monitor takes the test subject to a briefing room and explains what will happen during the session—what the test subject will be reading, what tasks will be performed, and how to record feedback. The monitor may also need to ask the test subject to sign a nondisclosure agreement to ensure confidentiality. The test monitor should assure the test subject that it is the work of the writer or writing team that is being tested, *not* the test subject's skill or efficiency in using the product. The test monitor should encourage comments and assure the test subject that any problem encountered will be taken as evidence of a problem with the document, not with the test subject. (Similarly, test monitors should ask test subjects to question anything they do not understand and assure them that it is not their fault if they cannot find information or complete a procedure.)

The test monitor then takes the test subject to the testing area and gives him or her time to become familiar with the room and equipment. The room should have the necessary equipment, a comfortable chair, the document being tested, and a chair for the test coordinator (see Figure 11.5). It must also have a help phone (with the help number prominently displayed) that the test subject can use for help if necessary. The author of the document being tested should be on the other end of the help line to answer questions (to simulate the support available to customers). The test monitor asks the test subject to write comments in the document and to elaborate on these comments as fully as possible. The test subject is sometimes given a list of symbols to use to represent frequently made

***Figure 11.5***   Ideal Testing Environment   (Photo illustrating process in use at NCR provided courtesy of NCR Corporation)

comments, such as "confusing," "helpful," "incomplete," and "needs an illustration."

Test monitors sometimes use a video camera to record the test sessions so that the evaluation team can review the tapes later if they find that they have missed something during the live sessions, such as the time required to complete a procedure. If cost is a problem, however, the video camera can be omitted.

During the test session, the test monitor prompts the test subject to "think out loud" so both will know what is occurring. The test monitor should be able to see what the test subject is reading and should keep notes about the test subject's comments and behavior (Davis and Jaye 177). The monitor participates only to suggest the help phone when it becomes apparent that the test subject is stuck; the monitor must resist any temptation to help the test subject find information or figure out how to perform a procedure.

When the test subject has completed all tasks, the test monitor takes the test subject to a debriefing area to complete a posttest questionnaire about the test experience and the document (see Figure 11.6). The questionnaire also asks the test subject to rate the document on its ease of use, organization, and the quality of writing and illustrations (Davis and Jaye 177). Meanwhile, the document's author restores the test room to its original state for the next test.

Both the test subject and the test monitor document any usability problems they encountered during the test and note positive experiences encountered while using the document as well.

### The Evaluation Phase

Three main types of information are collected during a usability test: observational logs, objective data, and subjective data. The type of data gathered during a usability test depends on what is going to be done with the data and how it is going to be analyzed.

*Observational logs* include the test monitor's observations, such as when and how long a test subject looked at the document or looked at the computer screen. The amount of detail in the log would depend on the purpose of the test. It would also include the author's help-line log. An observational log provides information not only about how a subject performs a task but also about what that subject is thinking while doing so.

*Objective data* includes how long tasks took, the number of times help was required, and the number of errors made. A time measure can help isolate difficult aspects of the task. For errors, both the number of errors that occurred and the time of their occurrence are important. One measure of the amount of help required is the number of calls for help. The number of errors and help calls attributable to different parts of the document shows where it needs to be revised for clarity.

*Subjective data* includes the test subjects' feelings and opinions about the document and the product. Subjective data reveals what test subjects found satisfactory and unsatisfactory about the document. It also enables test subjects to point out specific problems or make recommendations. Subjective data is gathered from interviews and questionnaires.

# Information Product Evaluation Questionnaire

## General Evaluation

1. Indicate your general opinion about the information product by circling the number that most closely corresponds to your feelings.

The information product is:

| Difficult to understand | 1 | 2 | 3 | 4 | 5 | 6 | 7 | Easy to understand |
|---|---|---|---|---|---|---|---|---|
| Poorly illustrated | 1 | 2 | 3 | 4 | 5 | 6 | 7 | Well illustrated |
| Poorly organized | 1 | 2 | 3 | 4 | 5 | 6 | 7 | Well organized |
| Poorly written | 1 | 2 | 3 | 4 | 5 | 6 | 7 | Well written |
| Poorly designed | 1 | 2 | 3 | 4 | 5 | 6 | 7 | Well designed |
| Confusing | 1 | 2 | 3 | 4 | 5 | 6 | 7 | Clear |

. . . . . . . . . . . . . . . . . . . . . . . . . . . . . . . . . . . . . . . . . . . . . . . . . . . . . . . . . .

## Accessing Information

1. When did you use the table of contents?

\_\_\_ Before searching for information

\_\_\_ After unsuccessfully searching for information

\_\_\_ Did not use the table of contents

. . . . . . . . . . . . . . . . . . . . . . . . . . . . . . . . . . . . . . . . . . . . . . . . . . . . . . . . . .

## Organization

1. Was the sequence of the information product appropriate?

Yes \_\_\_ No \_\_\_

If no, list the appropriate sequence:

. . . . . . . . . . . . . . . . . . . . . . . . . . . . . . . . . . . . . . . . . . . . . . . . . . . . . . . . . .

*Figure 11.6*  Posttest Questionnaire   (Courtesy of NCR Corporation)

The evaluation team should include the author of the document being tested, the test monitor who observed the tests, and the author's project leader. The test monitor and author prepare and distribute to every team member copies of the notes they compiled during the test. The team then meets to discuss the problems encountered during the tests and possible solutions to them.

During the analysis of data collected from the usability tests, the evaluation team may find unanticipated patterns of behavior that may indicate problems with the product itself or how the test subjects used it. If the data indicates that the test subjects consistently use the product in a way that is not reflected in the document, the document should be revised to reflect that way of using it. The kinds of problems that are commonly reported include poor organization, ineffective heads, inadequately explained procedures, undefined terms, ambiguity, unnecessary repetition, poor graphics, ineffective page layout, and poor indexing.

The result of the test analysis should be a list of problems and their frequency. The evaluation team should rank these as major or minor problems and develop a list of recommended ways to improve the document by resolving them.

The evaluation team then prepares a summary report that includes all usability problems, as well as anything positive they may have to report. Certain items in the report are essential:

- Information about the testing procedures
- Summaries of the collected data
- Major problems that affect the product, the user, or the document
- The evaluation team's assessment of the probable causes of problems and its recommendations for their solutions

The writer then prepares an action plan for implementing the solutions agreed on by the evaluation team (see Figure 11.7).

## CUSTOMER VALIDATION TESTS

Customer validation tests (also sometimes called field tests) are another means of evaluating user manuals and documentation. They are conducted at a pilot site of a real customer prior to release of the product. The subjects who use and test your document during a customer validation test are real customers in their own workplace—the people who will use the document after the product has been released—in contrast to usability test subjects, who merely *resemble* the people who will ultimately use the document. Customer validation tests, of course, require the agreement and participation of customers.

A good time for customer validation testing of a document is during the user acceptance phase of a product's introduction. If document development time is limited, you can actually use the initial release of your product as a customer validation test and then reissue the revised document as quickly as possible after the validation test.

## Action Plan

| Problem Number | Problem Description | Page Number | Frequency Encountered | Required Help Phone | Action to Be Taken | Problem Type Code |
|---|---|---|---|---|---|---|
| 1 | Wording in the manual and the labeling on the diskettes don't match | 2–2 | 1, 3, 4, 5, 9 | | Make sure the wording matches. | TT, TI |
| 2 | Did not know what to do with configuration menu | 2–5 | 1, 2, 5, 7, 9, 10, 11, 12 | 1, 2, 5, 7, 10 | On page 2–1 "Before Starting," tell the user that during the installation of diskette #4, a configuration menu will display. Indicate what type of configuration information the user needs. Also, on page 2–5, show the configuration main menu and tell the user to follow the instructions on the screen. | SE |
| 3 | Got "bad command or filename" message when participant entered Email1 | 4–1 | 1, 7 | 1 | None—page 4–1 tells the users to make sure that they are in the root directory. Both participants #1 & #7 were not using the manual for this procedure. | |
| 4 | Participant couldn't move cursor to previous line to edit text | 4–4 | 3 | 3 | • Make engineering aware of the problem and document the change. • If the problem can't be fixed, add a note on page 4–4 telling the user that he/she cannot edit a line once he/she moves to the next line. | PD |

### Problem Type Codes

Product Design Errors
PD —Product design error

Strategic Errors
StA —An audience is not defined
StT —A task or set of tasks is not explained
StW—Information is planned for the wrong audience
StI —Incorrect information products are identified
StM—Inappropriate media is chosen

Tactical Errors
TTY—Types
TT —Inconsistent terminology
TS —Style (jargon, passive voice, wordiness, etc.)
TG —Grammatical (punctuation, sentence structure, etc.)
TI —Inaccurate information
TU —Unclear/vague text or illustrations

Structural Errors
SO —Poor organization (material is presented out of sequence)
SE —Essential information/illustrations are missing
SEt—Extraneous information is included
SF —Ineffective format is chosen
SI —Illustrations and text are not on the same page or facing pages
SA —Ineffective access aids are used (user cannot locate the information)
ST —Tasks are not separated into comprehensible units

*Figure 11.7*  Usability Test Action Plan  (Courtesy of NCR Corporation)

During the customer validation test, you issue a working draft of your document for use by real customer employees in their workplace and then monitor their experience using the document as they perform their jobs. In some companies, writers must work through the company's support personnel to conduct customer validation tests at pilot sites rather than doing it themselves (Bartlett and Stuckey). As with reviews and usability tests, customer validation tests can further expose omissions of definitions, descriptions, or procedures. They can test whether the document distinguishes between important and secondary procedures, uses correct terminology, and reflects accurate and approved work practices. They also tend to generate user acceptance of the document because the users have helped develop it. Secondary audiences, such as supervisors, can also participate in the customer validation tests, of course.

Just as for reviews and usability tests, the customer validation test document should be as near to the final version as you can make it, and its title page should identify it as a customer validation document. It should be in final page format, with all graphics professionally rendered and in place. Pages should be numbered, and the document should be in a binder, with tabs between sections and with a table of contents, a glossary, and an index, as applicable.

You must brief the customer validation test participants. The briefing description given earlier for reviewers is equally applicable here, and you would add any specific considerations that apply only to the customer validation test. Your instructions to users must be clear and specific.

From the customer validation test, you will discover explanations that need further clarification, any procedural steps that need to be improved or added, any graphics that should be added, any material that is too technical, and any material that should be deleted. At the very least, your final document will be more user-oriented as a result of the customer validation test.

## CUSTOMER FEEDBACK

Direct customer feedback occurs after the document has been released for customer use and is used to determine how well the document is meeting customers' needs in the field. Customer feedback is commonly solicited in a number of ways.

In some few cases, federations of user groups have identified specific customers who are willing to log their use of documents for a specified period of time (such as three months), logging what they looked for, when they looked for it, and whether they found it (Bartlett and Stuckey).

*Questionnaires* or *response cards* that solicit comments from customers about the document's accuracy, usefulness, and appropriate level of detail can be included in the document (Gillihan and Herrin 117). This method is probably the least satisfactory, however, because of the small number of customers who respond.

*Telephone surveys* can provide an appropriate measure of the document's usability by asking specific questions about it. The greatest drawback here is that it may be difficult to find the appropriate person to answer your questions. Again, some user groups have cooperated in surveys on how to design the next generation of documentation.

A *customer help line* can be used and then monitored for the number and types of calls customers make to it. By noting the areas of the document where customers need help, technical writers can see what changes they should make when they revise that document or write a similar one.

Finally, *field research* requires the writer to visit the customers in their workplace to see how customers use the document. This procedure provides the most valid and useful information; however, it is also the most expensive.

*Checklist 11.1*

# Reviews

☐ Conduct the following reviews as appropriate
- Documentation plan review
- Peer review
- Edit review
- Technical review
- Management review
- Marketing review
- Legal review
- Service department review

☐ Prepare the document for review

☐ Brief reviewers on their responsibilities

☐ Prepare a master copy of reviewer comments

☐ Revise the document on the basis of these comments

*Checklist 11.2*

# Usability Test

☐ Analyze your audience or audiences and select test subjects on the basis of that analysis.

☐ Select technical writers to function as test coordinators.

☐ Prepare the test, the questionnaires, and nondisclosure forms.

☐ Reserve a briefing room and a test room.

☐ Reserve the necessary equipment.

☐ Evaluate the results of the test.

☐ Revise your document on the basis of results from the test.

---

## Checklist 11.3

# Customer Validation Test

☐ Prepare the document for the customer validation test.

☐ Gather the feedback from the customer validation test.

☐ Revise the document on the basis of comments from the customer validation test.

---

## Checklist 11.4

# Customer Feedback

☐ Consider various forms of feedback
  • Questionnaires or response cards
  • Telephone survey
  • Customer help line
  • Field research

☐ Revise the document on the basis of the customer feedback

---

### WORKS CITED

Bartlett, Michael, and William Stuckey (NCR Corp.). Personal interview. 5 May 1989.

Davis, Lauren, and Alecia Jaye. "Beginning a Documentation Usability Testing Program on a Shoestring." *Proceedings of the 34th International Technical Communications Conference* (1987): WE-175–WE-178.

Gillihan, Dana L., and Jennifer L. Herrin. "Evaluating Manuals." *Proceedings of the 35th International Technical Communications Conference* (1988): RET-116–RET-119.

Hartshorn, Roy W. "Building Rapport with Reviewers." *Proceedings of the 35th International Technical Communications Conference* (1988): WE-122–WE-124.

Maggiore, James A. "A Three-Step Approach for Conducting Formal Peer Reviews." *Proceedings of the 35th International Technical Communications Conference* (1988): MPD-84–MPD-87.

Meadows, Ed W. "Improving Your Documentation Review Cycles." *Proceedings of the 35th International Technical Communications Conference* (1988): MPD-78–MPD-80.

NCR Corporation. *Document Validation: A Practical Approach.* Videotape. Wichita, KS: NCR Corp., 1987.

Sherwood, Amy C., and Joan E. Cowan. "A System of Formal Peer Review for Documentation." *Proceedings of the 34th International Technical Communications Conference* (1987): MPD-70–MPD-73.

Zirinsky, Mark A. "Usability Testing of Documentation." *Proceedings of the 34th International Technical Communications Conference* (1987): WE-10–WE-12.

## FURTHER READING

Bolden, Mark, and Heather Chaboya. "Usability Testing and You." *Proceedings of the 35th International Technical Communications Conference* (1988): RET-96–RET-98.

Felker, Daniel B., et al. *Document Design: A Review of the Relevant Research.* Washington, DC: American Institutes for Research, 1980.

Hansen, Kathleen J. "Test Methods: Adopt or Write Your Own?" *Proceedings of the 34th International Technical Communications Conference* (1987): WE-181–WE-184.

Hubbard, Scott E., and Michael Willoughby. "Testing Documentation: A Practical Approach." *Proceedings of the 32nd International Technical Communications Conference* (1985): WE-41–WE-44.

Johnson, Lora A. "Development of a Quality Documentation Program." *Proceedings of the 34th International Technical Communications Conference* (1987): ATA-19–ATA-21.

Kraus, Dorie A. "Usability Testing for Lotus Documentation." *Proceedings of the 35th International Technical Communications Conference* (1988): RET-165–RET-167.

MacKenzie, Raymond N., and Laraine J. Gerdes. "The Role of Protocol Analysis in User Testing." *Proceedings of the 34th International Technical Communications Conference* (1987): WE-179–WE-180.

Miller, Deborah C. *How to Test Your Documentation.* Arlington, VA: Write Words, 1987.

Pakin, Sandra, and Associates. *Documentation Development Methodology.* Englewood Cliffs, NJ: Prentice, 1984.

Soderston, Candace. "The Usability Edit: A New Level." *Technical Communication* 32.1 (1985): 16–18.

Wagner, Carl B. "Quality Control Methods for IBM Computer Manuals." *Technical Writing and Communication* 10 (1980): 93–103.

Winbush, Barbara, and Glenda McDowell. "Testing: How to Increase the Usability of User Manuals." *Technical Communication* 27.4 (1980): 20–22.

Zirkler, Dieter J. "Evaluating Document Quality with BARS." *Proceedings of the 35th International Technical Communications Conference* (1988): WE-156–WE-157.

## Chapter 11: Review and Evaluation

**INTERVIEWER:** At what point in the document process do you seek reviews of a document?

**MARY:** I usually schedule two reviews (planned at the beginning of the project): one first-draft review, and one final review of the ready-for-publication draft. If a project is quite small, however, I may not schedule the final review.

**INTERVIEWER:** Do you personally ever seek any earlier, perhaps more informal reviews?

**MARY:** For a project the size of the International Banking System documentation, I would consult the peer consultant (another writer) assigned to review my project plans and prototype formats. The peer consultant would be my sounding board for early decisions regarding plans for the project.

For things like format and organization, it's better to get *early* feedback than to wait for the first-draft review. You don't want to write 15 documents, then have someone evaluate and suggest changes in their format, and then have to go back and change the format for all 15. It's better to write a prototype, have it critiqued, and then go on.

**INTERVIEWER:** Could you describe, in general, the reviews used in the preparation of the *Introduction to Using the International Banking System* manual?

**MARY:** Several system developers reviewed the material technically—that is, they verified that the drafts described the system accurately. Three user representatives reviewed drafts for readability and suitability: one actual regular user, one trainer, and one manager in the International Banking Department. All drafts were also copyedited by another writer in my department (the peer consultant). In addition, a writer in the Marketing Department edited the entire manual for organization, style, and format. [Figure C.12 shows the technical writing final review sign-off form used by reviewers of the *Introduction* manual.]

**INTERVIEWER:** Did you have any legal reviews?

**MARY:** Only the proprietary statement at the beginning of the manual needed to be reviewed by a legal expert.

**INTERVIEWER:** Who reviewed the manual technically?

**MARY:** The lead analyst on the project reviewed everything. Other programmers and analysts reviewed the parts that pertained to their areas of expertise. For example, a programmer responsible for the proper setup of the hardware reviewed the section on logon procedures.

# Technical Writing Final Review Sign-Off

Date: 10/25/88

To:    Jenny Acevedo, George Clark, Homer Farias, Marty Freier, Jean Heiss, Jim Lansing, Gary Maradik, Jim Murphy, Ralph Tatman

From:    Mary Mullins, Technical Writing  *MM*

Project #: CI80107

Project:   International Banking System

Manual(s) Affected: Introduction to Using the International Banking System

Material Attached: Basic Skills chapters. See reverse side for document list and notes.

## Instructions

1. Concentrate on the content of the material.

2. Mark all changes in red on this photocopy. Include initials next to changes if more than one reviewer marks the same draft.

3. Additional comments: Note questions in Using Menus chapter.

4. Sign this draft review form to indicate that you have seen the draft and are familiar with the technical writing policies described below.

Signature(s) _____ Date____/____/____

_____ ____/____/____

Return the draft with your changes by    Monday, Nov. 7th

To: [Mary Mullins], Technical Writer
    ISD, Administrative Services, 9-08

## Policy Statement

1. Format and style changes are up to the discretion of the Technical Writer as imposed by the Technical Writing Area's style guidelines.

2. If there are significant changes to this draft, you will receive a final draft for review. On the final draft, only minor changes should be made.

3. Any additions or changes received after the due date above may:
   - Not be included in this release.
   - Change the priority established by Technical Writing for this project.
   - Push back the project's deadline.

4. Any additions or changes outside this project's scope require a separate work request.

cc: G. Atkins

*Figure C.12*    Technical Writing Final Review Sign-off Form

INTERVIEWER: How about the management review?

MARY: A senior manager in the International Banking Department had the opportunity to review the manual but responded only to certain chapters. In some cases, this manager distributed pieces of the manual to people on his staff for their review.

INTERVIEWER: Do you have any general advice from your experience about identifying appropriate reviewers for a publication?

MARY: I've found that it's easy to get the wrong reviewers if you ask someone who doesn't really know the project for suggestions. In our company, we're supposed to get a list of reviewers from the systems project leader, but sometimes other writers in the department know the history of the manual better and can identify other important reviewers. For example, a writer who'd worked on a system before might say something like, "Sharon in Accounting has always had problems with our manuals. She would really like to review accounting procedures before they are published."

INTERVIEWER: How did you identify reviewers for the *Introduction* manual?

MARY: I started with Marty Freier, my main contact for the project. He was an operations manager in the International Banking Department who knew and would be training the future users of the manual.

INTERVIEWER: And he was able to suggest other reviewers to you?

MARY: Right. Marty's initial list of reviewers included the senior manager in the International Banking Department, the lead analyst, and himself.

I also requested review by real live users (the users worked in my own company, so actual users were more accessible than they are for some projects). I got real user reviews for most of the chapters.

INTERVIEWER: It sounds as if you cultivate reviewers very carefully. Once you have the reviewers, how do you prepare them so as to do a good job?

MARY: When I first identify potential reviewers, I ask whether they can do the review, give them an estimate of the amount of time needed to do the review, and generally outline the kind of review I expect from them. I tell technical reviewers, "Don't worry about spelling or style, but do worry about the content." I also ask user reviewers to ignore things like spelling but to let me know if something's confusing. I work out review expectations with the copyeditor, too.

INTERVIEWER: What problems have you encountered with reviewers?

MARY: There's the reviewer who's late, the one who doesn't respond to the deadline and doesn't inform me. And then there are reviewers who rewrite entire chapters without speaking to me first. They feel they have to take the manual, rewrite it, and then tell me to type it.

Then there's the reviewer who takes the review as an opportunity to change the system. A reviewer will tell me that the documentation is wrong, but I later learn that the system does work as described in the documentation and that the reviewer just didn't like the way the system was set up.

Once in a while I'll have two reviewers who take opposite stances on how the document should read. I generally ask them to discuss this difference and come up with a solution.

INTERVIEWER: Do you ever evaluate the usability of a document through either *validation* (having potential users use the document in their actual workplace) or *usability tests* (having test subjects similar to the intended users follow the publication's instructions)?

---

Manual Name _____ Date____/____/____

We would like to get your feedback on the adequacy of this manual. Your comments, both positive and negative, will influence the design and content of future texts. Thank you for your comments.

**BACKGROUND INFORMATION**

1. Name of Bank _____

2. Your Name _____

3. Job Responsibility _____

4. Length of Time in Current Position _____

5. Which of the following do you consider yourself to be?

       ☐ Experienced user of the system   ☐ Inexperienced user

**MANUAL EVALUATION**

| | | | |
|---|---|---|---|
| 1. Readability | ☐ Too simply written | ☐ Easy to understand | ☐ Too technical |
| 2. Thoroughness | ☐ Covers too little of subject | ☐ Complete in its coverage of the subject | ☐ Covers too much unrelated material |
| 3. Organization | ☐ Not in logical sequence | ☐ Sequenced logically | |
| 4. Referenceability | ☐ Topics difficult to find | ☐ Topics easy to find | |
| 5. Suitability | ☐ Not applicable to my job | ☐ Very applicable | ☐ Partially applicable |
| 6. Timeliness | ☐ Very out of date | ☐ Not at all out of date | ☐ Partially out of date |

**ADDITIONAL COMMENTS**

1. What additional subject areas should be included in this manual?

_____

_____

2. What would you suggest to improve this manual?

_____

_____

3. Would you recommend this manual to someone who needs to know this information?  ☐ Yes   ☐ No

If not, why? _____

_____

---

*Figure C.13*   Reader Comment Sheet

**MARY:** I've had the opportunity to do validation, but it doesn't happen on every project. The company has never participated in a usability test, but we've heard about them, listened to lectures on them, and thought about them.

**INTERVIEWER:** Could you describe the process of validation?

**MARY:** I'll give you an example. I asked one of the International Banking employees who uses a personal computer to try my procedure for logging on to the International Banking System from the personal computer. She tried it and said it worked just fine. Then she started talking about how sometimes the system didn't work right, and she described those problems in detail. I changed the procedure to address those problems.

**INTERVIEWER:** So you could have gone away thinking simply that it worked, but instead you teased some more information out of the user.

**MARY:** Right.

**INTERVIEWER:** Did you do any validation for the *Introduction* manual?

**MARY:** There was a form of validation in that that manual was published just as users were beginning to learn the system. We planned that initial learning time as a kind of test of the manual, and we expected to revise the manual after the first month or two in response to user feedback from that initial learning period.

**INTERVIEWER:** That brings me to another kind of validation: do you use "customer feedback" to improve your manuals? (Maybe in your case you'd call it "user feedback.")

**MARY:** We have a reader comment sheet that goes out with every manual [see Figure C.13]. Readers have the option of filling it in and returning it to us.

**INTERVIEWER:** How successful are you in getting feedback this way?

**MARY:** Statistically, we get a pretty low response rate, but the feedback is still useful. Though they may be few in number, the people who respond are the true readers, so their opinions really matter.

# 12

# Editing the Work of Others

The Editing Process

Working with Authors

Working with Other Document Specialists

Mechanics of Copymarking

Editing Online

Levels of Edit

Editing is the art of joining your knowledge of language, logic, graphics, and publication production with your powers of tact and diplomacy in the service of someone else's document. Professional editors usually put these combined skills to the service of a technical expert who is working on job-related writing such as a journal article, a report, a proposal, a feasibility study, or a service, procedure, or training manual. The author may be a scientist, an engineer, a systems analyst, an economist, a physician, or a sociologist. If you are a technical writer in an organization that encourages peer review among its writers, you may edit the work of your writing colleagues.

The editor acts as a mediator between the author—whether technical expert or writing colleague—and the audience for the document. As a mediator, the editor serves both. For the author, you as an editor bring an outsider's objective perspective and employ your editorial capabilities to improve the document so that it meets the author's intentions to the fullest extent. You work equally hard on the reader's behalf to ensure that the document contains a unified thread of meaning throughout, that its various parts are cohesively linked, and that its terminology, symbols, equations, graphics, and format conventions are accurate and consistent. Also central to the success of any editing project is the editor's relationship with the author, a topic discussed in some detail later in this chapter.

## THE EDITING PROCESS

Before beginning editing work on a manuscript, you must meet with its author to lay the groundwork for your mutual working relationship until the end of the project. Use this meeting to reach agreement on the following points about the manuscript:

- Purpose
- Audience
- Scope
- Final form (manual, report, journal article, conference paper)
- Special design requirements (loose-leaf binding, index tabs, oversize graphics)
- Level of edit required
- Deadline

Each of these requirements will affect the time you can devote to the manuscript, the types of changes you make to the text and graphics, and the amount of coordination necessary with other document specialists.

How will you know what is required of you as an editor when you begin working on a manuscript? Does the author expect you to look solely for errors in spelling, grammar, punctuation, capitalization, abbreviations, and numbers? Or are you expected to review for all these elements as well as for problems in the unity, coherence, and organization of the information? Will you be required to rewrite some passages and check the accuracy of any data? Should you ensure that the manuscript conforms to predefined format guidelines? All these options must be discussed and at least tentatively agreed on with the author before you begin work. (Of course, you may find after you start editing that the manuscript requires either more or less effort than the author thought originally.) In an attempt to answer these questions, editors and publications managers divide editorial tasks into a number of categories based on the level of edit required on a given manuscript. These categories are based on the recognition that some manuscripts require detailed, comprehensive editing while others need relatively little work. (Application of the levels of edit concept is discussed in depth later in this chapter.)

After reaching agreement with the author about the level of edit required, insist on receiving a double-spaced draft of the manuscript with wide margins at the top, bottom, and sides. You'll need the extra white space around the text to make queries and to mark your additions, deletions, substitutions, and reorganizations. Before making any marks on the manuscript, photocopy it, and make your editorial changes on that copy, saving the original in case you make a false start. (Online editing is discussed later in this chapter.)

The range of editing tasks described in the following discussion of the editing process runs the gamut from substantive editing for content and organization to "editorial style" editing for consistency in such elements as abbreviations, capitalization, numbers, equations, compound words, and symbols. Professional editors are expected to perform all these tasks successfully.

## The Organization Edit

Analyze the manuscript systematically. Begin with the large issues before proceeding to a line-by-line edit of the text. The best and quickest way to gain an overview of a publication is to review its organization. If it has a table of contents, review it. If it does not, create one from the topic headings within the manuscript. Review the organization carefully in light of your discussion of the purpose, audience, and scope of the document with the author. Do all headings fall within an easy-to-see hierarchy of meaning that allows you (and thus the intended reader) to follow the logic of the document?

If you edit documents such as proposals, customer manuals, journal articles, or technical reports that have a similar structure and contain standard sections, review the manuscript to ensure that all the typical parts are present. If you discover any obvious omissions, ask the author whether they are missing by accident or by design. If by accident, ask that the missing material be provided to you as soon as possible. Depending on the author's schedule, your grasp of the information, and your experience with similar documents, you may be expected to write some sections, perhaps the abstract, introduction, or executive summary, after you have edited the rest of the manuscript.

## The Content Edit

After you are satisfied that the manuscript contains all the essential parts and that they appear to be in the correct sequence, read the draft from start to finish critically for content and level of detail. During this reading, you will notice errors in grammar, punctuation, spelling, and mechanical conventions such as abbreviations. Some editors cannot resist the temptation to correct at least some of these errors during this reading, using the process as a warm-up to the assignment, even though they are after bigger quarry at this point. Other editors leave these corrections for subsequent readings, concentrating instead on content and reviewing everything in the light of what they know about the document's audience and purpose. Sometimes the purpose is stated in the introduction:

> This manual provides information on the care and use of the XXX Personal Computer and serves as a convenient reference source for the various options, supplies, and attachments available.

Briefer documents may contain nothing more formal in the way of a purpose statement than a title:

Operating Instructions for the
Model EZ-12 Remote Sensor Controller

If the document does not contain a purpose statement, create one based on your discussion with the author. Let this statement be your guide during the initial critical reading stage. Recognize, too, that some documents—often ones intended solely for in-house use—may not require an explicit purpose statement.

Review and then attempt to answer the questions on the following checklist during your first reading of the manuscript (adapted from Haness 15–16). The points addressed in this checklist, taken as a whole, may seem daunting. They simply make explicit the questions that all thoughtful readers ask of any piece of expository prose. As you gain editorial experience, asking them as you read will become second nature.

- Why was the document written?
- Does the purpose statement appear in the first or second paragraph of the opening? The introduction? The title? Is a purpose statement necessary?
- Is the document being written for a technician? The author's peers? Senior executives? Clients or customers?
- Are all the facts, details, and examples relevant to the stated purpose?
- Are the main points highlighted?
- Are the subordinate points related to the main points?
- Is each section distinct? Is its relationship to the overall purpose made clear?
- Does each section follow logically from the one that precedes it?
- Are there any contradictory statements of policy or fact anywhere?
- Do the descriptions and illustrations aid clarity? Are there enough of them?
- Does the text support and lead to the conclusions reached?
- For research or investigatory documents, are clear distinctions maintained for the sections on scope, methodology, findings, conclusions, and recommendations?
- Are recommendations adequately supported by the conclusions?
- Are any topics mentioned in the introduction that are not addressed in the conclusions?
- Do the conclusions raise topics not mentioned in the introduction or elsewhere in the text?

Note in the margins of the manuscript any problems you encounter in answering these questions. These marginal notes must be clear enough to enable you to recall the problem accurately in the postediting conference with the author.

The draft passage in Figure 12.1 has been edited using the appropriate principles from the checklist for content editing. The passage—about weather conditions in the Atlanta, Georgia, area—appears in a feasibility study written for the management of a brass foundry that is considering whether to open a new plant in Atlanta. Note that the editor reordered some information, deleted redundant or unnecessary material, and asked several questions of the author. The most significant query concerned a topic raised in the opening sentence, construction costs, that was not mentioned elsewhere. The editor also deleted the passage about the months that tropical cyclones affect Georgia. This information is not central to the main point about these storms: they rarely affect Atlanta. By

6.6 Weather

[not mentioned elsewhere]

Weather affects labor, (construction costs,) heating and cooling costs, and insurance costs. It can also affect the morale and health of workers. For comfort, the average temperature in Georgia is ideal at 64F. The winters are too cold further north and the summers too hot and long further south. In Atlanta, the average annual temperature is 62F., the average maximum in August is 89F, and the average minimum in January is 36F. The Atlanta area has average annual precipitation of 48 inches.

[graph of average temps for 12 mos. would help]

[move up]

Since foundry work is hot and needs summer temperatures that don't get too high, the mild temperatures are crucial to keeping the workers comfortable and productive.

If damaging storms are common to an area, insurance costs could be affected. Georgia has an expectancy of one full fledged hurricane in ten years. This does not limit the tropical storm damage in the State to once in ten years. In the past the exposed coastal areas suffered damage from less intense storms and the rains that accompanied them caused flooding. The tropical cyclones affect mostly the southeastern half of Georgia in the month of October. There is less effect in the northeast, including Atlanta, and less in August and September. Most of Georgia's tropical cyclones approach from the south or southwest and move across the State in a northeasterly direction. Having usually already passed over much land, their intensity is decreased from friction by the time they near Atlanta, and their source of energy, the (water,) has been cut off. Atlanta is not very much affected since it is in the top half of the state. It appears that storms don't often cause extensive damage in the Atlanta area. Insurance costs, therefore, should not be prohibitive.

[Gulf of Mexico?]

*Figure 12.1*    Content Edit of a Sample Passage

no means is the passage finished at this stage, however, because the editor concentrated exclusively on *what* was said rather than on *how well* it was said.

How you proceed from here depends on the severity of the problems you find. If the manuscript contains enough fundamental flaws (like those listed on the checklist) that further editing would not be practical, explain why in as much specific detail as possible, and return the draft to the author for revision. Otherwise, go on to the next critical reading.

## The Copy Edit

Now that you have read the manuscript once, you will go through it again, this time focusing on every paragraph, sentence, phrase, word, table, graphic, and symbol to determine how well each element contributes to fulfilling the document's purpose. In the first stage, you looked for what the manuscript said. This time through, you are looking for how well the manuscript says what it has to say. To that end, the following 11 areas must be examined during this reading.

### *Paragraphs*
- Does each paragraph develop a single idea?
- Does each develop the idea adequately?
- Does each contain a topic sentence?
- Are brief transitional paragraphs used when appropriate?

### *Sentences*
- Is each complete?
- Does each clause contain an identifiable subject?
- Are secondary ideas subordinate to primary ideas?
- Are coordinate ideas expressed in parallel words, phrases, and clauses?
- Do most sentences follow the subject-verb-object pattern?
- Do sentence type and construction vary occasionally to offset monotony?
- Are complex ideas stated in simple sentences as often as possible?
- Are simple ideas linked into complex sentences to offset choppiness?
- Are expletives (*it is*, *there are*) at the beginning of sentences kept to a minimum?

### *Transition*
- Are transitions adequate between sentences?
- Are transitions adequate between paragraphs?
- Are transitions adequate between sections?
- Are transitional expressions used within paragraphs to clarify the movement of ideas?

### *Voice*
- Is the subject highlighted properly by use of the active voice?
- Does either tact or an irrelevant agent or actor make the passive voice preferable?
- Does the writer shift voice inappropriately within sentences, paragraphs, or sections?

### Point of View

- If the writer is a participant or an observer, is the text written in the first person?
- If the text gives directions or instructions, are they written in the second person?
- Is the impersonal point of view used only when appropriate?
- Does the writer properly separate corporate or organizational opinion ("The staff concludes . . .") from personal opinion ("I conclude . . .")?
- Does the writer avoid self-references in the third person ("The writer believes . . .")?
- Is the point of view expressed consistently?
- For documents with multiple chapters or sections, each containing a separate perspective on the topic, is the point of view of each internally consistent?
- Are the nouns to which pronouns refer easy to identify?

### Parallel Structure

When they are intended to be parallel, are the following elements alike in grammatical structure as well as in function?

- Phrases and clauses
- Lists
- Topic headings in the text
- Topic headings in the table of contents

### Consistency

- Is terminology used consistently? (Has the author avoided using more than one term for the same idea or item?)
- Has the author avoided using the same term for more than one idea or object?
- Are abbreviations, acronyms, and symbols used only when appropriate? Are they consistent throughout the manuscript?
- Are verb tenses (past, present, future) consistent and appropriate?

### Word Choice and Diction

- Is the terminology appropriate to the intended audience?
- Is there a potential for confusion because terms commonly understood in one context are used in a different sense in the document (e.g., *affect* used as a noun in the field of psychology when it is used as a verb in other contexts)?
- Would functional definitions of specialized terms enhance communication?
- Would a glossary do likewise?
- Are usage choices in words (*irregardless*) and constructions (*don't got*) expressed in standard English?
- Do all descriptive terms convey the intended connotations?
- Is explicit or implicit sexist language present in pronouns, titles, or examples?

### Wordiness and Repetition

- Is the same idea repeated in different words unnecessarily?
- Are any phrases, clauses, or sentences redundant?
- Is repetition that enhances emphasis, expresses transition, or promotes consistency effective?

### Punctuation

- Does punctuation promote understanding and help the reader interpret the writer's meaning?

### Graphics and Tables

- Can any parts of the text be better supported or even replaced by a table or a graphic?
- Will adding graphics enhance communications to a multinational readership?
- Does the text introduce each table and figure adequately?
- Is each table and figure placed as soon as possible following its first mention in the text?
- Does the text reinforce or otherwise clarify the information in each table and figure?
- Does each have a concise, descriptive title?
- Are terms used consistently in tables, graphics, and the text?

(Additional guidance on the use of tables and graphics appears in Chapter 7.)

Figure 12.2 shows the passage from Figure 12.1 edited word for word to make the text coherent, unified, and consistent throughout.[1] Following the first and second edits, the retyped passage reads as in Figure 12.3. Although this version reads better than the original and is more coherently organized, the editor's questions must be answered before the text can be considered fully edited.

## The Editorial Style Edit

The third, and usually last, critical review of the entire manuscript is made for "editorial style" and usually requires several more readings through the manuscript. Editorial style refers to the conventions editors apply to eliminate inconsistencies in abbreviations, capitalization, numbers, equations, compound words, symbols, spelling, and other recurrent features of the manuscript. During these read-throughs, also check for whether all sections, subsections, tables, graphs, equations, bibliographic references, and the like are present, accurate, and in the right sequence. For editors with access to the appropriate equipment, computer software is now available that "analyzes" the editorial style of a piece of prose.

The following guidelines provide an overview of the range of document components to which editorial style applies. The list cannot be definitive because

---

[1]Copyediting symbols and conventions are shown in Figure 12.4.

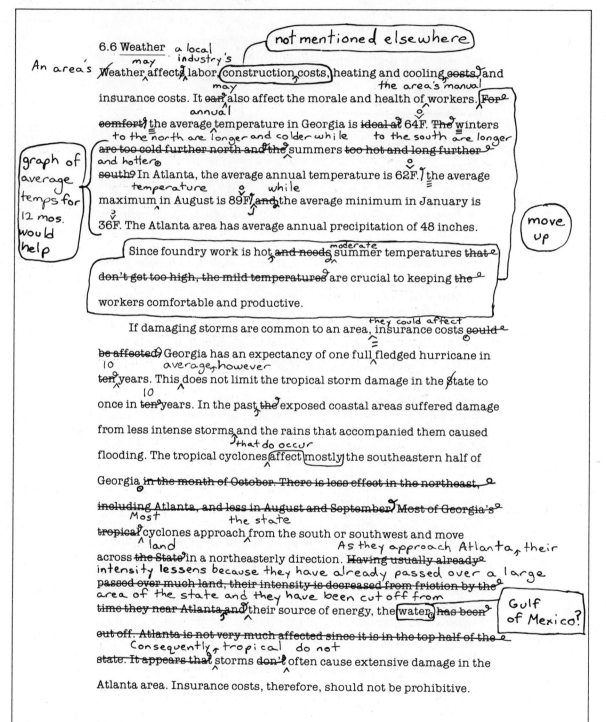

The marked-up manuscript reads, with editorial annotations:

6.6 Weather

*not mentioned elsewhere*

An area's Weather affects a local industry's labor, (construction, costs,) heating and cooling costs, and the area's manual insurance costs. It can also affect the morale and health of workers. For annual comfort, the average temperature in Georgia is ideal at 64F. The winters to the north are longer and colder while are too cold further north and the summers to the south are longer and hotter too hot and long further south. In Atlanta, the average annual temperature is 62F. the average temperature maximum in August is 89F while and the average minimum in January is 36F. The Atlanta area has average annual precipitation of 48 inches.

*graph of average temps for 12 mos. would help*

Since foundry work is hot and needs moderate summer temperatures that don't get too high, the mild temperatures are crucial to keeping the workers comfortable and productive.

*move up*

If damaging storms are common to an area, insurance costs could be affected. they could affect Georgia has an expectancy of one full fledged hurricane in ten 10 years average, however. This does not limit the tropical storm damage in the state to once in ten 10 years. In the past, the exposed coastal areas suffered damage from less intense storms and the rains that accompanied them caused flooding. The tropical cyclones affect mostly that do occur the southeastern half of Georgia in the month of October. There is less effect in the northeast, including Atlanta, and less in August and September. Most of Georgia's Most tropical cyclones approach from the south or southwest and move the state across the State land in a northeasterly direction. Having usually already As they approach Atlanta, their intensity lessens because they have already passed over a large passed over much land, their intensity is decreased from friction by the area of the state and they have been cut off from time they near Atlanta, and their source of energy, the water has been Gulf of Mexico? out off. Atlanta is not very much affected since it is in the top half of the state. It appears that Consequently, tropical storms don't do not often cause extensive damage in the Atlanta area. Insurance costs, therefore, should not be prohibitive.

**Figure 12.2**  Copy Edit of Sample Passage

### 6.6 Weather

An area's weather may affect local industry's labor, construction, heating and cooling, and insurance costs. It may also affect the morale and health of the area's manual workers. Since foundry work is hot, moderate summer temperatures are crucial to keeping workers comfortable and productive. The average annual temperature in Georgia is 64° F. Winters to the north are longer and colder while summers to the south are longer and hotter. In Atlanta, the average annual temperature is 62° F. The average maximum temperature in August is 89° F, while the average minimum in January is 36° F. The Atlanta area has an average annual precipitation of 48 inches.

If damaging storms are common to an area, they could affect insurance costs. Georgia has an expectancy of one full-fledged hurricane in 10 years. This average, however, does not limit the potential for tropical storm damage in the state to once in 10 years. In the past, exposed coastal areas suffered damage from less intense storms and the rains that accompanied them caused flooding. The tropical cyclones that do occur mostly affect the southeastern half of Georgia. Most cyclones approach the state from the south or southwest and move across land in a northeasterly direction. As they approach Atlanta, their intensity lessens because they have already passed over a large area of the state and they have been cut off from their source of energy, the water. Consequently, tropical storms do not often cause extensive damage in the Atlanta area. Insurance costs, therefore, should not be prohibitive.

*Gulf of Mexico?*

*Author: Construction costs mentioned in opening not discussed elsewhere.*

**Figure 12.3**  Final Draft of Edited Passage for Review by Author

of the breadth of specialized fields to which these guidelines might apply—mathematics, chemistry, psychology, biology and medicine, and economics, to name a few. Each such field has discipline-specific style conventions for its journals and other types of documents. Chapter 14 discusses how corporations, professional associations, government agencies, and other organizations standardize these editorial style guidelines for their documents.

### Text Elements

- Are all sections and subsections present, and are they in the correct sequence?
- Are cross-references accurate, especially those that refer to page, figure, and table numbers?
- Are entries in glossaries, indexes, and lists of symbols in correct alphabetical sequence?
- Does each heading in the table of contents match, word for word, the heading in the text?
- Are page numbers in the table of contents accurate?
- Are page numbers in the lists of tables and figures accurate?
- Are page numbers in the subject index accurate?

### Graphics and Tables

- Are all graphics and tables present and in the correct sequence?
- Does each have a number and a caption?
- Are callouts accurate and legible?
- Is each referenced in the list of figures or tables?
- Do those borrowed from other sources have a credit line?
- Has a written release been received for all copyrighted materials?[2]

### Lists in Text

- Are lists consistently formatted throughout the publication?
  - Are they indented or flush left?
  - Are they preceded by bullets, numbers, or letters?
  - Are they consistent in capitalization and in end punctuation?

### Equations

- Are equations numbered in sequence?
- Are symbols and numbers consistent in type style and size?
- Are power-of-10 notations (exponents) referred to and formatted consistently in text and in tables?
- Is there adequate white space on the page around equations?
- Do multiline equations break at operational signs, parentheses, or brackets?

### References and Footnotes

- Are all bibliographic references complete and accurate?
- Are they consistent in style?

---

[2]See Chapter 5 for a sample copyright release form letter.

After you finish marking the manuscript, *always* read through it one more time. On a heavily edited manuscript, you are sure to find at least one major and several minor mistakes that you have overlooked on each page, some of which will be of your making. Eliminate them before the author sees the manuscript. If these mistakes are seen by the author, their cumulative effect could diminish your credibility. Finally, photocopy the entire manuscript for safekeeping. You thereby avoid the potential disaster of having all your hard work wasted if you mail or hand over the manuscript to the author and it gets lost. You also need to keep a copy of your changes handy should the author phone you to discuss them.

In some organizations, editors mark (or "spec") the manuscript for the typesetter following review and final approval by the author. Typemarking entails specifying how all elements in a document are differentiated typographically. Typical features include these:

- Size and style of typefaces throughout (text, headings, footnotes, figure captions)
- Number of text columns on a page
- Height and width of each column
- Use of justified or ragged right-hand margins
- Size of all margins
- Location of headers, footers, and page numbers
- Placement and size of graphics
- Positioning of tables and lists

Typemarking may be done in a variety of ways, depending on the kind of typesetting system used. A detailed discussion of traditional and electronic typesetting techniques appears in Chapter 13.

## WORKING WITH AUTHORS

Following your final review of the manuscript, meet with the author to go over your questions, comments, and suggested changes. This meeting will last longer than the first meeting in which you interviewed the author about the intended purpose, audience, and scope of the manuscript. During this meeting, keep in mind that the work you are commenting on in many cases represents the basis on which the author's professional reputation will be judged by both peers and superiors. The likelihood is high that you will have an attentive, even critical listener during this review session. Recognize too that some authors will have an inherent antipathy toward anyone who's not a peer in their area of expertise making changes to their work. The sooner you establish that your purpose is not to criticize but to help the author realize his or her intentions to the fullest extent possible, the sooner you will gain the author's respect and cooperation.

Where to begin? This question has an appearance and a substance component. Begin with appearances. The manuscript will more than likely look a mess, so first consider the author's reaction to it. Explain that many of the marks simply indicate the need for mechanical or editorial style changes and that they do not reflect on the quality of the writing. You can further offset the untidy look of the manuscript by retyping long handwritten or especially messy passages before your meeting. The retyped passages will not only make a better first impression but will also be much easier for the author to review. Elsewhere, and for the same reasons, make your editorial marks as neat and clear as possible. (We will return to the topic of copymarking shortly.)

Begin by discussing the major content and organizational problems. When tackling substantive issues, recognize that content is the author's province. "An editor's job is to shape the *expression* of an author's thoughts, not the thoughts themselves" (Plotnik 32). Rather than making assertions about content problems, ask questions. What sounds like a mistake or an inconsistency may well be a common characteristic of the topic that the intended audience is quite familiar with. By allowing the author to answer your questions, you save yourself the potential embarrassment of appearing to challenge what the author knows best. Being respectful doesn't mean that you should shrink from asking hard questions or suggesting substantive changes in organization or format. Nor should you be so timid that you don't ask questions because they will reveal how little you know. Any author can tolerate a few seemingly stupid questions if even one of them uncovers an error or an inconsistency that would have gone unrecognized if you had not challenged it. If authors were infallible, they would not need technical peer review or editing. Finally, be aware that some authors will reject even your most helpful suggestions. If you have communicated the reasons for your suggested changes as convincingly as you can, be prepared to let go and move on. You have fulfilled your responsibility, and you have the consolation of knowing that the author bears final responsibility for the document.

As your review with the author moves away from content and organization problems toward tone, point of view, diction, and style, distinguish between required and recommended changes. Represent the readers in these matters to the extent that they would likely be confused by the author's original text. Anyone would be confused by inconsistent terminology, graphics dropped into the text without a context for them, shifting or contradictory points of view, an inadequate introduction, or units of measurement that switch from the metric system to the avoirdupois system and back again throughout the publication. Some authors may think that your changes are merely discretionary and represent things only "editor types" care about or that because these matters don't directly pertain to the technical content, one person's opinion is as good as anyone else's. Be tactful but firm as you go over these points. These types of changes are well within your professional competence, so hold firm on them when you know that the reader will be confused otherwise.

Just as important, learn when to compromise. If you indiscriminately hold firm on every point as though it were of equal consequence, you will alienate all but the most thick-skinned of authors. Foolish consistency is every bit as debilitating a practice in editing as it is elsewhere. Recognize, for example, that not every sentence of every document can or should be expressed in the active voice.

Likewise, despite your preference to the contrary, you may not always be able to position every graphic so that it immediately follows its first mention in the text. These are not minor points, but they often require compromise. With minor points, your hold should be much less tenacious. Time spent changing every *utilize* to *use*, switching restrictive *which* to *that* and nonrestrictive *that* to *which*, or "correcting" split infinitives adds little of value to the content and makes the edited manuscript look messier than it would otherwise. Moreover, these are stylistic choices about which respected usage guides themselves differ. When no style is specified, or when the standard guides offer conflicting advice, follow the author's expressed preference, making it consistent throughout the manuscript. You will gain the esteem of your author clients when they recognize in you an advocate for the substance of their message rather than a purist trying to uphold an illusory law of "correctness." Don't be correct but irrelevant like the purist in the famous Ogden Nash poem.

> I give you now Professor Twist,
> A conscientious scientist.
> Trustees exclaimed, "He never bungles!"
> And sent him off to distant jungles.
> Camped on a tropic riverside,
> One day he missed his loving bride.
> She had, the guide informed him later,
> Been eaten by an alligator.
> Professor Twist could not help but smile.
> "You mean," he said, "a crocodile."

You bear another important responsibility to your author-clients. In the course of your editorial work, you may find yourself the willing or unwilling recipient of a personal or professional confidence that you are asked not to betray. You may also be made privy to confidential or proprietary information about the inner workings of your organization that is not intended for disclosure elsewhere. Unless these confidences or disclosures involve criminal wrongdoing, keep them to yourself. If you are perceived as someone who betrays a confidence, regardless of whether you were explicitly asked not to betray it, you could be removed from sensitive projects or even fired.

## WORKING WITH OTHER DOCUMENT SPECIALISTS

During and following your edit, you will work closely with other document specialists. Word processing operators, graphic designers, and typesetters may all participate to one degree or another in helping you create the camera-ready version of the manuscript that is finally submitted for printing. In this capacity, you act as the intermediary for the author in carrying out the agreed-on changes and as the client for the other specialists. To obtain the best service possible, maintain an attitude of active cooperation with these specialists. Learn as much as you can about what they do, how long their tasks take, and the limitations under which they work. This knowledge will allow you to profit from their advice—thus making you more efficient—and to cooperate with them in working around technical and schedule problems. For example, you may be able to save time by obtaining, from an existing document, graphics that require only minor modifications by your organization's graphics staff. Or you may be able to supply a computer diskette of the final manuscript for the typesetters which they can then "read" in to their electronic composition equipment without having to retype the text word by word. Your cooperation may take more mundane but still important forms as well, like making all your instructions legible and unambiguous when you submit work to these specialists and by being available to answer any of their questions. When submitting graphics work, for example, adhere to the guidelines detailed in Chapter 7.

## MECHANICS OF COPYMARKING

Professional editors go about the mechanics of manually marking copy in a variety of ways best suited to the requirements of their tasks and the bent of their personalities. (Editing on a computer will be discussed in the following section.) No consensus has emerged for the techniques that work most efficiently in all circumstances, although some widely applicable guidelines do exist.

Editors communicate with authors, typists, and artists in a symbolic shorthand called copyediting marks, listed in Figure 12.4. These marks are almost identical to those used by proofreaders. Their differences reflect the fact that editors work on draft manuscripts (that is, manuscripts before they are typeset) whereas proofreaders work on manuscripts after they have been typeset. Errors may occur in a typeset text that either cannot occur or are unimportant in the original draft manuscript—for example, a wrong font or poor proportional spacing in a column. The last three marks shown in Figure 12.4 are used solely by proofreaders, although they use all the other copyediting marks, as well.

When marking copy, be certain that the symbols you use are understood by the author and by your fellow document professionals. If you have any doubts about that, draw up a one-page "style sheet" for them that includes each pertinent symbol and an example of how it is used.

Another universally agreed-to principle of copymarking is that you write or print legibly. Common courtesy and the need to communicate your message clearly and unambiguously make neatness imperative. For the same reasons, type long passages you add and any other passage that is especially sloppy.

Some editors handwrite all their changes to the manuscript and editorial queries to the author directly on the manuscript. Other editors mark their additions, deletions, substitutions, and reorganizations on the manuscript but reserve their queries to the author for gummed or self-adhesive "flags" that they attach to the edge of the appropriate page. (Editorial queries ask the author to

| Mark in margin | Instructions | Mark on manuscript | Corrected type |
|---|---|---|---|
| ℓ | Delete | the ~~lawyer's~~ bible | the bible |
| *lawyer's* | Insert | The ∧bible | The lawyer's bible |
| (stet) | Let stand | the ~~lawyer's~~ bible | the lawyer's bible |
| (cap) | Capitalize | the bible | the Bible |
| (lc) | Make lowercase | the Law | the law |
| (ital) | Italicize | the lawyer's bible | the *lawyer's* bible |
| (tr) | Transpose | the bible lawyer's | the lawyer's bible |
| ⌣ | Close space | the Bi ble | the Bible |
| (sp) | Spell out | ②bibles | two bibles |
| # | Insert space | TheBible | The Bible |
| ¶ | Start paragraph | ¶ The lawyer's... | The lawyer's... |
| (run in) | No paragraph | ...marks. Below is a... | ...marks. Below is a... |
| (sc) | Set in small capitals | The bible | The BIBLE |
| (rom) | Set in roman type | The (bible) | The bible |
| (bf) | Set in boldface | The bible | The **bible** |
| (lf) | Set in lightface | The (bible) | The bible |
| ⊙ | Insert period | The lawyers have their own bible∧ | The lawyers have their own bible. |
| | Insert comma | However∧we cannot... | However, we cannot... |
| | Insert hyphens | half∧and∧half | half-and-half |
| | Insert colon | We need the following∧ | We need the following: |
| | Insert semicolon | Use the law∧don't... | Use the law; don't... |
| | Insert apostrophe | John∨s law book | John's law book |
| | Insert quotation marks | The ∨law∨ is law. | The "law" is law. |
| | Insert parentheses | John's ∧law∧book | John's (law) book |
| | Insert brackets | John∧1920-1962∧went... | John [1920-1962] went... |
| | Insert en dash | 1920∧1962 | 1920–1962 |
| | Insert em dash | Our goal∧victory | Our goal—victory |
| | Insert superior type | 3∨=9 | $3^2=9$ |
| | Insert inferior type | H∧SO₄ | $H_2SO_4$ |
| | Insert asterisk | The law∨ | The law* |
| | Insert dagger | The law∧ | The law † |
| | Insert double dagger | The bible∧ | The bible ‡ |
| | Insert section symbol | ∧Research | § Research |
| (X) | Broken letter; dirty letter or space | law∧yer∧ | lawyer |
| (wf) | Wrong font | (lawyer) | lawyer |
| (eq) | Unequal word spacing | equal ∨space ∨in | equal space in |

*Figure 12.4* Copyediting and Proofreading Marks

resolve apparent contradictions, clarify a point, supply missing information, or verify that passages rewritten by the editor have not changed the author's intended meaning.) Personal preference and sometimes organizational policy govern these options. Each has its advocates and detractors. The flags are easy for the author to spot, but they can also fall off easily. They also make photocopying the manuscript nearly impossible. Conversely, queries and changes written on the page are convenient for the editor but can make the page look very messy and may confuse the author. (Most editors circle all queries on the manuscript to distinguish them from additions and revisions to the text.) One other liability of writing on the page is that writing near the edges tends to get lopped off when the pages are photocopied. To correct for this problem, set the photocopier to reduce the page by 6 or 7 percent.

Editorial opinion is also divided over whether to mark the manuscript in pen, pencil, or felt-tip marker and which color ink to use. Colored ink is easy to see, will not smudge, and photocopies well. It is, however, difficult to correct. No matter whether you use ink eradicator (such as household bleach), correction fluid, or self-adhesive correction tape or simply cross out errors, you run the risk of making a mess. If you favor colored ink, avoid red: not only does it look strident, but it may remind authors unpleasantly of the marks made on school compositions. Do not use blue either because it will barely reproduce when photocopied. Avoid felt-tip pens also. They lose their points very quickly, and the marks they make are even more difficult to correct than those of ink pens. Many editors favor mechanical pencils with black or brown lead. These pencils do not need sharpening, and the marks they make are easy to erase and to read and photocopy well if the original is written distinctly.

## EDITING ONLINE

With the increasing availability of computers in the workplace, many editors now have the option of editing on the computer screen rather than on a paper ("hard") copy. Editors and writers in some organizations, in fact, conduct all their business online, not producing a paper copy until the manuscript is in final form ready for printing (Coletta and Oately).

Despite some unquestioned advantages, online editing challenges editors at the most basic level at which they work: marking copy. The editor's copymarking permits the author to review editorial changes against the background of the original text. However, the proofreading symbols and copymarking conventions used by editors working with paper copy do not have interchangeable or convenient counterparts for editors working online. To solve this problem, editors have devised a variety of substitutions for copyediting marks.

To show additions, for example, they put the new text in **all boldface** or in *italics*, [place it in brackets], underline it, place an *i* (for "insert") before and after it, or place a vertical change bar ( | ) in the margin next to the appropriate passage.

To show deletions, they may strike through the original text with hyphens (·) or invoke a strikeout mode, thus allowing the author to see the original text and the editor's intended changes.

To make editorial queries, they may place their questions in parentheses or brackets; put them in boldface, italics, or all capital letters; or devise some other means that does not overlap an existing convention.

This type of copymarking also introduces potential ethical dilemmas about the degree to which editors "flag" changes for the author's attention. Done day after day, this process becomes tedious and could tempt the editor to make "routine" changes to the punctuation, spelling, and phrasing of the manuscript without marking them for the author's review. The author could easily miss these changes unless he or she makes a word-for-word comparison between the original of the electronic manuscript, now likely gone, and the edited version. To make these changes silently is wrong both ethically and practically. Would you want someone changing your text without telling you? You could also introduce errors inadvertently, a likelihood that increases in direct proportion to the editor's lack of familiarity with the subject matter. Editors must avoid making such changes on their own until they are knowledgeable and experienced enough to have gained the trust of those whose work they edit.

Online editing has several other distinct disadvantages. The size of the computer screen limits the amount of text that the editor can view at one time. Comparing information in other parts of a manuscript with the information on the screen can be difficult because the editor must scroll forward or backward, hoping to locate salient passages. Sheets of paper, by contrast, can be spread out and shuffled, making the passages sought easier to locate. Working online can also be fatiguing and tedious, thus increasing the possibility that the editor will both make and miss typographical errors. Moreover, certain types of information do not lend themselves to online editing. Tables, graphics, equations, and other nontext materials must usually be edited in hard copy.

These liabilities are acknowledged by an editorial consulting firm that recently publicized a decline in requests for the online editing services it offers. The firm noted that "clients said they found it easier to review a paper copy of editing changes, pass the manuscript back to [us] for final corrections, and receive a clean printout and disk of the edited manuscript" ("Calls" 8). The potential advantages that online editing offers cannot compensate at this time for its numerous liabilities.

## LEVELS OF EDIT

The range of editing tasks described in this chapter—organization, content, copy, and style edits— rests on the premise that each manuscript will require the same level of editorial effort and that editors have no time constraints to inhibit their quest to produce flawless documents. Experience argues otherwise, of course. Editors in most settings often do not have as much time as they would like to spend revising and polishing their projects. Moreover, not all documents require the same degree of effort. The document standards for an organization's in-house documents are often far less stringent than its standards for documents created for customers and clients. Recognizing that the level of effort between assignments varies, editorial managers describe the tasks editors perform in a variety of ways. Such descriptive categories also help to establish a rational basis for scheduling work, describing options to clients, and quantifying costs (Van Buren and Buehler 3).

The four stages of editing described here are sometimes collapsed into fewer categories. One editorial consulting firm divides the editing done for its clients into two broad categories: copyediting and substantive editing. *Copyediting* "includes reviewing a . . . manuscript for spelling, grammar, punctuation, and consistency. Copyediting can include checking completeness, accuracy, and formats of tables, bibliographies, and footnotes. It does not include rewriting and reorganizing" (Boston 71). *Substantive* editing is described as

> including copyediting, rewriting, reorganizing, writing transitions, writing chapter or section summaries, eliminating wordiness, reviewing content for accuracy and logic, and ensuring the proper tone and approach to the intended audience. Substantive editing can also include helping plan and outline publications and consulting with authors and publishers. (71)

In *The Elements of Editing*, Arthur Plotnik notes that "when told to 'edit' a manuscript, no two editors in the world will go about it the same way" (11). He then identifies several levels or "depths" at which a manuscript can be edited, depending on the time available: light, medium, and heavy. *Light editing* involves correcting errors in spelling and grammar; revising punctuation, numbers, capitalization, and the like according to an agreed-to standard of style; resolving any stylistic inconsistencies; and checking statements that seem absurd or libelous. *Medium editing* includes all the light editing tasks as well as reworking some prose to make it more active and concrete; reorganizing paragraphs as needed to make a strong opening and to enhance logic; and spot-checking a few sources for accuracy. *Heavy editing* encompasses light and medium editing as well as rewriting some weak passages, cutting material that interferes with momentum, providing transitions where needed, checking as many facts as possible, and conferring with the author (41).

The best-known levels-of-edit concept was developed by Robert Van Buren and Mary Fran Buehler at the Jet Propulsion Laboratory of the California Institute of Technology. Their goal in developing these categories "was directed not toward defining terminology as such, but toward specifically identifying the domains of editorial activity and establishing boundaries between them" (Buehler 12). They divide editorial tasks into five "levels" that encompass nine "types" of editing. Their system, described in *The Levels of Edit*, is structured as shown in Figure 12.5. The editors who master this range of responsibilities, regardless of how they are categorized, and apply them tactfully and objectively will gain the respect of their author-clients and a deserved sense of self-satisfaction.

### Fifth Level

*Coordination edit* concerns the editor's administrative responsibilities for the production of a publication: scheduling, budgeting, planning, meeting, explaining, and interpreting.

*Policy edit* involves the enforcement of the organization's guidelines on the kind of information that may not be included in a publication, such as proprietary, classified, or company confidential, and the use of legal or policy disclaimers about the information that is included.

### Fourth Level

*Integrity edit* requires ensuring that all elements of the manuscript referred to in the text are there, properly identified, and in the proper sequence. These elements include tables, figures, equations, references, footnotes, appendixes, page numbers, and all text sections and their titles.

*Screening edit* is a minimal language review for subject-verb agreement, spelling, and complete and ungarbled sentences, as well as for unreproducible or inappropriate graphics (e.g., a hand-drawn graph in an otherwise professionally produced publication).

### Third Level

*Copy clarification edit* requires that the editor unambiguously mark the manuscript with clear instructions for the typist, typesetter, and graphics staff.

*Format edit* involves marking the manuscript for the typesetter with instructions for the physical design and layout of the information on the page: for example, the type size and style for the text, headings, and running heads and feet and the number, height, and width of text columns.

### Second Level

*Mechanical style edit* is a review of the manuscript to ensure consistency in spelling, capitalization, hyphenation, acronyms, abbreviations, numbers, and use of compound terms.

*Language edit* incorporates the screening edit and entails the full range of text components, including grammar, punctuation, usage, syntax, and verb tenses.

### First Level

*Substantive edit* entails a thorough review of the contents to correct any problems caused by lack of unity and coherence, poor organization, discrepancies in meaning, needless redundancy, and the presence of inappropriate material.

*Figure 12.5*   Van Buren and Buehler's Levels of Edit

*Checklist 12.1*

## First Meeting with the Author

☐ What is the purpose of the document?

☐ Who is the audience for the document?

☐ What is the scope of the information covered in the document?

☐ What kind of document is it?
  • Manual
  • Report
  • Journal article
  • Conference paper

☐ Will the document have any special design requirements?
  • Loose-leaf binding
  • Index tabs
  • Oversize graphics

☐ What level of edit will be necessary?
  • Coordination edit
  • Policy edit
  • Integrity edit
  • Screening edit
  • Copy clarification edit
  • Format edit
  • Mechanical style edit
  • Language edit
  • Substantive edit

☐ What is the deadline for the document?

---

*Checklist 12.2*

## The Editing Process

### The Organization Edit

☐ Does the document have a table of contents?

☐ Are headings organized into a logical hierarchy?

☐ Are all conventional parts present for this particular type of document?
  • Abstract
  • Table of contents
  • Executive summary

- Introduction
- Methodology
- Conclusions

## The Content Edit

☐ Does the document have or need a purpose statement?

☐ Is the document aimed at the right audience?
- Technicians
- Senior executives
- Clients or customers

☐ Are facts, details, and examples relevant?

☐ Are major points highlighted?

☐ Are subordinate points relevant to major points?

☐ Does each section serve a distinct purpose?

☐ Are there proper transitions between sections?

☐ Are there any contradictory statements of fact or policy?

☐ Do descriptions and visuals aid clarity?

☐ Does the text support the conclusions reached?

☐ Are findings, conclusions, and recommendations clearly distinguished from one another?

☐ Are all topics raised in the introduction addressed in the conclusions?

☐ Are all topics addressed in the conclusions mentioned in the introduction?

## The Copy Edit

☐ Do the following elements contribute effectively to fulfilling the document's purpose?
- Paragraphs
- Sentences
- Transitional devices
- Point of view
- Parallel elements
- Terminology
- Abbreviations
- Symbols
- Verb tenses
- Punctuation
- Graphics and tables

### The Editorial Style Edit

☐ Are the following components of the document consistent and accurate?
- Text elements
- Graphics and tables
- Lists
- Equations
- References and footnotes

### Typemarking

☐ Have the following components been accurately marked for typesetting?
- Typeface size and style
- Number and size of text columns
- Location of headers, footers, and page numbers
- Size and placement of graphics
- Position of tables and lists

---

## WORKS CITED

Boston, Bruce O., ed. *Stet! Tricks of the Trade for Writers and Editors*. Alexandria, VA: Editorial Experts, 1986.

Buehler, Mary Fran. "Defining Terms in Technical Editing: The Levels of Edit as a Model." *Technical Communication* 28.4 (1984): 10–15.

"Calls for On-Line Editing Decline." *Editorial Eye* 12.2 (1988): 8.

Coletta, Barbara, and Lorraine Oately. "Online Editing: Is It Worth the Trouble?" *Proceedings of the 34th International Technical Communications Conference* (1987): WE-59–WE-62.

Haness, Joel. "How to Critique a Document." *IEEE Transactions on Professional Communication* 26 (1983): 15–17.

Nash, Ogden. "The Purist." *Selected Verse of Ogden Nash*. New York: Modern Library, 1945. 229.

Plotnik, Arthur. *The Elements of Editing*. New York: Macmillan, 1982.

Van Buren, Robert, and Mary Fran Buehler. *The Levels of Edit*. 2nd ed. Pasadena, CA: Jet Propulsion Laboratory, 1980.

## FURTHER READING

Amsden, Dorothy Corner, and Scott P. Sanders. "Developing Taste and Judgment: Correctness and the Technical Editor." *Technical Writing Teacher* 12 (1985): 111–114.

Bates, Jefferson. "Writers and Editors: Can't We Be Friends?" *Editorial Eye* 11.10 (1988): 1–3.

Bennett, John B. *Editing for Engineers*. New York: Wiley-Interscience, 1970.

Bush, Don. "Content Editing: An Opportunity for Growth." *Technical Communication* 28.4 (1981): 15–18.

Chaffee, Patricia. "Human Engineering and Technical Writing." *Technical Writing Teacher* 10 (1983): 130–133.

Cheney, Patrick, and David Schleicher. "Redesigning Technical Reports: A Rhetorical Editing Method." *Journal of Technical Writing and Communication* 14 (1984): 317–337.

Cheney, Theodore A. R. *Getting the Words Right: How to Revise, Edit, and Rewrite.* Cincinnati: Writer's Digest, 1983.

*The Chicago Manual of Style.* 13th ed. Chicago: U of Chicago, 1982.

Clements, Wallace, and Robert G. Waite. *Guide for Beginning Technical Editors.* Livermore, CA: Lawrence Livermore Laboratories, 1979.

Cook, John M. "The Technical Writing Student as Editor." *Technical Writing Teacher* 10 (1983): 114–117.

Cox, Alberta L. "Copy Editing: The Final Word." *Technical Communication* 28.4 (1981): 18–20.

Dukes, Eva. "Some Authors I Have Known." *Technical Communication* 28.4 (1981): 27–30.

Farkas, David K. "Online Editing and Document Review." *Technical Communication* 34 (1987): 180–183.

Grodsky, Susan. "Indexing Technical Communications: What, When, and How." *Technical Communication* 32.2 (1985): 26–30.

Holder, Laurel N., and Stephen P. O'Neill. "Who's in Control: Editorial Authority and the Author." *Technical Communication* 34 (1987): 19–23.

Jack, Judith. "Teaching Analytical Editing." *Technical Communication* 31.1 (1984): 9–11.

Judd, Karen. *Copyediting: A Practical Guide.* Los Altos, CA: Kaufmann, 1982.

Kellner, R. S. "A Necessary and Natural Sequel: Technical Editing." *Journal of Technical Writing and Communication* 12 (1982): 25–33.

Lutz, Jean A. "Attitude toward the Editing Process: Theory, Research, and Pedagogy." *Journal of Technical Writing and Communication* 16 (1986): 157–165.

Mann, Michele H. "How to Edit the Passive Writer's Work." *Technical Communication* 32.3 (1985): 14–15.

Michaelson, Herbert B. "How to Review an Engineering Manuscript." *Engineering Education* 5 (1985): 688–692.

Moy, Connie. "Editing the Prima Donna." *Editorial Eye* Aug. 1985: 1–3.

Paxon, William. "A Survival Kit for Editors." *Technical Communication* 28.4 (1981): 3.

Power, Ruth M. "Who Needs a Technical Editor?" *IEEE Transactions on Professional Communication* 24 (1981): 139–140.

Putnam, Constance E. "Myths about Editing." *Technical Communication* 32.3 (1985): 17–20.

Rainey, Kenneth T. "Technical Editing at the Oak Ridge National Laboratory." *Journal of Technical Writing and Communication* 18 (1988): 175–181.

Rude, Carolyn D. "Word Processing in the Technical Editing Class." *Journal of Technical Writing and Communication* 15 (1985): 181–190.

Scroggins, Mary J. "In Search of Editorial Absolutes." *INTERCOM* Aug.-Sept. 1988: 7.

Skillen, Marjorie E., and Robert M. Gray. *Words into Type.* 3rd ed. Englewood Cliffs, NJ: Prentice, 1974.

Soderston, Candace. "The Usability Edit: A New Level." *Technical Communication* 32.1 (1985): 16–18.

U.S. Government Printing Office. *Style Manual.* Washington, DC: GPO, 1984.

Whalen, Elizabeth. "The Author-Editor Relationship: Observations and Suggestions." *INTERCOM* Apr. (1988): 8–9.

Zook, Lola M. "Editing and the Editor: Views and Values." *Technical Communication* 28.4 (1981): 5–9.

———. "Technical Editors Look at Technical Editing." *Technical Communication* 30.3 (1983): 20–26.

### Chapter 12: Editing the Work of Others

Besides her peer editor's review, Mary felt that the *Introduction to Using the International Banking System* manual should also be edited by someone from First Wisconsin's Marketing Division. She enlisted the services of Judy Sheldon, writer and editor in the Corporate Communications Department of the Marketing Division (see the First Wisconsin Organization Chart in Figure C.1 on page 16). Judy joins Mary for this portion of the Case History to discuss the editor's role.

INTERVIEWER: Judy, how do you normally get an editing assignment?

JUDY: From my supervisor, who hands out all the assignments.

INTERVIEWER: How did you get the assignment to edit the *Introduction* manual?

JUDY: My supervisor got a memo from the supervisor in Technical Writing asking for some editing help for a high-profile project. My supervisor assigned the project to me.

MARY: We were concerned that this manual be well received by its anxious and skeptical readers. We wanted a professional editor to edit the manual—in addition to having the manual undergo the peer editing that is a normal part of our department's document cycle—so that the manual would have the highest possible quality.

INTERVIEWER: What was the sequence of contacts between the two of you?

JUDY: After my manager agreed to have me do the editing, Mary called me to describe the nature of the edit that she wanted and to discuss the audience and purpose of the manual and of the library of manuals planned for the system. She explained why her area wanted me to edit the manual, adding that she wanted a substantive edit that looked at organization, style, conciseness, and appropriateness of format.

When the draft of the manual arrived, it was accompanied by a memo stating the specific items Mary wanted me to look at in the edit [see Figure C.14]. Mary also sent copies of the style sheet for the manual [see Figure C.17 on pages 362–365], information on the audience and purpose of the manual, and her department's standard list of copyediting marks.

After I edited the material, I returned the marked-up draft to Mary. A few days later we met to resolve a few remaining questions about my comments.

INTERVIEWER: Mary, what was your attitude when you gave your writing to Judy? Did you feel like a student submitting writing to an instructor, fearful of corrections? What sort of attitude did you have?

MARY: I think there's always an element of fearful attentiveness, especially on a job that involves asking another department to edit. But one of the things you agree to when you're a writer is that you *want* to

To: Judy Sheldon, Corporate Communications, Marketing

From: Mary Mullins, Technical Writing, Information Services.  *MM*

Date: December 1st, 1988

Subject: Editing for <u>Introduction to Using the International Banking System</u>

Here is the <u>Introduction to Using the International Banking System</u> for your review. I was very happy to hear your supervisor approved your assignment to the project, since this is the first manual in a high-profile project and we want it to have the best possible edit. I know your work is excellent.

<u>Requirements for Edit</u>

As I mentioned on the phone, we usually copyedit each other's work within the Technical Writing Department. Another technical writer will be reviewing the manual while you give it a more substantive edit. Please consider the following items:

- Organization.

  Does the manual's organization make sense? Within each chapter, is the information presented in an organized fashion? Please pay special attention to the oganization of the System Standards chapter.

- Style.

  Is it clear? Please circle any confusing language, especially complex constructions. As I mentioned, the readers are familiar with international banking terminology, but many are not native English speakers. In addition, employees are anxious about the new system since it will be reducing the need for a large clerical staff.

- Conciseness.

  I would like to reduce the size of the manual from 60 pages to 50 pages, if possible. Can you suggest changes to make it more concise?

- Format.

  The manual should reflect the format decisions listed in the style sheet for the manual (enclosed).

<u>Reference Card</u>

In addition to editing the manual, could you also look over the reference card that comes with the manual? I'm planning to reformat the card and I'd like suggestions for making it look less cluttered.

<u>December 15th Return Date</u>

We are in a crunch to publish the manual by January 31st. Although I realize you have other projects, could you possibly edit the manual by December 15th? If so, I can be assured of meeting the January target date for publication.

I have signed out Conference Room H for the morning of December 15th (from 9 to 10:30). If you can make the December 15th deadline, we can meet then to discuss your comments.


cc: G. Atkins

*Figure C.14*  Cover Memo from Writer to Editor

hear feedback on your work. Even though you might have some anxiety about being judged, you have to separate yourself from the work and recognize that it's the work that is being judged, not you, and that the edit will improve your work.

INTERVIEWER: Judy, what tactics do you use to avoid offending a writer?

JUDY: I try to make sure every issue I raise is something that needs questioning. For example, I try not to pick on a point that's just an arbitrary choice of words, where I just happen to like one phrasing and the writer happens to like another.

I also try to present my point in an open-ended way—to say, "Is this what you meant?" or "Would it be better to say this instead?" rather than just saying, "This is wrong."

INTERVIEWER: Generally, how do you begin the editing process? What do you do first?

JUDY: The first thing I do is read through the whole document to get an overview of what it is and how its parts fit together. I do this because I want to make sure that things seem to be in the right order, that subheadings all have the same relationship to the main topic, that sort of thing. I just look at the bigger items first.

When I read through the first time, I'm normally reading as though I were trying to perform the task that's being explained. I try to use that document as it is intended to be used. Later, I'll read through it again just to catch things like typos and punctuation errors.

I usually end up reading things two, three, or four times before I get down to the detail level. Sometimes it takes that long to analyze the structure of a piece. For example, it took me a few readings of the *Introduction* manual before I got to the point where I decided that pages 2 and 3 should be switched.

INTERVIEWER: Does a writer ever choose not to follow your advice?

JUDY: Sure, that happens sometimes. A writer doesn't necessarily follow editing suggestions to the letter. Editing comments are just pointers that something may be wrong; they're suggestions, not definitive corrections. The writer may come up with an entirely different solution for the problem.

INTERVIEWER: How do you note these suggestions?

JUDY: Normally, I make a copy of the document and write in pencil my first time through, then make final marks in red. It depends on what the writer wants. Often I'll work with a preliminary copy and then get a fresh copy to put my final comments on neatly. At other times I'll write in pencil and then just erase a lot of it. I guess I don't always work in the same way.

INTERVIEWER: I'd like to ask you a few questions about your editorial comments on the first five pages of the "Introduction" chapter of the manual. [See Figure C.15.] Why did you suggest that Mary reverse the order of presentation by putting the list of chapters in the manual first

*Topic heading? Should begin on second line, run across page, end with underlined, bold-tab, and have triple spacing after.*

## Introduction

*€IBS≠ — (used on next page for Maintenance IBS Guide)*

This manual provides an overview of the entire International Banking System and describes basic skills needed to use the system. We hope you find it useful as a guide to your new system.

**Basic Services**

As you know, an international banking department provides many different services--often as many services as a domestic bank. As a result, in your work you may specialize in any of the following areas.

*The International Banking System can help you do your work in any of these areas. (Or some statement of the relationship)*

| | |
|---|---|
| Letters of Credit | Instruments issued on behalf of a buyer that give the buyer the financial backing of the issuing bank. |
| Loans *(Can you avoid repeating?)* | Loans made to businesses and other banks for the financing of inventory purchases and trade, including foreign loans that require specialized knowledge of legal requirements for lending in different countries. |
| Savings | Deposit instruments such as open and term time deposits, certificates of deposit, and interbank placements. |
| Current Accounts | Running accounts between our bank and other banks that reflect the movement between the banks of cash, loan reserves, and settlements. |
| Foreign Exchange Trading | The buying and selling of currencies in relation to either U.S. dollars or other foreign currencies. |
| Payments | The receipt of payments made via SWIFT, the international bank-to-bank wire service. |
| Financial Control | Cash management, individual and global credit risk analysis, and financial control. |

*Figure C.15*   Edited Pages from "Introduction"

**Related Manuals**

*[handwritten: one or more of these]*

This manual describes *[handwritten: the]* basic skills everyone needs to use the system. In addition to this manual, however, you will receive other manuals that describe how to do work in your own specialty. *[handwritten circled: cap]*

### For Daily Users

*Entering Letters of Credit.* *[handwritten: Tells]* How to set up and monitor letters of credit. *[handwritten circled: lc]*

*Entering Deposits and Loans.* *[handwritten: Tells]* How to set up deposit accounts and enter business transactions. *[handwritten circled: lc]*

*Entering Current Accounts.* *[handwritten: Tells]* How to set up settlement, interbank placement, and disbursement accounts. *[handwritten circled: lc]*

*Entering Payments.* *[handwritten: Tells]* How to enter and receive payments. *[handwritten circled: lc]*

*[handwritten: Entering]* *Foreign Exchange* *[handwritten: Contracts]* *[handwritten: Tells]* How to enter foreign exchange contracts. *[handwritten circled: lc]*

*Using General Inquiry.* *[handwritten: Tells]* How to view and act on previously entered transactions and incoming wires.. *[handwritten circled: lc]*

*[handwritten left margin: I prefer your other idea of using a side margin caption for third-level headings too. The way it is you need another space above to separate enough. (The other way would save space too.)]*

### For Credit Managers

*Interpreting Reports.* Describes standard reports in detail and shows how to request customized reports.

### For Special Employees

*[handwritten: Make others parallel by adding verbs — or delete verbs here?]*

*Entering Customer Records.* *[handwritten: Tells]* How to enter customer information, including customer credit lines. *[handwritten circled: lc]*

*IBS Maintenance Guide.* *[handwritten: Tells]* How to perform routine tasks such as changing accounting dates and system rates. *[handwritten circled: lc]*

### For Operations Managers

*[handwritten circled: cap]* *Startup Guide.* *[handwritten: Describes]* Options your office can choose from in setting up an office; procedures for setting up an office. *[handwritten circled: lc]*

*Conversion Guide.* *[handwritten: Describes]* Strategies and procedures for converting existing customer and account information for use in the new system. *[handwritten circled: lc]*

*[handwritten left margin: Should be chapter name, not topic name (on all pages) (Or are they the same here?)]*

*Figure C.15*   *Continued*

*I'd switch pages 2+3. Page 1 describes this manual & so does 3; pages 2, 4 +5 describe the whole set of manuals.* ⟶2

## Introduction

| | |
|---|---|
| **Chapters in This Manual** | Chapters in this manual include: |

| | |
|---|---|
| System Overview | An overview of the entire system, including samples showing how the system performs business tasks such as processing payments. |
| Using Passwords | *Procedures for ~ing ing* ~~How to~~ establish or change your personal password. |
| Logging On and Off | *Procedures for accessing* ~~How to access~~ the system and ~~how~~ ~~to~~ end a session with the system. |
| System Standards | The 'rules of the road' for entering amounts, dates, rates and other information. |
| Using Menus | *Procedures for ing* ~~How to~~ use menus, screens from which you can select from a list of available options. |
| Glossary | *Definitions of* Banking and data processing terms that appear in these manuals. |
| Index | Chapter names and page numbers for specific information on a topic. |

Each chapter includes a table of contents and a chapter summary.

| | |
|---|---|
| **Quick Reference Card** | The front pocket of this manual contains a card showing how to use the *function keys* that appear on the keyboard you will use for your work. You may find it useful when reading the chapters on Using Menus or Logging On and Off to refer to the quick reference card. |

*Figure C.15 Continued*

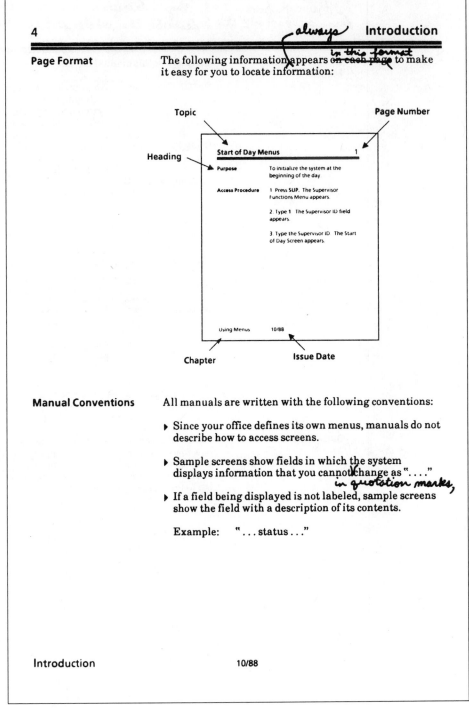

**Page Format**

The following information *~~appears on each page~~* to make it easy for you to locate information:

Topic

Page Number

Heading

Start of Day Menus      1

Purpose      To initialize the system at the beginning of the day.

Access Procedure      1. Press SUP. The Supervisor Functions Menu appears.

2. Type 1. The Supervisor ID field appears.

3. Type the Supervisor ID. The Start of Day Screen appears.

Using Menus      10/88

Chapter

Issue Date

**Manual Conventions**

All manuals are written with the following conventions:

▸ Since your office defines its own menus, manuals do not describe how to access screens.

▸ Sample screens show fields in which the system displays information that you cannot change as "...." *in quotation marks,*

▸ If a field being displayed is not labeled, sample screens show the field with a description of its contents.

Example: "...status..."

Introduction      10/88

*Figure C.15*    *Continued*

**Manual Conventions**
(continued)

For field descriptions, the following conventions apply:

Fields not marked *optional* or *conditional* are required. For example:

Employee Name  Enter your name as it is registered on the system.

Fields are marked *optional* if they can be entered or left blank under any circumstance. For example:

*? These look out of place to me if not capitalized*

Their Ref.  (optional) Enter the number, if known, by which the customer refers to the account. 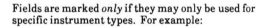 *cap*

Fields are marked *conditional* if they may be either optional or required, depending on the circumstance. For example:

Offset Portfolio  (conditional: required for account types *cap* that belong to the commercial portfolio; optional otherwise) If this account type *tr* uses an offset portfolio, enter the offset portfolio that usually funds the instrument.

Fields are marked *only* if they may only be used for specific instrument types. For example:

Option Date  (foreign exchange contracts only) Indicate *cap* whether the maturity date can be a range of dates.

**Your Comments**

We welcome your comments on the format, organization, and content of this manual. Please complete the Readers' Comment Card in the pocket of your manual and forward it to First Wisconsin.

*Figure C.15*  *Continued*

and the description of related manuals after it? [See Judy's comment at the top of original page 3.]

JUDY: Because the entire set of manuals was described on pages 2, 4, and 5 of the original version, with a description of this particular manual on page 3. I reversed the order of pages 2 and 3 so that readers would learn first about the contents of this particular manual and then several items of information about the entire set of manuals.

MARY: I was especially concerned about the organization of this document. I felt that something was wrong, but I couldn't put my finger on it. Judy caught it.

INTERVIEWER: What other things did you suggest, and why?

JUDY: My main suggestion was to make items more parallel. For instance, the list of chapter descriptions (original page 3) originally included some noun phrases and some verb phrases. I suggested changing them to make them all noun phrases.

Other changes were just minor and picky: for example, I changed the chapter name from "Using Passwords" to "Using a Password." Mary called the chapter "Using Passwords" from her perspective as a writer. I thought of it from the reader's point of view: each reader would only have one password.

INTERVIEWER: I notice you recommended changing the format of the minor headings in the "Related Manuals" section (original page 2). What did you think Mary should do with these?

JUDY: I thought the side-margin headings (such as "Entering Letters of Credit") in the body text should have been pulled farther over to the left. Either that or another space should have been added above the underlined headings (the labels for groupings of related manuals, such as "For Daily Users"), because the lines under these headings seem to cut them off from the information below them, which is the information they label.

MARY: I really had trouble with the minor headings. When I added the extra space, the material no longer fit on the page. I took out the line and created another side margin within the body text, but I needed some way to label the groupings of related manuals. I tried putting the grouping labels in italics, but that didn't really work, either.

JUDY: So in the final version, you pulled the label for the grouping out to the far left margin.

MARY: Right. But to make it clear that these labels do not have the same heading level as "Related Manuals," I put them in italic type [see the final version of this page in Figure C.16].

INTERVIEWER: So you weren't able to follow Judy's suggestion of adding more space above grouping labels, but you came up with a third alternative that she hadn't thought of.

MARY: Yes. That's one of the things that happens when someone edits your work. You get a comment, and you start thinking about these kinds

# Introduction 1

**Purpose of manual**

To provide an overview of the entire International Banking System (IBS) and describe the basic skills needed to use the system. We hope you find this manual useful as a guide to your system.

**Basic services**

As you know, an international banking department provides many different services–often as many services as a domestic bank. The International Banking System assists your work in any of these service areas:

*Letters of credit*

Instruments issued on behalf of a buyer that give the buyer the financial backing of the issuing bank.

*Loans*

Credit extended to businesses and other banks for the financing of inventory purchases and trade.

*Savings*

Deposit instruments such as open and term time deposits, certificates of deposit, and interbank placements.

*Current accounts*

Running accounts that reflect the movement of cash, loan reserves, and settlements between banks.

*Foreign exchange trading*

The buying and selling of currencies in relation to either U.S. dollars or other foreign currencies.

*Payments*

The receipt of payments made via SWIFT, the international bank-to-bank wire service.

*Financial control*

Cash management, individual and global credit risk analysis, and financial control.

▶

*Figure C.16* Final Pages from "Introduction"

| | |
|---|---|
| **Chapters in this manual** | Although other manuals will tell you how to use the system to do work in your area of expertise, this manual describes the basic skills in using the system that everyone must have. |
| | Each chapter includes a table of contents and a chapter summary. The chapters are: |
| *System Overview* | An overview, designed for managers, of the entire system. The overview includes descriptions of the system's components, benefits, and security features, as well as task flows for typical transactions such as payments. |
| *Using a Password* | Procedures for establishing or changing your personal password. |
| *Logging On and Off* | Procedures for beginning and ending a session. |
| *System Standards* | The "rules of the road" for entering amounts, dates, rates, and other information. |
| *Using Menus* | Procedures for using *menus,* screens from which you can select available options. |
| *Glossary* | Definitions of banking and data processing terms that appear in these manuals. |
| *Quick reference card* | The front pocket of this manual contains a card that lists function keys on Tandem terminals and on personal computers. When reading the chapters "Using Menus" or "Logging On and Off," you may find it useful to refer to the Quick Reference Card. |

**Figure C.16**   *Continued*

**Related manuals**

In addition to this manual, you will receive one or more of the following manuals, depending on your area of expertise:

*For daily users*

| | |
|---|---|
| Entering Letters of Credit | Tells how to set up and monitor letters of credit. |
| Entering Deposits and Loans | Tells how to set up and manage deposit and loan accounts. |
| Entering Current Accounts | Tells how to set up and manage settlement, interbank placement, and disbursement accounts. |
| Entering Payments | Tells how to enter and receive payments. |
| Entering Foreign Exchange Contracts | Tells how to set up and monitor foreign exchange contracts. |
| Using General Inquiry | Tells how to view and act on previously entered transactions and incoming wires. |

*For credit managers*

| | |
|---|---|
| Interpreting Reports | Describes standard reports in detail and shows how to request customized reports. |

*For those who maintain system records*

| | |
|---|---|
| Entering Customer Records | Tells how to enter customer information, including customer credit lines. |
| IBS Maintenance Guide | Tells how to perform routine tasks such as changing accounting dates and system rates. |

▶

*Figure C.16*   *Continued*

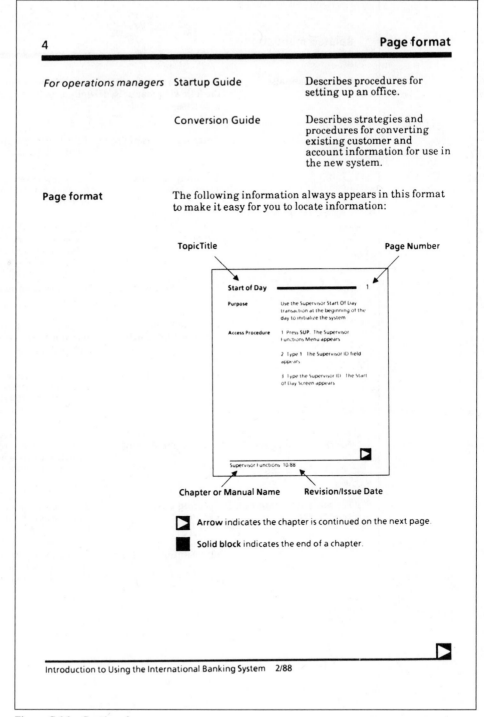

*For operations managers*    Startup Guide         Describes procedures for setting up an office.

                              Conversion Guide      Describes strategies and procedures for converting existing customer and account information for use in the new system.

**Page format**              The following information always appears in this format to make it easy for you to locate information:

TopicTitle                                          Page Number

Start of Day                    1

Purpose        Use the Supervisor Start Of Day transaction at the beginning of the day to initialize the system

Access Procedure    1 Press SUP. The Supervisor Functions Menu appears

                        2 Type 1. The Supervisor ID field appears

                        3 Type the Supervisor ID. The Start of Day Screen appears

Supervisor Functions 10/88

Chapter or Manual Name      Revision/Issue Date

▶ **Arrow** indicates the chapter is continued on the next page.

■ **Solid block** indicates the end of a chapter.

*Figure C.16*    *Continued*

312

| | |
|---|---|
| **Manual conventions** | All International Banking System manuals are written with the following conventions: |
| *No access procedures* | Since your office defines its own menus and uses them to access individual screens, manuals do not describe how to access individual screens. |
| *Fields* | ▸ Fields not marked *optional, conditional,* or *only* are required. |

  EMPLOYEE NAME      Enter your name as it is registered
                    on the system.

▸ Fields are marked *optional* if they can be entered or left blank under any circumstance.

  THEIR REF.         (Optional) Enter the number, if
                    known, by which the customer
                    refers to the account.

▸ Fields are marked *conditional* if they may be either optional or required, depending on the circumstance.

  OFFSET PORTFOLIO   (Conditional: required for account
                    types that belong to the commercial
                    portfolio; otherwise, optional) If
                    this account type uses an offset
                    portfolio, enter the offset portfolio
                    that usually funds the instrument.

▸ Fields are marked *only* if they may only be used for specific instrument types.

  OPTION DATE        (Foreign exchange contracts only)
                    Indicate whether the maturity date
                    can be a range of dates.

▸ Sample screens show fields that cannot be changed in quotation marks: " . . . . . ."

▸ If such fields are not labeled, sample screens show the fields with descriptions of their contents.

  ▷ **example:** " . . . status . . ."

*Figure C.16*   *Continued*

*Keys*

▸ Keys are shown in **boldface** as they are defined on the International Banking System.

▷ **example:** Press **POST**.

Refer to your Quick Reference Card for a listing of key names as they appear on your keyboard.

**Your comments**

We welcome your comments on the format, organization, and content of this manual. If you have any suggestions as to how this manual could be made more useful to you, please send us your ideas on the enclosed Reader's Comment Card.

**Figure C.16** *Continued*

314

of things and you realize, "Yes, that heading really was a problem. I do have to do something about it." Then you might have to experiment to find out what exactly will solve the problem.

INTERVIEWER: Another item you considered, Judy, was the page format description on page 4.

JUDY: Actually, I just nudged a little, and then Mary came up with an alternate solution. The only thing I corrected was the statement in the page format description that read, "The following information appears on each page." The statement preceded an illustration of headings and footers in the document. But the statement wasn't really true: the information itself did not appear on every page, it appeared in the same *format* on every page.

INTERVIEWER: What was your reaction, Mary, when you got back Judy's marked-up draft?

MARY: My first reaction was, "Gee, she didn't mark it up that much," which was a relief because I did have some anxiety about sending the draft to an editor in another department.

INTERVIEWER: Were you at all offended by any of her comments?

MARY: Oh, no! I thought she was very polite in the tone of her comments. For example, she circled the word *loan*, which I had used to begin the definition of the term *commercial loans*. It's well known that you're not supposed to begin a definition by repeating part of the term, but to make her point, Judy simply circled both instances of the word *loan* and asked, "Can you avoid repeating?" Her style was succinct and nonjudgmental.

The experiences of editing someone else's work *and* of getting someone's comments on your own work teach you what's appropriate and what's best, both for improving writing and for explaining writing problems to another person. In the end, there is such a spirit of cooperation in the endeavor that you get over any kind of fear.

It's worse to hold back, really. As a copyeditor, if you don't let the writer know the writing is unclear because you're trying to save the writer's feelings, you're not really helping that person out. Writing generally gets a final quality check by the department manager, so if you haven't helped improve the writer's work, you make the writer look bad to the manager!

INTERVIEWER: Judy, how did you determine how polished the document needed to be?

JUDY: In this particular case, Mary was very concerned that this document be as polished as possible. Because it was a high-profile project, she and her department were concerned that it be as good as it could be.

MARY: A lot of people were going to read this document. It was going to be highly visible.

JUDY: For some projects, there is a limited amount of time that you can spend on a given document. That's probably the main limiting factor—how much time you have. If time were not an issue, you could go on forever, but you would reach a point of diminishing returns. Generally, I try to get everything as good as I can possibly get it. But sometimes the deadline limits that.

INTERVIEWER: So for some projects, you might get a clue that the project merits a more thorough edit?

JUDY: Yes. I went over the *Introduction* manual five times, but I went over a less important draft that I was working on at the same time only twice. My manager lets me know how much time I have and how important a project is.

INTERVIEWER: What sort of deadline did you work under for this project?

JUDY: For this particular project, I had two weeks to edit a 60-page document, which would be a generous amount of time if I had no other projects. But I had a lot of other projects going, so I had to request that much time.

INTERVIEWER: How tight are your deadlines, generally?

JUDY: Again, if you think in terms of a single project, the deadlines are generous. However, we may have to edit five or six drafts within the same period of time.

For example, right now, I'm editing six different projects, with new projects coming all the time. Four of the projects have a due date of June 1. Since it's only mid-January now, that sounds like a lot of time, but some of the projects are great big manuals, and for others I'm doing writing in addition to editing. So really, I will have to work hard to meet those deadlines.

INTERVIEWER: How do you estimate the time it will take you to edit a document?

JUDY: This is the toughest question there is. There are seminars all the time on this question, and the only answer seems to be to keep a history of how long similar projects have taken in the past and use that history to estimate current projects.

INTERVIEWER: What resources are most useful to you as an editor?

JUDY: I use the company's style guide, but since I've used it so much, I don't need to refer to it often because I automatically incorporate that information. The *Chicago Manual of Style* and the *Handbook of Technical Writing* are also excellent references, but again, since I'm so familiar with these references, I now need to refer to them only if something really unusual comes up.

INTERVIEWER: What charactertistics should an individual possess to be a good technical editor?

JUDY: I guess probably carefulness, pickiness. A real concern for all the details. In my own experience, the more I know about a topic, the easier it is to edit documentation that describes that topic. Also, the more familiar I become with editing and with references for editors, the easier it becomes for me to edit.

INTERVIEWER: Would you say that good writing skills make for a better editor?

**JUDY:** It doesn't necessarily follow. A good writer is not necessarily a good editor, and a good editor doesn't really have to be a good writer either, but a good editor must be able to recognize good writing or be able to recognize problems in writing.

**INTERVIEWER:** Let me ask you something. You're a writer and an editor. You're functioning in both worlds. What do you see as the difference when you approach writing versus approaching editing? Do you see different parts of your mind at work?

**JUDY:** Boy, they're such different types of tasks. When you are writing something, you are of course trying to get it to the best point you possibly can. As a result, you finally approach the point where you are *editing* your own work as you write.

All writers have to be, in a way, their own editors. But there comes a point when you simply can't edit your own work anymore and you need somebody else to look at it.

**MARY:** You know, it's always kind of irritating to me when I get through editing my own work and I give it to other people and they find more problems. I think, "How did I miss that?" But that's what happens. After a while, you're just too close to the material; you generated it, you thought about it, and you've edited it as much as you can. But somebody outside can bring a fresh perspective.

**INTERVIEWER:** I have a large, philosophical question for you, Mary. Considering the changes an editor might make and that the original writer might accept, whose writing emerges from this process?

**MARY:** Since I have the final say in what changes I will or will not make, I guess it's my writing. It's definitely been enhanced by Judy's edit, but I don't feel as though it's her work. Of course, in some companies, the work is written by a technical person and edited by a professional editor. The writer may not have the authority to reject editing suggestions in a situation like that.

# 13 Document Production

Typewritten Copy
Typeset Copy
   Traditional Typesetting
   Electronic Publishing
Printing

Document production—the process of turning your final manuscript into a printed document—will be affected by the design decisions made when the document was being planned. Its size, page layout, number and types of graphics, color, binding, number of copies required, and similar decisions affect the time, the degree of professional expertise, and the costs required to produce and print the manuscript. Producing the printed version of your manuscript can occur only after the contents have undergone the final reviews described in Chapter 11. You must then work closely with other document specialists to convert the final draft into a printed version of the manuscript. A close-knit working relationship between you and the production specialists is essential because the quality, timeliness, and production costs of your document depend on how the project is scheduled, organized, and coordinated. Keep everyone involved—editor, graphics specialist, typesetter, printer, distribution personnel—informed about the purpose of the document and its proposed and actual schedule throughout the production cycle.

The writer's role in preparing a manuscript for printing is changing in many organizations. With the increasing use of electronic publishing, writers and editors are taking a larger role in the production process. They may write or edit at microcomputers, code the text with typesetting instructions, and communicate it to electronic publishing systems. (Typesetting coding will be explained shortly.) Writers may create some or all of their documents directly on electronic publishing workstations. The resulting typeset manuscript, known as camera-ready copy, can be sent directly to the printer. The expression *camera-ready copy* refers to the process by which a photographic negative of the manuscript is used to make plates for the printing press. The plates are usually made of metal, plastic, or heavy paper. They contain the text and visual images that are transferred to paper during the printing process. (Figure 13.10 shows how the photographic negative is used to create the plate.) If you are or will be directly involved in the production process using these new tools, the advice on layout and design in Chapter 11 will be of great benefit to you, as will the information on electronic publishing in this chapter.

In other settings, writers and editors belong to a structured documents team that includes a production manager, graphics specialists, word processors, and sometimes typesetters. The writers or editors then coordinate manuscript preparation tasks with the production manager, who makes up and manages a production schedule for the document.

## TYPEWRITTEN COPY

Camera-ready copy can be created from a typeset or typewritten manuscript. The latter can be sent directly to the printer in the form that emerges from the word processing printer. Most word processors offer a limited range of typefaces, fonts, styles, and sizes. However, recent developments in word processing software for personal computers used with laser printers make sophisticated typesetting options increasingly available. Some organizations, however, still create many documents with a "typewriter" look that is adequate for most in-house documents. These documents are sometimes enhanced with typeset headings, icons, or other devices to augment the limited type styles and sizes available, as in the sample page from an in-house employee booklet shown in Figure 13.1.

These types of manuscripts are often photoreduced 10 to 15 percent before printing to improve the sharpness of the typed images. Such reduction also permits more words per page, thus reducing the number of pages per document, which in turn reduces printing costs. Do not reduce such copy by more than 15 percent, however, because it will be too difficult to read.

Before this kind of document goes to the printer, ensure not only that the contents are accurate but also that the document meets your organization's standards.

- Are the text and white space balanced on each page?
- Are headings properly differentiated from the text?
- Are parallel text headings identical in typeface and style?

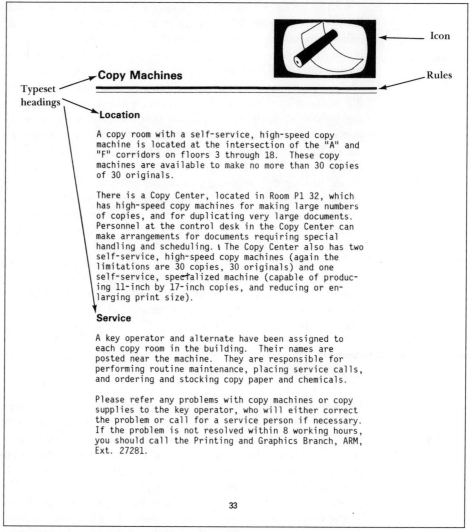

**Figure 13.1** Sample Page with Typewritten Text

- Does the left margin begin at the same place on each page?
- Are headers, footers, and page numbers located correctly and consistently?
- Is the text single- or double-spaced uniformly?
- Have widows and orphans been eliminated?
- Are all tables and figures titled and numbered appropriately?
- Have "windows" (blank spaces) been left for photographs or other artwork?
- Are cropping and sizing instructions for photographs clearly marked for the printer?
- Are printing instructions for screens, color, and other design elements properly marked?

Once this review is completed to your satisfaction, the manuscript can be sent to the printer.

---

## TYPESET COPY

Typesetting offers a myriad of options that must be chosen and combined carefully for optimal effect, as the guidelines in Chapter 8 make clear. Choosing among these options is the province of the document designer—a role traditionally held by graphic designers but one that you may be called on to fulfill at some point in your career.

The actual mechanics of typesetting continue to undergo significant changes because of the application of computer technology to this field.

> New developments in typesetting occur every few months mainly because typesetting is probably the part of the printing industry that most lends itself to computerization, so that its speed of development is linked to the rapid developments in the computer industry. (Bann 70)

A detailed exposition of these developments is beyond the scope of this chapter. (The technical details of how photocomposition and electronic publishing systems work are described in many of the books and articles listed under "Further Reading" at the end of this chapter.) The following discussion will focus instead on the main outlines of two major typesetting processes that coexist today for the preparation of camera-ready copy: *traditional typesetting* (also called *composition*, *phototypesetting*, or *photocomposition*) and *electronic publishing*. These two processes differ in technology, page makeup and graphics capabilities, quality of the printed image they create (its optical resolution), and number of steps required to produce a camera-ready manuscript.

### Traditional Typesetting

Traditional typesetting converts a final manuscript into composed, camera-ready copy in the series of steps depicted in Figure 13.2. The final manuscript can be submitted to the typesetter in different forms: on paper, on computer disk, or communicated electronically through a modem. (A *modem*, short for *modulator-demodulator*, converts electronic signals for transmission over telephone lines from a sender to a receiver. Such transmissions are a form of telecommunications.) If you are using an outside commercial typesetter, agree in advance about how you will submit the manuscript to ensure that the form in which it is sent and the equipment that receives it are compatible. If you submit paper, the text may either be retyped by the typesetter or scanned into the typesetting system via an optical character recognition (OCR) device. The OCR device "scans the lines of a typewritten manuscript and records the characters in digital form on magnetic tape for use in driving a typesetting system" (*Chicago* 634). Scanning eliminates the need for the typesetter to retype the entire manuscript—text, tables,

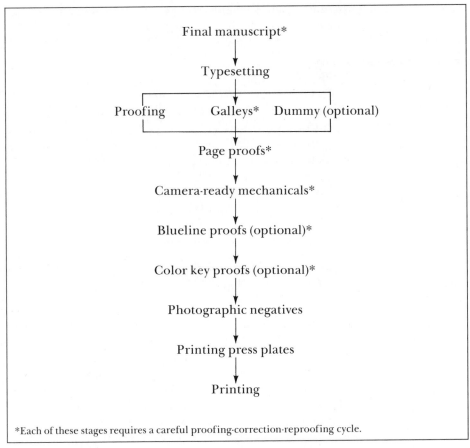

*Figure 13.2*   Typical Stages in the Traditional Typesetting Process

equations—character for character. Adhere to the following guidelines when submitting copy for scanning:

- Learn from the typesetter ahead of time which typefaces the scanner can read (some require a special OCR font).
- Type the copy double-spaced on one side of white bond paper.
- Use 1-inch margins on all sides.
- Keep the copy clean and free of last-minute handwritten editorial changes (the scanner will be unable to "read" them).

Providing the manuscript to the typesetter in a form that does not require that it be retyped saves time and money; above all, it ensures accuracy. The typesetter can also return galley proofs overnight, at a cost savings of from 30 to 60 percent compared with the costs of retyping the text from paper copy. If you provide a disk or telecommunicate the manuscript to the typesetter, you should also send a paper copy and retain a backup copy of the disk in case the copy sent is lost or damaged. To benefit from these savings, submit the manuscript coded with the appropriate typesetting commands. These commands—a sequence of letters and numbers in brackets like that shown in the next paragraph—instruct the typesetting equipment software to create the desired typeface, point size, leading, line length, column justification, and hyphenation standards.

As a general rule, typesetting commands precede the words or characters to be typeset. For example, to instruct the typesetting equipment to set a block of text in 10-point Helvetica bold on 11 points of leading, by a line length of 14 picas, ragged right margin, the appropriate commands would be entered immediately before the words to be set, as follows:

[t89][p10][l11][ml4][rr]Text . . .

Helvetica bold is identified as **t89**.

Although typesetting commands vary among systems, they generally consist of three basic elements: brackets, a mnemonic, and the "argument." Commands must be enclosed in brackets in order for the system to recognize them as commands. A mnemonic is one or more letters that represent the keyboard shorthand for the command. For example, *t* in certain systems is the mnemonic for the typeface command; *m* represents *measure*, the printer's term for line length; and *p* introduces the point size. The argument portion of the command refers to the number that follows the mnemonic. In the command **[p10]**, **10** is the argument that specifies the point size.

Figure 13.3 shows a sample of text from a journal article with the typesetting codes highlighted in boldface. Note that it contains numerous codes that control all facets of the text and spacing. When the disk containing the coded text is inserted into the typesetting equipment, its software "reads" the commands and prints out typeset galleys. The finished printed page, of which the sample is a part, is shown in Figure 13.4.

Submit tables and other nontext material clipped together in separate batches because these are generally typeset separately. Retain a copy of everything sent and a backup file on disk so that you can answer any questions the typesetter may have.

### Galleys and Dummies

After the text has been typeset, the typesetter usually returns two sets of galleys. Proofread one set of galleys against the original manuscript, and, if necessary, use the other to create a "dummy"—a mock-up of the document. Proof the galleys not only for typographical errors but for any missing or out-of-sequence information as well. Such errors are easily and inexpensively changed at this stage of the production process. As the manuscript gets closer to the printing press, correcting major mistakes, like missing or out-of-order text, becomes increasingly expensive.

After a proofing-correction-reproofing cycle, return the corrected galleys and a dummy of the document to the typesetter. A dummy is a page-sized version of the whole document for the typesetter to follow when creating the page proofs.

```
[f50] 88-1020 jof;l
marjof international file 23 transmitted 1/24/89 repro;l;l


[p22,l24,m27.5,fl0,rr,ah4,t231]Taungya Plantings in Puerto Rico;l

[el40,p16,l20,m27.5,t197]Assessing the growth of mahogany;l and maria stands.;l

[el24,p11,l12,t231]By Peter L. Weaver;l

[el11,p9,l11,m13,fl0,xr,ah3,t227][2,231]!drop!As the USDA Forest Service became more active
in international forestry in the early 1980s, various approaches to tropical wood production
became better known to American foresters. These approaches included enrichment planting,
agroforestry, and taungya plantations.;ts.;ts.;ts.;l
.......................................................................................
[fl1s]When good agricultural practices are complemented by the judicious
use of trees, agroforestry can supply wood, foodstuffs, and animal products from a single
management unit--a farm, a small community, or a portion of a watershed (Weaver 1979). Recent
studies in the American tropics have documented numerous agroforestry systems and have quan-
tified production on some properties (De las Salas 1979, Organizaci;a1on para Estudios
Tropicales and Centro Agron;a1omico Tropical de Invetigac;a1ion y Ense;a5nanza 1986).;ts.;ts
.;ts.;l
.......................................................................................
[el11,p9,l11,m13,ah4,cr,t231]Taungya Plantations in Puerto Rico;l

[el11,p9,l11,xr,ah3,t227]Taungya plantings began 50 years ago in Puerto Rico, and the USDA
Forest Service played a role in their development. In the late 1930s, maria [t229] (Cal-
ophylluum calaba [t227]L.[t229]) [t227]and mahogany [t229](Swietenia macrophylla [t227]King
[t229]), [t227]two commercial species (Longwood 1961, Little and Wadsworth 1964), were es-
tablished on the Luquillo Experimental Forest and in the Commonwealth of Puerto Rico's Rio
Abajo Forest Reserve.;ts.;ts.;ts.;l
```

*Figure 13.3*     Typical Embedded Typesetting Codes

It is made by a graphics specialist, who cuts and pastes it together from a set of galleys. The dummy contains all elements of the document laid out in the proper sequence: columns of text, photocopies of artwork (rather than the original artwork), and tables. Dummies are indispensable for large, complex projects, although they are often not needed for smaller, more routine documents. The common practice of publication staffers and their experience with the typesetter, as well as the complexity of the document, will determine whether a dummy is necessary.

## Taungya Plantings in Puerto Rico

*Assessing the growth of mahogany and maria stands.*

By Peter L. Weaver

As the USDA Forest Service became more active in international forestry in the early 1980s, various approaches to tropical wood production became better known to American foresters. These approaches included enrichment planting, agroforestry, and taungya plantations.

Enrichment planting introduces valuable timber species into degraded woodlands or forest without eliminating useful trees. At least 163 tree species in 12 or more neotropical countries have been established using several different techniques of enrichment planting (Weaver 1987). Line planting, a technique to convert degraded forests into valuable plantations at maturity, is probably the most successful enrichment method. Plantings are usually made in cleared strips 2 meters wide and spaced at 10-meter intervals throughout the woodland. These strips simulate small openings in which many timber species regenerate naturally, and the wide spacings reduce establishment costs. Because much of the surrounding woodland is left intact when the seedlings are planted, environmental damage is minimized, particularly on

single management unit—a farm, a small community, or a portion of a watershed (Weaver 1979). Recent studies in the American tropics have documented numerous agroforestry systems and have quantified production on some properties (De las Salas 1979, Organización para Estudios Tropicales and Centro Agronómico Tropical de Investigación y Enseñanza 1986). Agroforestry systems often simulate the natural forest in appearance, remain productive throughout the year, resist plagues and infestations better than monocultures because of plant diversity, and minimize soil erosion. Tree cover modifies the microclimate, and minerals are recycled through natural processes. Despite its numerous benefits, however, agroforestry is not a universal substitute for intensive agriculture or forestry. Trees compete for light and water, and unless properly managed, they can reduce marketable products. Agroforestry is best seen as one way to keep sloping terrain in permanent production or to rehabilitate lands previously degraded by poor agricultural practices.

"Taungya," a Burmese word meaning a cultivated hill plot, is a tropical agroforestry system with a long history. Timber trees are planted at some point the slash-and-burn agricultural cycle on public lands loaned to

tree species either before, along with, or after the sowing of the agricultural crop." In 1856, taungya plantations of teak were established on migratory plots in Burma. The same system is known by 25 different names in various countries (King 1968).

### Taungya Plantations in Puerto Rico

Taungya plantings began 50 years ago in Puerto Rico, and the USDA Forest Service played a role in their development. In the late 1930s, maria (*Calophyllum calaba* L.) and mahogany (*Swietenia macrophylla* King), two commercial species (Longwood 1961, Little and Wadsworth 1964), were established on the Luquillo Experimental Forest and in the Commonwealth of Puerto Rico's Rio Abajo Forest Reserve. Farmers planted tree seedlings among subsistence crops—usually corn, beans, bananas, plantains, sweet potatoes, yautias, dasheens, and cassava.

The program, implemented in conjunction with Civilian Conservation Corps projects, operated from the mid-1930s through the late 1940s as the government was purchasing land to add to public holdings. When the land was occupied, permits for use of the land were issued until the government decided to remove the inhabitants. Farm parcels averaged 5 hectares and were designated on a map.

The program's purpose was gradually to convert agricultural land into timberland and simultaneously to provide gainful employment during hardship. Houses were timber species, catchments wer first few maintai ern

**Figure 13.4** Typeset Version of Coded Text in Figure 13.3

### Page Proofs

Following instructions on the dummy, a graphic specialist, called a paste-up artist, cuts the galleys into column-width strips and affixes them to "boards," which the artist then photocopies and returns to you as page proofs. Figure 13.5 shows a three-column page proof laid out on a board. Page proofs contain all the elements that the finished page will: the text (one, two, or more columns), page numbers, running heads and feet, tables (which have been typeset and proofed independently of the text), and either artwork or windows where the artwork will be affixed before printing. In Figure 13.5, space has been left for a complex line drawing that straddles all three columns. Note that the header and the page number are outside the box for the text and artwork but within the page image

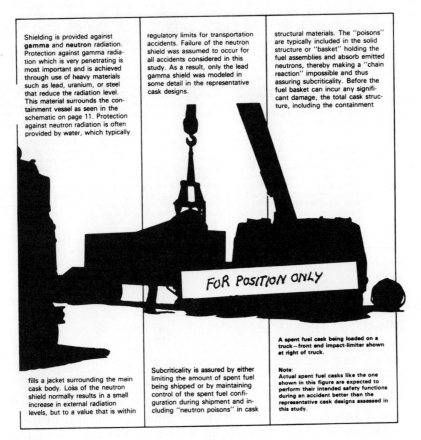

**CASK CHARACTERISTICS
AND RESPONSES**

Shielding is provided against **gamma** and **neutron** radiation. Protection against gamma radiation which is very penetrating is most important and is achieved through use of heavy materials such as lead, uranium, or steel that reduce the radiation level. This material surrounds the containment vessel as seen in the schematic on page 11. Protection against neutron radiation is often provided by water, which typically fills a jacket surrounding the main cask body. Loss of the neutron shield normally results in a small increase in external radiation levels, but to a value that is within

regulatory limits for transportation accidents. Failure of the neutron shield was assumed to occur for all accidents considered in this study. As a result, only the lead gamma shield was modeled in some detail in the representative cask designs.

Subcriticality is assured by either limiting the amount of spent fuel being shipped or by maintaining control of the spent fuel configuration during shipment and including "neutron poisons" in cask

structural materials. The "poisons" are typically included in the solid structure or "basket" holding the fuel assemblies and absorb emitted neutrons, thereby making a "chain reaction" impossible and thus assuring subcriticality. Before the fuel basket can incur any significant damage, the total cask structure, including the containment

FOR POSITION ONLY

A spent fuel cask being loaded on a truck—front end impact-limiter shown at right of truck.

Note:
Actual spent fuel casks like the one shown in this figure are expected to perform their intended safety functions during an accident better than the representative cask designs assessed in this study.

12

*Figure 13.5*    Pasted-up Page Proof

area, which is marked at the four corners by horizontal and vertical lines called *trim lines*. Note also that rather than containing the original artwork—a truck trailer being loaded with a shipping cask—the page proof contains an outline of the artwork, marked "For Position Only."

The page proofs may or may not contain halftone or color illustrations. (The special preparation required for these visuals will be discussed shortly.) If you prepared a dummy, compare the page proofs with it carefully. Major errors and last-minute changes to the text should have been corrected during the galley proof stage. At this point, review the page proofs to verify all of the following:

- All elements are present and in the correct sequence.
- Page margins are consistent, line lengths are accurate, column depth is even, graphics are correctly placed, and all other elements are correctly aligned.
- "Continued" lines in tables and elsewhere are placed correctly.
- The sequence of titles and page numbers in the table of contents is correct.
- The placement of captions and credit lines is correct.
- Any widows and orphans have been deleted.

### Pages to Mechanicals

Return page proofs with errors noted to the typesetter, retaining a copy so that you can answer any questions that the typesetter may have. The typesetter makes final corrections on the boards, photocopies this version, and returns it and the original boards to you for final proofing. The error-free boards, known as camera-ready mechanicals, are then sent to the printer, where they are made into photographic negatives and plates for the printing press, a process shown in Figure 13.10.

### Blueline Proofs

You may sometimes request that the printer prepare a final set of proofs for complex manuscripts. Called *blueline proofs*, they are made when the printer exposes the photo negatives to photosensitive paper. The bluelines show how the finished manuscript will appear when printed and provide you with the last chance to proofread and correct errors before the document is printed. Check each page carefully to ensure that the printer has followed every instruction and that the instructions were correct. Examine them especially carefully for these aspects:

- Proper alignment of elements on side-by-side pages
- Flopped or upside-down photographs
- Obscured or missing page numbers
- Consistency or intensity of screens
- The fit of each graphic in its window
- Any broken or imperfect type
- Printer's markings inadvertently left on any copy

If the document requires color, request that the printer provide a color proof copy for you to check. (Printers usually charge extra for a color proof.)

Once you have signed the blueline, you have given permission to print "as is" or to print with the "corrections as marked." If you made major changes, request a second set of bluelines after the changes have been made. Although you should examine the blueline carefully, do not make last-minute *editorial* changes to the text or the artwork. Corrections made at this late date are very expensive.

### Halftones and Color Separations

Photographs and other visual elements, some or all of which may be in color, require special processing prior to printing. Although the prepress processing for halftones and color separations is done by graphics specialists and printers, writers and editors should be aware of what their preparation entails. This knowledge not only affords an appreciation of the skills involved but also helps explain the time and expense that these tasks require and their impact on your production schedule and budget.

Compare a black-and-white photograph with the text and other graphics components in a document. The letters, numbers, and symbols that make up the text are solid black images on a white background (the paper), as are line drawings, schematics, graphs, tables, and charts. These materials are collectively known as *line copy* or, in the case of visuals, *line art*. A photograph, by contrast, is made up of blacks, whites, and varying shades of gray. Because of these shadings or gradations from black to white, photographs are known as *continuous-tone copy*. Printing presses cannot reproduce these shadings accurately as is, so continuous-tone copy must be converted into a pattern of black and white dots before it can be printed. The dots vary in size and closeness, depending on the lightness or darkness of the original image. Figure 13.6 illustrates this effect. When printed, these dots give the illusion of the original continuous tones. This pattern, called a *halftone screen*, is created when the printer places a grid-pattern screen (which can vary in degree of fineness) between the original glossy photograph and the lens of the camera used in making plates for the printing press.

The screens vary in degree of coarseness or fineness (from 55 to 150 lines per square inch) to accommodate the type of paper, printing plate (metal, plastic, or heavy paper), and printing press, all of which have been selected to produce the quality necessary for the project. The naked eye sees or "integrates" these patterns as the blacks, whites, and grays of the original photograph. But the dot pattern is easily visible in any printed photograph under a 10X magnifying glass. Newspaper-quality paper and ink require coarse screens, whereas magazines "use a smoother paper and are able to use finer screens . . . and thus obtain finer, more detailed halftones" (Craig 74).

Color images require even more preparation before they are ready for the printing press. They, too, are like halftones in that the image prepared for the printing press is made by photographing the color image through a color filter that divides it into a dot pattern. Black-and-white photographs require one halftone screen, whereas color images require four separate films. Four-color process printing, using yellow, blue (cyan), red (magenta), and black inks in varying proportions, reproduces all the colors of the spectrum. The halftone screens are used to create halftone printing plates that, when combined during

55-line screen

85-line screen

120-line screen

150-line screen

*Figure 13.6*　Four Screened Photos of Varying Quality

printing, simulate the original color image from which they were reproduced. "Just as an ordinary halftone gives an illusion of continuous tone in black and white and shades of gray, so process color printing gives an illusion of continuous tone in natural colors." Under a magnifying glass, however, "you will see that the image is composed of tiny dots of pure primary colors" (*Chicago* 637–638). The unaided eye, unable to distinguish among the individual dots, interprets the blended pattern as natural colors.

## Electronic Publishing

Electronic publishing systems use computer technology for the input, composition, page makeup, and output of finished, camera-ready pages. As used in this chapter, *electronic publishing* refers to computerized publishing systems whose sole purpose is to produce typeset, camera-ready manuscripts for the printing press. In addition to their software, large-scale processing capacity, and host of dedicated peripheral equipment for communications and printing, these systems use big-screen computer terminals, known as workstations, for online page layout and design. Desktop publishing systems, by contrast, use page layout software on general-purpose microcomputers (that is, personal desktop computers) to output typeset pages on a desktop laser printer.

Electronic publishing systems permit the operator to preview, edit, and lay out full pages on high-resolution "what you see is what you get" (WYSIWYG) workstation screens. Most of these systems can also incorporate line art and halftones from other sources through a process called *digitizing* or can leave windows (space on the page) for artwork they produce. Some systems include highly sophisticated graphics capabilities that allow graphic designers to create images that can be merged online with text during page makeup or printed and used independently.

These systems, regardless of their scope and capacity, are gaining widespread use because of the painstaking manual work they eliminate compared with traditional typesetting, a process outlined in Figure 13.2. Traditional typesetting first produces text in one-column galley sheets of the author's manuscript. After they are proofed and corrected, the galleys require several other time-consuming manual operations before they qualify as camera-ready copy. Graphics specialists must cut all page makeup components from the galleys—columns of text, headers and footers, tables, footnotes, graphics captions, page numbers—and paste them onto page-size proof boards for subsequent proofing and printing. The cut-and-paste process must be repeated for any corrections or other last-minute changes, especially those that necessitate the movement of tables, graphics, or blocks of text.

Electronic publishing systems almost completely eliminate time-consuming cutting and pasting by allowing the operator to manipulate the page makeup components on the WYSIWYG screen. (These screens show how a full-size document page will actually appear in print.) Once each page conforms to the document's format requirements, the entire manuscript is printed on a laser printer as a camera-ready master copy. Electronic publishing systems can drive laser printers or typesetters as output devices. The major difference between them is the resolution quality, measured in dots per inch (dpi), of the image they create: devices below 600 dpi are classified as printers, and those above 600 dpi are considered typesetters. Typesetter resolution generally ranges from 1,200 to 2,400 dpi. For comparison, the typical dot-matrix printer has a resolution of approximately 70 dpi. A resolution of 300 dpi is adequate for the text, tables, and line art of much technical documentation. Compare the images produced at 150 dpi and 300 dpi, respectively, in Figures 7.19 and 7.20 in Chapter 7.

### Corporate Documents and Electronic Publishing Systems

*Corporate publishing* refers to the publishing requirements of businesses, government agencies, educational institutions, professional associations, and other organizations for which publishing is a *sideline* rather than the organization's primary purpose, as with newspaper, magazine, or book publishers. These organizations create such items as sales brochures, price lists, forms, reports, regulations, newsletters, proposals, journals, directories, parts lists, and training, service, and maintenance manuals. According to how they are *produced* prior to printing (i.e., typeset and laid out), these documents fall into one of two categories: those with *uniform formats*, regardless of content, and those with *variable formats*.

In documents with variable formats—brochures, booklets, newsletters, magazines—the position and style of text columns and graphics components for each page are determined by a graphic designer's layout decisions rather than a set of predefined format guidelines that are uniformly applied throughout the document. Accordingly, these documents are called *layout-intensive*. Figure 13.5 shows a page from such a document. In that booklet, each page was laid out to differ in appearance from every other page, although certain unifying elements, like typeface, color, and three columns per page, combine to tie the overall design of the document together. The "look" of these documents is almost always planned and executed by graphic designers because the task requires layout and design skills, although writers usually write and edit the text.

Uniform-format documents, by contrast, comprise the largest share produced by technical writers. This category, often called *technical documentation*, contains product support literature that explains a procedure or describes a piece of equipment; it includes "product installation guides, training manuals, operations guides, or reference, support, or maintenance manuals," and may even encompass proposals and product specifications. These documents share a broad range of characteristics (Seybold 3):

- They feature regular rules for page makeup.
- They result from the collaboration of different people.
- They go through successive revisions and updates after initial publication.
- They frequently include graphics (especially line art), tables, and equations.
- They may be quite long—100 or more pages.
- They may be composed according to rigorous specifications, such as those of the Department of Defense.

Even though the individual pages in a uniform-format document are not exact mirror images of one another because some contain graphics, tables, and equations and others do not, the text (columns) and other page elements (headers, footers, page numbers, borders) are uniform from page to page throughout.

For the production of uniform-format documents on electronic publishing systems, form and content can be considered separately: once the format has been established, the system applies it uniformly, regardless of the content. This distinction is useful for thinking about the capabilities of electronic publishing systems as long as we recognize that even documents with uniform formats become layout-intensive as operators make final decisions about where to end pages and position graphics during final page makeup on the screen. It is also desirable to have a graphic designer create the original design specifications on electronic publishing systems for uniform-format documents. After the system has been programmed with these format guidelines, other typical system users—writers, editors, secretaries, word processing operators—can be trained to adjust page components so that they conform to the predefined format specifications.

We will focus on corporate-level rather than desktop systems since the former are most often used to produce technical documentation. The flowchart in Figure 13.7 shows the typical production process for such a system. These systems process documents in three distinct stages: input, composition and pagination, and output.

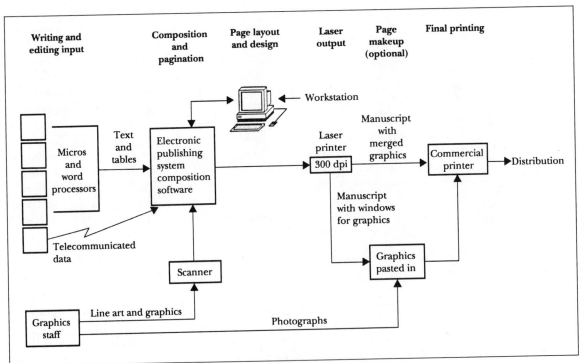

*Figure 13.7*    Publication Processing for a Typical Electronic Publishing System

## System Input

Text from word processors and personal computers can be entered directly into the system on disks or communicated via modems from remote locations. System keyboards are used for making last-minute changes to the text and for creating equations online rather than for typing original text.

For documents with uniform formats, the systems are programmed with the salient format specifications. Called a *format table*, *style sheet*, or *template*, the program translates precoded text and tabular matter entering the system into the document's predefined typeface, point sizes, and other typographic elements. As with traditional typesetting, writers, editors, and others usually code the text with typesetting specifications before submitting their disks for input to the system. After the disk is read into the system, an operator makes a series of final adjustments with a mouse on a WYSIWYG screen so that the page conforms in every particular to the predefined format specifications. For layout-intensive documents, however, the operator—usually a graphics specialist—creates the optimal layout on the screen after a series of trials and errors by using the system's composition, graphics, and layout options.

These systems can also incorporate existing hard-copy graphics via a scanner, generate business graphics from statistical data, and permit graphics specialists to create line art, graphs, and a host of other images online. The process by which hard-copy graphics are scanned into the system is illustrated in Figure 13.8.

Once scanned into the system, the graphics can be enlarged, reduced, cropped, duplicated, "cut and pasted" selectively from one figure to the next, and ultimately merged with text and copyfitted for the correct page layout before being output on the system printer. Not all graphics can be effectively scanned in, however. The resolution for some scanners is too low to capture the continuous tones in photographs. Instead, copies of photographs are pasted to the page in a

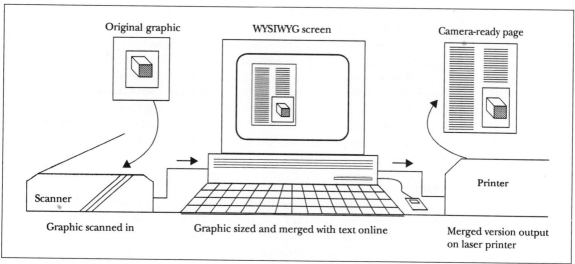

*Figure 13.8*    Graphic Scanner for an Electronic Publishing System

window sized to the photograph's dimensions. The section on system output explains the special handling that photographs require. Figure 13.7 shows the separate track they take in most electronic publishing systems.

### Composition and Pagination

Most systems offer the following range of composition and pagination options:

- A selection of serif and sans serif typeface families
- A range of styles within each typeface family—italic, bold, small capitals, roman
- A range of type sizes—typically, 6 to 36 points
- A variety of mathematical and other symbols
- Special programs for creating and aligning multiline mathematical equations
- Proportional spacing between letters (including kerning), between words, and between lines (leading)
- Automatic and operator-controlled hyphenation and justification
- Horizontal ("landscape") and vertical ("portrait") orientation of text, tables, and graphics
- Multicolumn formats
- Precise control of line lengths and column depths
- Precise control of space (windows) for graphics
- Control of headers, footers, widows, and orphans
- Creation and sizing of variable-thickness rules for tables, boxes for graphics windows, and borders for page design

Operators choose from among these options by pointing a mouse-controlled cursor at icons and "pop-up" menus on a WYSIWYG screen. The four overlapping menus in Figure 13.9 show how an operator chooses from among the typeface options available to select 12-point Modern. The choices selected from each menu are shown in reverse screen, indicated on the figure by the shaded band. Pop-up menus simplify operator interactions because only those menus needed, of the hundreds available, are displayed in a given situation.

Among the most useful system features is the capability to create a table of contents or a subject index from precoded or "tagged" elements in text. Once so marked, all numbered features in a manuscript—sections, figures, tables, footnotes, equations, references, lists—can be renumbered automatically in response to a change in their sequence. Likewise, as on-screen revisions are made to text, graphics, or tables, the composition software automatically "ripples" pagination changes throughout the manuscript, updating changes to the table of contents and subject index in the process.

### Output

After the text has been set in the appropriate typeface, merged with the graphics components, and laid out on a WYSIWYG screen, it is ready to be output on plain paper as camera-ready copy. The resulting output can go to the printer as is. The printer photographs each page and from the resulting photonegative produces printing plates for offset printing, as shown in Figure 13.10.

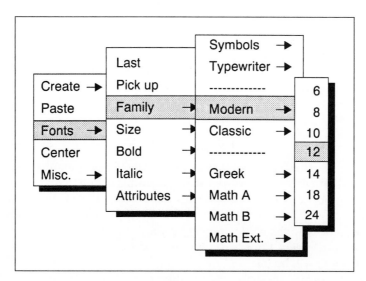

**Figure 13.9**
Sample Pop-up Menus

As a rule, black-and-white and color halftones are handled differently. The manuscript pages for them are output with windows that are sized to the correct dimensions for the artwork. Both the pages and halftones are sent to the printer with photocopies of the halftones affixed in the windows and marked FPO ("for

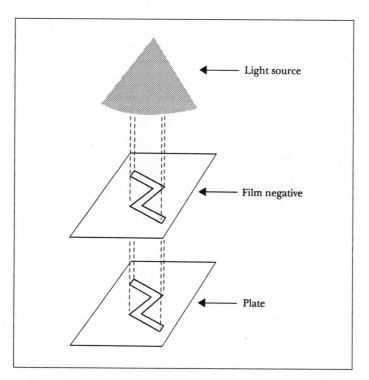

**Figure 13.10**
Creating a Printing Press Plate from a Photonegative

position only"), as in Figure 13.5. If the halftones need to be cropped or sized, they are marked with the appropriate instructions before being sent. The printer makes halftone screens of them and adds ("strips in") the negatives in the correct windows as part of the plate-making process. This process is repeated for each color image times the number of colors required, usually four.

## PRINTING

The time, effort, and expense of producing the best manuscript possible will be compromised or even wasted if the printing is botched because of a mix-up in communications between you and the printer. If you do not have an in-house print shop, choose a commercial printer experienced in handling your type of manuscripts and capable of meeting your schedules, with whom you can communicate effectively. Otherwise, mistakes will be made, schedules will be delayed, and money will be lost.

To ensure the best job possible, provide the printer with explicit instructions for how the manuscript should be printed. Complex manuscripts require more detailed instructions than simple ones. Many document departments prepare dummies to accompany complex manuscripts. The dummy may be a rough sketch showing the format and sequence of each page or a comprehensive mock-up, called a *comp*, showing each element precisely as it should appear in the finished document. Not all jobs require a dummy, however. For relatively straightforward manuscripts comprising mostly text and few graphics, you may only need to fill out a reproduction assembly sheet, like that shown in Figure 13.11. This form specifies the precise location of each page in the manuscript, whether it's a front (F), or right-hand, page or a back (B), or left-hand, page. In accord with a centuries-old printer's convention, all right-hand pages are odd-numbered and all left-hand pages are even-numbered. This convention applies to front-matter pages numbered with lowercase roman numerals as well. The form contains blocks that identify the pages where halftones and line art appear (pages 3-2 and 3-5) and indicates the figure number of each. The halftones will be screened by the printer and assembled (stripped in) for making printing plates. Typically, any reduction required for halftones or other graphics will be marked on the camera-ready manuscript. The form also indicates the location and special markings required for the covers. The form shown specifies that a mailing indicia be printed on cover 4—the outside back cover—so that the document can be used as a self-mailer. Note also that once the left-right (front-back) pattern is established within a section, the location of each individual page need not be identified. The dash in the "Page Number" column next to back (B) pages indicates that they're to be blank. Page 1-10, for example, is blank so that Section 2 can begin on page 2-1, a right-hand page.

For long documents with complicated page numbers, such as service manuals where single pages and entire sections are updated and replaced routinely, it is usually easier to number each page of the camera-ready manuscript manually in pencil in 1, 2, 3 sequence. This practice is called "ring-folioing" because the

| REPRODUCTION ASSEMBLY SHEET | | | | | |
|---|---|---|---|---|---|

**TITLE:** Operating Units Status Report  **JOB NUMBER:** 9- 109

F = FRONT     B = BACK     INDICATE FOLD-INS, BLEED PAGES & DIE CUT TABS

|   | PAGE NUMBER | HALF-TONE | LINE | STRIP-IN | SCREENS | SPECIAL INSTRUCTIONS |
|---|---|---|---|---|---|---|
| F | Cover 1 | | | | | |
| B | Cover 2 | | | | | |
| F | Title pg. | | | | | |
| B | ——— | | | | | |
| F | Statement of Purpose | | | | | |
| B | ——— | | | | | |
| F | i | | | | | |
| B | ii | | | | | |
| F | iii | | | | | |
| B | iv | | | | | |
| F | ✓ | | | | | |
| B | ——— | | | | | |
| F | SECTION 1 — Current Data Summaries | | | | | |
| B | 1-2 | | | | | |
| F | 1-3 | | | | | |
| B | Thru | | | | | |
| F | 1-9 | | | | | |
| B | ——— | | | | | |
| F | SECTION 2 — Operating Unit Data | | | | | |
| B | 2-2 | | | | | |
| F | Thru | | | | | |
| B | 2-248 | | | | | |
| F | SECTION 3 — Appendix | | | | | |
| B | 3-2 | | 3.1 | ✓ | | Shoot at 100% |
| F | 3-3 | | | | | |
| B | 3-4 | | | | | |
| F | 3-5 | 3.2 | | ✓ | | Shoot at 85% |
| B | 3-6 | | | | | |
| F | Cover 3 — blank | | | | | |
| B | Cover 4 — mailing indicia | | | | | |

***Figure 13.11***   Sample Reproduction Assembly Sheet

handwritten numbers—page numbers are called *folios* in printing jargon—are circled ("ringed") in pencil. That way, should the manuscript be dropped accidentally at the printer's or should pages otherwise get shuffled out of order, the printer can reassemble them in the correct order.

In addition to submitting either a reproduction assembly sheet or a ring-folioed manuscript, you must also submit a list of specifications to cover additional requirements or ones not marked on the manuscript itself. These should include some or all of the following kinds of information.

*Administrative Information*
- Title of job
- Person to contact and phone number
- Name and address of your firm
- Date manuscript was sent to the printer
- Date due back from the printer
- Whether job is new, an exact reprint, or a revision to an existing job
- Number of copies required
- Where the copies are to be delivered

*Format Information*
- Page trim size (8½ by 11 inches, 6 by 9 inches, etc.)
- Number and diameter of holes if punched for a ring binder
- Cover and spine
  - Self-cover
  - Separate cover (type of stock, color of stock and ink)
  - Size of cover and spine inserts for ring binders
- Divider tabs (originator provides printer with artwork for tabs)
  - Type and color of tab stock
  - Whether plastic laminate is needed on the tabs
  - Size of the tab inserts

*Paper*
- Weight
- Color
- Finish (glossy or matte)

*Binding*
- Wire-stitched
  - Saddle wire
  - Side wire
- Glued or perfect
- Thread-sewn
- Mechanical-bound
  - Three-hole-punched
  - Spiral
  - Plastic comb
- Other

*Proof Copies Needed*
- Bluelines
- Color proofs
- Other

*Artwork*
- Number of halftones
- Number and percentage of screens and tints
- Other

Detailed and unambiguous instructions to the printer are essential if you expect your document to be printed as you wish, on time, and within budget.

## *Checklist 13.1*

## Checking Galley Proofs

☐  Have galleys been proofread against the original manuscript?

☐  Are all text and tabular elements present and in the correct sequence?

☐  Will creating a dummy be necessary?

## *Checklist 13.2*

## Checking Page Proofs

☐  Are all elements present and in the correct sequence?

☐  Are page margins consistent?

☐  Are line lengths accurate?

☐  Are columns the same depth on each page?

☐  Are graphics (or boxes for them) correctly positioned?

☐  Is the sequence of titles and page numbers correct in the table of contents?

☐  Are captions and credit lines positioned correctly?

☐  Are "continued" lines (in tables and elsewhere) positioned properly?

☐  Have widows and orphans been deleted?

*Checklist 13.3*

## Checking Blueline Proof Copy

☐ Are parallel elements on facing pages properly aligned?

☐ Are photographs positioned and oriented correctly?

☐ Are all page numbers present and visible?

☐ Are screens correct and of uniform intensity?

☐ Does each graphic fit into its allotted space?

☐ Is any type broken or imperfect?

☐ Have any printer's marks been left on copy inadvertently?

*Checklist 13.4*

## Pre-Print Instructions

☐ Will a dummy or comp be necessary?

☐ Has a reproduction assembly sheet been completed?

☐ Has the printer been provided with information about the following aspects of the project?
- Administrative details
- Format
- Paper
- Binding
- Artwork
- Proof copies

### WORKS CITED

Bann, David. *The Print Production Handbook*. Cincinnati: North Light, 1985.

*The Chicago Manual of Style*. 13th ed. Chicago: U of Chicago, 1982.

Craig, James. *Production for the Graphic Designer*. New York: Watson, 1974.

Seybold, John. "Technical Documentation Systems." *Seybold Report on Publishing Systems* 17.7 (1987): 3–6.

### FURTHER READING

Allen, Robert. "In-House Publishing: From Concept to Print." *Modern Office Technology* Nov. 1985: 56–62.

Beach, Mark, Steve Sherpo, and Ken Russon. *Getting It Printed*. Portland, OR: Coast to Coast, 1986.

Brady, Philip. *Using Type Right*. Cincinnati: North Light, 1988.

Bruno, Michael H., ed. *Pocket Pal: A Graphic Arts Production Handbook*. 14th ed. Memphis: International Paper, 1989.

De Ford, Bob. "A Non-technical Guide to Technical Documentation." *Electronic Publishing and Printing* Feb.-Mar. 1986: 36–41.

Doebler, Paul D. "Electronic Publishing: How to Make It Happen." *Electronic Publishing and Printing* Feb.-Mar. 1986: 27–34.

Firman, Anthony H. *An Introduction to Technical Publishing*. Pembroke, MA: Firman, 1983.

Hackos, Joann T. "Redefining Corporate Design Standards for Desktop Publishing." *Technical Communication* 35 (1988): 288–291.

Kalmbach, James. "Reconceiving the Page: A Short History of Desktop Publishing." *Technical Communication* 35 (1988): 277–281.

Kleper, Michael L. *The Illustrated Handbook of Desktop Publishing and Typesetting*. Blue Ridge Summit, PA: Tab, 1987.

Nelson, Charles. "Help with Help: A Selected Bibliography of Desktop Publishing Resources." *Technical Communication* 35 (1988): 313–317.

Pickens, Judy E. *The Copy-to-Press Handbook: Preparing Words and Art for Print*. New York: Wiley, 1985.

Romano, Frank J. "In-House Printing: Where Does It Fit In?" *Electronic Publishing and Printing* Feb.-Mar. 1986: 22–26.

_____. *The Typencyclopedia: A User's Guide to Better Typography*. New York: Bowker, 1984.

Rubens, Philip. "Desktop Publishing: Technology and the Technical Communicator." *Technical Communication* 35 (1988): 296–299.

Sabatine, Natalie V. "How to Standardize Documents Using a Desktop Publishing System." *Proceedings of the 36th International Technical Communications Conference* (1989): RT-136.

Seybold, John, and Fritz Dressler. *Publishing from the Desktop*. New York: Bantam, 1987.

Standera, Oldrich. *The Electronic Era of Publishing: An Overview of Concepts, Technologies and Methods*. New York: Elsevier, 1987.

Sullivan, Patricia. "Desktop Publishing: A Powerful Tool for Advanced Composition Courses." *College Composition and Communication* 39 (1988): 344–347.

Wise, Mary R. "Is Electronic Publishing for You?" *Proceedings of the 33rd International Technical Communications Conference* (1986): ATA-1–ATA-3.

### Chapter 13: Document Production

**INTERVIEWER:** Did you work with any other document specialists in producing the *Introduction to Using the International Banking System* manual?

**MARY:** I worked with staff members in the Reprographics Department when I wanted to get the manual reproduced. The Reprographics Department also reproduced the reference card that was included with the manual. The divider tabs for the manual were ordered through our Purchasing Department, which in turn ordered them from a vendor.

**INTERVIEWER:** What about the binder cover?

**MARY:** The binder cover was chosen because it would be inexpensive, since we really didn't know if, down the line, there were going to be changes in the system that would require a new manual. The design was simple and was produced in-house. I did, however, have to get user agreement on the design and legal permission to use the logo, and inside I had to add the co-owner's name to the proprietary statement in the beginning of the manual. So there *were* some aspects of design that I had to get checked out or approved by others.

**INTERVIEWER:** Did you produce the finished manuscript on an electronic publishing system, or was it typeset by an outside vendor?

**MARY:** We use an in-house electronic publishing system that is tied to a mainframe computer.

**INTERVIEWER:** Does your electronic publishing system have both text and graphics capabilities?

**MARY:** Yes. The system provides access to multiple typefaces and sizes and permits multiple columns per page. It has a "what you see is what you get" (WYSIWYG) screen. It also allows us to display either an entire page on the screen or two adjacent pages side by side to compare them for layout symmetry.

**INTERVIEWER:** Were you able to produce all the graphics in the *Introduction* manual on your electronic publishing system?

**MARY:** Yes. It allowed us to create the frames around the sample computer screens as well as the frames around the tables. Our system contains a vector-oriented graphics package that permits us to manipulate graphics primitives to create organization charts, flowcharts, and the other typical graphics requirements for our manuals.

**INTERVIEWER:** So far you've talked about *creating* graphics online. Can you also *scan* existing graphics into the system?

**MARY:** Yes, we can now. A scanner was added to the system recently that allows for line drawings, charts, and even maps to be scanned in, brought up on a writer's screen, and sized online to fit into an existing page.

**INTERVIEWER:** Is the system capable of storing the format specifications for the *Introduction* manual and related manuals?

**MARY:** Yes. The format was established by a committee before we began writing, stored on the system as a template, and then used by all writers. Establishing specifications ahead of time like this not only makes the writer's job easier (because there are fewer decisions to make), but it also gives our manuals a uniform, readily identifiable look for our customers.

**INTERVIEWER:** How do you select the appropriate options when you're working online?

**MARY:** I used to use a mouse-controlled cursor and "click" the right choices from among multiple menus on the screen. But now I use style keys. By striking a prescribed key, I give the system specific format and typesetting instructions for headings, text, margins, and so forth. Striking a single key for each typesetting and format component saves time compared with the old method of using the cursor and a series of menus to give the system the same instructions.

**INTERVIEWER:** Does the system permit the automatic creation of a table of contents from headings scattered throughout the text?

**MARY:** Yes. It also permits us to create subject indexes, but we seldom need these for our manuals.

**INTERVIEWER:** What kind of output device did you use for the finished manuscript?

**MARY:** The final text pages were output on a 300-dots-per-inch laser printer.

**INTERVIEWER:** So there were no galleys, then, only page proof (which matched what was on the WYSIWYG screen) that was output on a laser printer?

**MARY:** Right?

**INTERVIEWER:** Was the manual printed commercially or within your own company?

**MARY:** The manual was photocopied by our in-house Reprographics Department. There was no cutting and pasting involved. We produce so few copies—typically under 100—that going to a commercial offset printer would not be economical, particularly since we do not use photographs or other elements that would require offset printing.

# 14

# Creating Document Standards

Definition of Standards
Purposes of Standards
How to Establish Standards
Possible Issues to Standardize
Warnings and Liability

This chapter describes the process of creating document standards (and a manual to contain them) that best fit your needs and capabilities and the needs of your audience. It describes the purposes of standards, tells how to establish standards, and lists some issues you should consider standardizing. The chapter also provides an overview of one of the most frequently standardized features, the use of warnings.

## DEFINITION OF STANDARDS

A document standard is a rule or model that is followed consistently. For instance, a document standard might prescribe that first- and second-level headings must always appear in all capital letters, with first-level headings centered and second-level headings flush left. Any writing project, from an academic paper to a ten-volume set of manuals, needs standards to some extent. The extent depends on the project's purpose, audience, and complexity; the number of writers working on it; and the capabilities of the electronic publishing or typesetting system used to produce it.[1]

A standard differs from a guideline, which merely advises, or from a convention, which is simply common practice. These terms are often used interchangeably, but the distinction is that standards are enforced whenever possible. Enforcing standards is not always possible, however, because rules always have exceptions and situations change, and standards must allow for those exceptions and changes. Sometimes standards can at best fit only the majority of situations.

The decisions made in creating standards must be based on good reasoning, thorough research, and a full understanding of the alternatives. Knowing the reasons behind standards allows you to decide when they must be enforced and when they cannot.

Standards have also gained in importance with the increasing availability of electronic publishing systems. Because of those advances in technology, today's technical writers are able—and increasingly expected—to create entire documents, from the beginning outline to the final published product. Many decisions that formerly belonged to editors, artists, page designers, printers, and other experts are now in the hands of the writers. Unless they have mastered the requisite design principles, writers must rely on others—or on standards—for guidance. Carefully developed document standards help writers to create consistently attractive and readable documents.

Document standards allow any number of writers to create any number of any document types, all with the same appearance. The best way to implement document standards is to publish them in a manual. Standards manuals may be called "style guides," but they can cover matters beyond the merely stylistic.

## PURPOSES OF STANDARDS

Document standards have four important purposes:

- To ensure consistency
- To enhance the organization's image
- To save writers time
- To ensure quality

First, standards ensure consistency within documents and throughout a group of related documents. Readers recognize when the specific elements of a document are treated in a standard way and come to expect that they will receive similar treatment throughout. So the use of standards helps fulfill reader expectations. The more broadly these standards apply, the better the communication that results. If the same standards can be used across all documents related to a product or a system—or, preferably, across all documents of a company or an organization—the readers have that much less to learn. For example, if headings

---

[1]This book follows the current standards (or "house style") of St. Martin's Press for such issues as hyphenation, capitalization, and optional punctuation.

are treated the same throughout all documents, readers can always tell the relative importance of the topic they are reading.

Another purpose of standards is to enhance the image of the organization publishing the documents. Brochures or manuals with carefully developed, consistent standards reflect positively on the organization. Such documents can even convey an organization's "personality." For example, some organizations' documents might adopt a casual tone and use an informal approach, while others could maintain a rigorous writing style and project a more formal image. Most organizations concern themselves with projecting a positive image that is consistent and appropriate. When all published documents do not match that image, readers may believe that the organization is not unified. Applying the same standards to all documents, therefore, ensures that the image presented is unified and appropriate.

Document standards also save time for writers and make their jobs easier. With established standards, writers do not have to make decisions independently on all the details of a project. Standards are especially helpful for collaborators. Collaborators can work independently of one another with the assurance that their final product will be consistent throughout. The more writers there are, the more important standards become. By reducing choices, standards especially help new writers because deciding on the details is more difficult for them and because they have so many other new things to learn. But in order to make the writers' jobs easier, *the standards chosen must be easy to use*. Forcing writers to use standards that are awkward or time-consuming defeats this purpose.

Finally, standards can also help ensure quality in documents. Quality has received much attention in the workplace, where it sometimes means that a product meets clearly defined requirements. Standards can be used to set those requirements, thus ensuring that the document is accepted as a high-quality product.

## HOW TO ESTABLISH STANDARDS

As discussed earlier, one of the principal reasons for establishing document standards is to ensure consistency. But consistency has to be tempered with reason, and knowing why certain standards are preferable in certain situations allows you to determine the degree of consistency you can achieve. The best course is to know as much as possible about rules, conventions, and research findings on many of the subjects covered in this book so that you can make informed decisions. Before establishing standards, you must balance this knowledge with what you know of your organization's typical audience and purpose, your text processing equipment, your time or budget constraints, and any other factors that might affect your work.

Some things cannot or should not be required in a standard, for various reasons. For example, you might not have a choice of typefaces or type sizes because of the word or text processing system you use. Further, you might need to make trade-offs, such as considering the usefulness or helpfulness of a design standard versus its cost efficiency or the ease of producing and revising it. Color graphics, for example, command attention but are expensive and take longer than black-and-white graphics to produce. As suggested in Chapter 8, the document design can depend on the constraints of the schedule and budget and the flexibility of your printing method.

On the job, managerial support also determines the kinds of standards you can establish. Before beginning to prepare a standards manual, be certain that you have your management's backing. Creating organizational document standards requires a great deal of time and effort, and your management must approve of the plan to create such standards—both to allow you the time for it and to enforce the standards once they are created. (Your management must also be involved in determining whether the standards you devise are appropriate for the company.)

## Determine What Kind of Standards You Need

Begin the process by determining your needs, based on such issues as these:

- Your typical document types
- Your typical audience
- Any special requirements of the audience, task, or location
- Your typesetting and printing capabilities
- The number of writers in your organization and their rate of turnover

While you determine your needs, you should look at a range of authoritative sources for guidance about which features you might want to standardize—what specific standards you might prescribe.

- U.S. Government Printing Office *Style Manual*
- *The Chicago Manual of Style*
- *Handbook of Technical Writing* (Brusaw, Alred, and Oliu)
- Technical dictionary for your specialty
- English-language dictionary

To begin, you need to answer some key questions. What standards do members of your group use? What needs standardizing? What works well and could easily be made into a standard? What doesn't work and should not be standardized?

Next, find out what similar organizations are doing. Attend technical writing workshops, conferences, seminars, and meetings. Ask to see others' standards manuals. Examine the kinds of features they're standardizing and how they present their solutions.

Three options exist for creating a standards manual. Which option you choose depends on how detailed you need to be and how unusual your circumstances are.

1. Rely entirely on an existing book, such as *The MLA Style Manual* (Achtert and Gibaldi), or set of books for your standards.

2. Create a standards manual to cover only issues *not* found in existing books and refer readers to those books that your manual supplements.
3. Create a standards manual covering all facets of the document process.

The last two options cost more because they involve creating a manual of some sort, but they enable you to solve your own unique problems. The second option (covering only issues not found elsewhere) is more efficient than the third. By using an existing book (or set of books) as a first resource, you save the time and expense of reproducing basic information. Then you need only record solutions to your organization's unique problems to ensure that you have covered every possible standardization need.

The next step is to research the issues themselves. List all the issues you need to account for and the possible alternatives for each. Keep in mind that the more research you do, the more issues and alternatives you will discover. Determine the reasons for choosing one alternative over another. Knowing the reasons behind your preferences allows you to decide when they can or cannot be enforced—or when you should establish a standard, a guideline, or nothing at all.

Some of the decisions you have to make will have no clear-cut best choice; one alternative might be preferable for one reason and another alternative for another reason. For example, research might indicate that boldfaced type highlights for emphasis better than italic (see Chapter 8). But italic might work better for emphasis if your standards already require the substantial use of boldface for a purpose other than emphasis, such as the use of boldface to cross-reference parts of a document. Other issues might not have been researched, or some might have an equal number of authorities supporting each side of the question. As a result, many of your decisions must be somewhat arbitrary. If future research should show sufficient reason for deciding one way or the other, allow yourself the flexibility to change your standards.

## Collaborating to Prepare Standards

If a writing group has been assigned to establish standards, you should break into several teams to work on different matters, according to team members' areas of expertise or interest. However, some method must be established for making final decisions, whether individual or collective (see Chapter 3).

Be sure to test all the alternatives you select to make sure that they are easy to use and that they produce the results you expect. Keep a record of all decisions you make so that you don't end up discussing the same matter again. Begin drafting your standards manual as soon as you begin discussing issues and making decisions.

When working with a group, be sure that everybody involved makes the commitment both to work on creating the standards and to follow them once they are established—even if a member should be outvoted and disagree with certain decisions. Some writers might resist change; others might complain of having too many restrictions and not enough room for creativity. Recognize that virtually everyone will balk at severe restrictions that have no clear advantage over current practices. You need to make sure that any standards you establish are in fact necessary and are perceived as such.

All writers need to understand the advantages of using standards, both to the organization and to themselves. Getting writers involved in creating the standards can help in two ways: (1) writers more willingly follow standards that they have helped create and when they understand that standards function as a tool to help them perform their jobs, not as means of controlling them, and (2) writers will become more knowledgeable about the reasons behind the standards and thus better equipped to do their jobs.

Finally, keep in mind the most important rule in creating standards: be consistent. But at the same time realize that as methods and tools advance, your standards must grow with them. Your standards manual must always evolve.

## POSSIBLE ISSUES TO STANDARDIZE

You should consider standardizing the following: document type, organizational sections of documents and their sequence, writing style, editorial style, access aids, page format, graphic devices, and in-house policies and procedures.

### Type of Document

You can standardize the type of document you need to create for each situation. For example, you might decide that each new product of a certain type requires an advertising brochure, a tutorial manual, and a reference card. Typical document types might include the following:

- Proposals
- Annual reports
- Feasibility reports
- Status reports
- Articles
- Brochures
- Reference cards

- Functional design documents
- Manuals
  - Reference
  - Task-oriented
  - Tutorial
  - Combination
- Technical specifications

### Sections and Sequence

You might standardize overall document formats so that each type contains certain sections in a specific order and so that those sections always contain certain items and information. A standard for computer-related manuals, for example, might include the following requirements:

- "The preface should contain an explanation of any conventions used in the document, guidelines for using the document, statements describing the purpose and intended audience of the document, and a list of any related documents to which readers can refer for more information."

- "The introduction should contain an overview of the main product, service, or system, as well as overviews of any subsystems involved with the main topic."
- "The body should contain any input instructions, processing descriptions, operating instructions, and output descriptions. Its organization should be task-oriented for a task-related manual, descriptive for a reference manual, and benefits-related for a marketing document."
- "The appendixes should contain lists of commands, lists of error messages, a glossary, or any reference-type material that is too large to fit in the text without interrupting it."

## Writing Style

You might standardize the writing style for each type of document and the appropriate tone for each audience. Particularly for technical documents, you can define the appropriate knowledge level for each of your different audiences so that you can be sure to give them the right amount of information. For manuals, you might specifically recommend text paragraphs for descriptions but provide alternatives for instructions, such as numbered steps or "cookbook-style" sentences (Grimm 49).

## Editorial Style

Your standards manual could also include specific editorial style guidelines. If you intend to cover only the issues not found in a desk or specialty dictionary or in a general style manual, deal with the issues specific to your needs that are not covered in your other sources. These might include the following:

- *Special terms.* Spelling, capitalization, abbreviation, and hyphenation of special terms used in your subject area or in your organization. For the workplace, that category might include product or service names, procedures, and other terminology unique to your company.
- *Definitions.* Consider creating a glossary of all technical terms to ensure that they are always used correctly.
- *Acronyms and symbols.* Standardize their use and form common to your subject area and organization.
- *Word usage.* For software documentation, decide, for example, whether to use *data is* or *data are* and whether to restrict the use of nonverbs like *impact* as verbs. Also consider the use of computer terms like *interface* in noncomputer areas. Decide whether or not to permit the use of contractions. You can also include reminders of such problems as nominalizations, redundancy, and noun strings.
- *Perceived discriminatory language.* You can replace gender-specific titles with neutral ones (chair, mail carrier, police officer) and create a list of ways to work around the indiscriminate use of masculine or feminine pronouns (Christian 89–93).

- *Punctuation.* Consider whether to use, for example, spaces around dashes and one space or two after a colon. Decide also whether to use a comma after short introductory phrases, in approximate dates (e.g., *June, 1991,* or *June 1991*), and in four-digit numbers.
- *Numbers.* Some style guides recommend spelling out numbers from one to ten in text; others say to spell out to one hundred. Determine which standard to adopt and any situations that might warrant making exceptions, such as precise measurements or codes.

## Access Aids

The next level of standards involves access aids, which enable readers to find the information they need quickly and efficiently.

- Design a standard cover and binding for each type of document, making sure that they fit the purpose of the document.
- If you use ring binders, set up standards for the type of binder (size, pockets, etc.). Design readable spine labels and tabs to be used for all major topics as well.
- Set up procedures to ensure that the table of contents and index mention all topics likely to be sought in them. You might want to include a table of contents for each section of a large document.
- Decide whether you want a title page and how you want it formatted. Also consider including a title page for each section of a large document.
- Create a standard hierarchy of headings for all major and minor divisions, especially for large documents. Consider also whether you want to limit headings to what will fit on one line. And consider whether you want to mandate the content and wording of the headings so that they consistently and clearly indicate what material the divisions cover.
- Decide how to format lists. Begin by mandating that they have parallel structure. Establish standards for consistent indention, spacing, and punctuation, too. Also determine the types of markers to use to set off each list item: bullets, other symbols, numbers, letters, or nothing (see also Chapter 8). You should establish a hierarchy of markers for lists, such as the following:
  - First-level item
    - Second-level item
    - Another second-level item
      - Third-level item
      - Another third-level item
  - Another first-level item

## Page Format

Next consider page format standards. Each type of document that your organization produces might require its own page-formatting standards, or at least variations of the primary standards. Consider standardizing the following issues, many of which are detailed in Chapter 8.

- Determine a page size that is appropriate for the users, the document's purpose, and the setting in which it will be used. Consider whether to use any landscape pages for reports or any wide illustrations and determine any format changes that will be necessary to accommodate them.
- Define a margin size that is adequate to accommodate whatever kind of binding you use and to provide adequate framing on each page.
- Define the use of white space on a page to avoid a cluttered appearance.
- Determine a readable line length to fit your subject matter, readers, and format.
- Decide on the number of text columns to use if your text processing system is capable of multicolumn formats.
- Decide where to place headings and how much space to leave around them.
- Decide on the use, wording, and placement of running heads and feet.
- Choose the format and location of your page numbers. Depending on the size of the document, decide whether to use a simple numbering scheme (1, 2, 3, etc.), a two-part scheme that includes the section number (1-1, 1-2, 1-3, etc.), or a more complicated, military-type scheme that includes more levels (1.1.1, 1.1.2, 1.1.3, etc.).
- Decide whether to use right-justified or ragged right text columns.
- Establish a standard for spacing of paragraphs and lists and for indention; be aware that too many levels of indents can be distracting.
- Choose typefaces to use for text, for headings, and in graphics. Consider whether you want to limit the number of typefaces to be used per page, per document, in all related documents, or even in all documents for the organization. Create sample pages for the manual to illustrate your choices.
- Establish the use of type sizes and leading for text, headings, covers, and other document elements.
- Standardize highlighting techniques (such as **boldface**, *italic*, and <u>underline</u>) because there are so many options and so many items that you might want to cause to stand out from the text. For software documentation, for example, consider which items you should highlight and how to handle them:
  - Names of products, modules, files, records, fields, screens, reports, forms, and codes
  - Programming terms (standard might require writers to use them exactly as they are used in the program)
  - Variables (by convention, $x$ represents a character value and $n$ represents a numeric one)
  - Values that the reader must enter (standard might prohibit quotation marks around these for emphasis because readers might be unsure whether to keyboard the quotation marks or not)
  - Keyboard keys (standard could require the same design as on the keyboard)
  - Headings, headers, footers, and figure captions

## Graphic Devices

Standardize the placement and treatment of graphics and examples. In addition to the following suggestions, review the guidelines in Chapters 7 and 8.

- Consider a standard range of sizes for your graphics. Determine the spacing and borders that go around illustrations and what kind of labels (or callouts), captions, and numbering to use. Also determine the standards for highlighting, shading, and color and for typefaces and type sizes in graphics, labels, and captions.
- Set standards that will keep illustrations simple, sufficient, clear, and accurate. For example, you might require the use of line drawings rather than photographs because they are generally easier to reproduce and to understand.
- Determine the appropriate types of illustrations for your standard uses. (For example, use tables to show values, bar graphs to show variations, and bar or pie graphs to show parts of a whole.)
- Establish standard revision codes so as to mark easily any changes you make to documents that are continually updated. The conventional treatment is to place vertical bars in the margin to the left of all changed text. Some writers also place revision codes in the table of contents for extra help. You can also add a revision date to any changed pages in a header or footer or in parentheses in text following the change.
- Standardize how warnings, which are discussed later in this chapter, will be displayed.
- Specify how to present procedures and descriptions. To make the material illustrated as realistic as possible, use actual data in all samples, and use a single source for the samples so that they always look the same. You should also define which elements to include in each type of description (of screens, reports, etc.): purpose, user, frequency of distribution, request or access procedure, important features, and so on. Figure 14.1 illustrates a sample report description as it might appear in a standards manual. For your descriptions, you might define the methods used to point out individual items needing explanation.
  - Text next to each item
  - Arrows between the items and their descriptions
  - Callouts (pairs of numbers or letters that match each item to its description in the text)

  For callouts, you need to use similar graphics for both sides of the reference—at the callout and at the text—to make it clear that they belong together, as shown in Figure 14.1. You might want to set a limit for the number of callouts that should appear on one page and decide that any examples requiring more than that limit be broken into smaller portions.
- Establish a standard format for your table of contents and index. The conventional layout for a table of contents has headings listed on the left side of the page with dotted lines (leaders) leading to the page numbers, which are right-justified. Indexes are customarily arranged in two columns.

## Report Forms
Production Reports

(1)

Production for January

| (2) Department | (3) Part# | (4) Needed | (5) Estimated | (6) Completed | (7) Difference | (8) Phase |
|---|---|---|---|---|---|---|
| 100 | A304 | 300 | 350 | 347 | +47 | 1 |
| | A305 | 4 | 5 | 4 | - | 1 |
| | B789 | 75 | 80 | 75 | - | 3 |
| 200 | D904 | 1 | 1 | 1 | - | 3 |
| | H612 | 7 | 8 | 10 | +2 | 5 |
| | J375 | | 123 | | - | |

**Figure 3.1 Standard Report Form**

This report shows the number of parts produced by each department for the month and the number estimated to be completed. It also compares the number produced to the number needed and shows which phase of the project the parts will be used in. The report is distributed to department managers within one week of the month reported on. Descriptions of the individual items on the report follow; the numbers at the descriptions match the numbers on the report.

(1) The title of the report, showing the month for which these figures apply.

(2) The department producing the parts. For more information on departments, see the listing in the ***Human Resources Handbook.***

(3) The part number. Part numbers are listed with their descriptions in Appendix 3.

(4) The number of parts needed for the month.

(5) The number of parts estimated to be completed for the month.

(6) The number of parts actually completed for the month.

(7) The difference between the number of parts completed and the number needed for the month.

(8) The phase of the project for which the parts are needed, as shown in Appendix 2.

**Documentation Standards (Internal) 3.2**

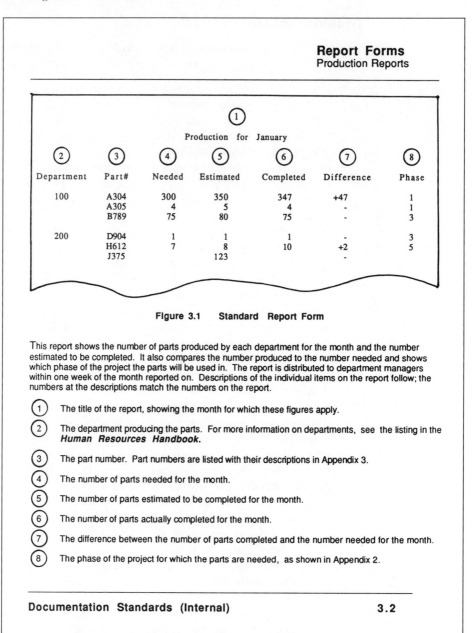

*Figure 14.1* Page from a Standards Manual Describing a Report Form

Consider making the typography of the headings in the table of contents and index consistent with headings in the document.

In general, it is best to show typical pages and examples to ensure that the users of standards manuals understand how to implement the standards you require.

## In-House Policies and Procedures

The last section in your standards manual might be a list of your organization's in-house policies and procedures. Here are some examples of items that might fit under this heading.

- *Personnel policies.* List job descriptions, performance reviews, instructions for administrative tasks such as filling out travel forms and time forms, and the like.
- *Descriptions of work-related resources.* Describe your word processing system and its capabilities (spelling checker, thesaurus, etc.) and your production procedures and equipment.
- *The document process.* Describe the normal method of preparing documents, illustrated with a flowchart: scheduling, planning (creating a project plan, performing task and audience analyses), researching, and so on.
- *Subject index.* Explain the method for creating a subject index if your text processing system does not generate one automatically.
- *Editing and review process.* Detail the normal editing and reviewing process (with a flowchart) and any testing procedures you use: usability testing for manuals or methods of analyzing the text itself, such as readability formulas.
- *Printing and production processes.* Describe those that might be different for each type of document and the standard terminology used.
- *Maintenance and distribution.* Outline procedures for maintaining and distributing your documents. Consider creating standard cover letters for both new and revised documents.
- *Quality control.* Include quality checklists and ways to measure productivity.
- *Bibliography.* List the contents of your current library, if small enough.
- *Exhibits.* Provide samples of standard covers, title pages, spine labels, page layouts, callouts, illustrations, descriptions, and any other items you commonly use.

Figure 14.2 shows a sample page from a standards manual describing the prescribed use of illustrations.

## WARNINGS AND LIABILITY

One of the most important and most common features to standardize is the use of warnings. Schoff and Robinson, specialists in the use of warnings, observe:

> The rise in the consumer movement and the increase in product liability suits in recent years have placed a greater responsibility on manufacturers to make their products safe and to provide both proper instructions for their use and proper warnings of hazards. (93)

Warnings involve both the text and the visual design of a document. Since product liability is a complex legal and technical area, you must work closely with safety experts and your organization's attorneys when you are preparing a man-

---

**Creating   Illustrations**
Graphic Standards

---

## Section  4:    Creating  Illustrations

In addition to the text in your documents, consider including the following illustrations:

o Record layouts          o Flow charts
o Report descriptions     o Diagrams
o Command explanations     o Tables
o Screens                 o Graphs
o Error messages          o Maps

Appendix 3 of this manual, "Models," contains a model of each of these illustration types and lists names of files you can copy to create them.  Also review manuals related to your topic for more specific details.

### Graphic  Standards

**Captions**

Provide a caption for each illustration.  The system numbers them automatically. Leave two blank lines between the illustration and its caption.  Try to keep the caption to less than one line and treat it as a title, with initial caps for all words except short prepositions, coordinating conjunctions, and articles unless they begin the caption.  If the figure requires additional explanation, either use a sentence for the caption or add a period to the caption and add a sentence after it.

**Paging**

Put all illustration on the left page with any necessary explanation on the right page facing it.  If an illustration takes more than one page, write "Continued" on the bottom of all continuing pages and  include a short caption on all pages.

Keep all illustrations as close as possible to their references in the text.

---

**Documentation  Standards  (Internal)**                    **4.1**

---

*Figure 14.2*    Standards Manual Page Describing the Use of Illustrations

ual that involves issues of safety and liability. And you need to know some of the basics to talk with the experts.

Product liability laws require a manufacturer to warn potential users of (1) dangers in the *normal use* of the product and (2) dangers in the *foreseeable misuse* of the product. A manufacturer need not warn of *open and obvious* dangers. Hence instructions for an electric knife would not need to warn the user not to use the knife for shaving a beard—an open and obvious danger that is also neither "normal use" nor "foreseeable." Instructions for an electric knife would need to warn users to take care when holding food to be sliced, since a slip of the hand could result in injury—a danger in normal use. The same instructions would also need to warn users to disconnect the power when cleaning the blades since the user could receive a shock if the knife were immersed in water while plugged in—a danger that is foreseeable.

Even if the danger is open and obvious, the manufacturer may have a duty to warn users who may not be aware of the extent or degree of danger. Schoff and Robinson give the example that a person using tile adhesive labeled "flammable" would probably refrain from smoking while using the product but might well fail to recognize the danger posed by stovetop pilot lights (96). If the likelihood of injury is serious, the manufacturer is also required to display a warning on the product itself.

## Language of Warnings

The front matter of a manual should include any disclaimers that a manufacturer's legal counsel may deem appropriate. A manual should be dated and tell exactly what model of the product it covers and what earlier documents, if any, it replaces.

In general, to avoid liability problems, all instructions for the proper use of a product should be *clear*, *readable*, and *understandable*—the goals of much of the advice throughout this book. However, readers must also be warned specifically of both dangers that they might expect and ones that they might not. Remember, a danger that is obvious to you as a technical writer may not be obvious to the user. Schoff and Robinson give the example that many people are

> unaware that burning charcoal emits carbon monoxide gas, which can be deadly with inadequate ventilation. Anyone concerned with the manufacture of charcoal briquettes surely knows this [certainly a technical writer who works in the industry], yet every year the news includes reports of people dying while using a charcoal grill in a trailer or closed garage. (95)

Instructions must not only warn of all risks and hazards, but they must also warn *adequately*. An adequate warning must do three things:

- Identify the hazard and the gravity of the risk
- Give the likely results of ignoring the warning
- Describe how to avoid the hazard and thus injury

To ensure that a warning is adequate, the language in it must be clear and specific, even vivid:

*Change:* Failure to disengage the blades may result in bodily harm.
   *To:* If you do not disengage the blades, they can amputate your fingers.

Avoid words that are open to interpretation or need further defining: *proper*, *excessive*, *frequently*, *often*, *seldom*, *may*, *might*, *could*, *recommended*, *occasionally*. If potential users comprise a diverse group, you must consider their age, expertise, familiarity, and level of literacy.[2]

## Visual Symbols and Signal Words

Make sure that warnings stand out from the text and are easily readable. Warnings should not blend in with the instructions. Use an open, uncrowded format. A clear border of heavy rules or white space adds visual emphasis to warnings. Use symbols and icons in the text to reinforce warnings, as illustrated in Figure 14.3. Use line drawings of products that depict the physical sources of hazards and, if possible, the nature of the hazards. Don't show a picture of what *not* to do unless you put an *X* through the picture. Don't use cartoons for safety warnings (Schoff and Robinson 107):

- They dilute warnings.
- They trivialize.
- They are difficult to design simply.
- They may imply that you are "talking down" to readers.

Warning words, symbols, icons, and labels are increasingly standardized, although industries do vary. Certain signal words in boldface and their corresponding colors are becoming standards:

**DANGER** (red) = hazard or unsafe practice that *will* result in severe injury or death

**WARNING** (orange) = *could* result in severe injury or death

**CAUTION** (yellow) = could result in *minor* injury or property damage

**NOTICE** (blue) = information unrelated to safety

*Figure 14.3*    Typical Warning Icons with Text

---

[2]Schoff and Robinson cite a court case (*Hubbard Hall Chemical Co. v. Silverman*) in which the court held that a written warning was not adequate because it failed to provide for illiterate users (95).

**Figure 14.4**
Hazard Alert
Symbol and Word

The hazard alert signal—an international standard—should appear with the signal word, as shown in Figure 14.4.

It is important to coordinate with artists and safety personnel the warnings in the manual with those affixed to the product. Although many more warnings may appear in the text than labels on the product, every DANGER-level warning in the text must have a corresponding label on the product. Don't overuse the DANGER designation; if you do, its impact is diluted. Further, if you *must* use an "inordinate number of DANGER warnings, it may indicate that the product is unreasonably dangerous and should be redesigned" (Schoff and Robinson 104).

Many sources recommend *Product Safety Sign and Label Systems* (Santa Clara, CA: FMC Corp., 1990) as a reference for warning symbols and icons. The bibliography of Ballard and Rode-Perkins (88) lists many useful works on warnings and icons.

*Checklist 14.1*

## Creating Document Standards

☐ Determine the kind of standards needed
  • Use an existing source or sources (manuals, handbooks, etc.)
  • Create a standards manual covering only those issues *not* found elsewhere
  • Create a standards manual covering *all* facets of the document process

☐ Research relevant standards issues

☐ Consider issues to standardize
  • Type of document needed for various situations
  • Sections and sequence
  • Writing style
  • Editorial style
  • Access aids
  • Page format
  • Graphic devices
  • Policies and procedures
  • Use of warnings
    ○ Language
    ○ Symbols

## WORKS CITED

Achtert, Walter, S., and Joseph Gibaldi. *The MLA Style Manual.* New York: Modern Language Association of America, 1985.

Ballard, James C., and Susan L. Rode-Perkins. "How to Create Mass Media Print Warnings." *Technical Communication* 34 (1987): 84–89.

Brusaw, Charles T., Gerald J. Alred, and Walter E. Oliu. *Handbook of Technical Writing.* 3rd ed. New York: St. Martin's, 1987.

*The Chicago Manual of Style.* 13th ed. Chicago: U of Chicago, 1982.

Christian, Barbara. "Doing without the Generic He/Man in Technical Communication." *Journal of Technical Writing and Communication* 16 (1986): 87–98.

Grimm, Susan, J. *How to Write Computer Manuals for Users.* Belmont, CA: Lifetime Learning, 1982.

Schoff, Gretchen H., and Patricia A. Robinson. *Writing and Designing Operator Manuals.* Belmont, CA: Lifetime Learning, 1984.

U. S. Government Printing Office. *Style Manual.* Washington, DC: GPO, 1984.

## FURTHER READING

Bandes, Hanna. "Defining and Controlling Documentation Quality, Part II." *Technical Communication* 33 (1986): 69–71.

Batik, Albert L. "The Future of Standards." *ASTM Standardization News* Aug. 1986: 36–39.

Brockmann, R. John. *Writing Better Computer User Documentation: From Paper to Online.* New York: Wiley, 1986.

Buehler, Mary Fran. "Defining Quality: It Is Not the Same as Goodness." *Proceedings of the 34th International Technical Communication Conference* (1987): MPD-37–MPD-39.

Howell, John Bruce. *Style Manuals of the English-speaking World.* Phoenix: Oryx, 1984.

Nesmith, Achsah. "A Long, Arduous March toward Standardization." *Smithsonian* Mar. 1985: 176–194.

Pakin, Sandra, and Associates, Inc. *Document Development Methodology.* Englewood Cliffs, NJ: Prentice, 1984.

Southard, Sherry G. "Practical Considerations in Formatting Manuals." *Technical Communication* 35 (1988): 173–178.

Wilson, Catherine Mason. "Product Liability and User Manuals." *Proceedings of the 34th International Technical Communication Conference* (1987): WE-68–WE-71.

## The Case History

### Chapter 14: Creating Document Standards

As Mary has mentioned throughout the Case History, many design and format decisions—or document standards—for the *Introduction to Using the International Banking System* manual were dictated by the style sheet for the International Banking System manuals. These standards address such issues as choice of binding and paper, format of headings, rules for tables and screen samples, and spelling and phrasing conventions. The IBS style sheet is reproduced here in Figure C.17.

## Style Sheet for All IBS Manuals_____

| | |
|---|---|
| **Binder** | 1 inch white binder with inside pocket and slip-in front and spine labels. |
| **Label** | To be designed by Technical Writing and typeset in Reprographics. Do not include the company logo. |
| **Paper** | 8½ x 11 inch white bond, regular weight |
| **Page Margins** | Left, right, and bottom margins: 1 inch. Top margin: 1.33 inches to accommodate running header. |
| **Line Height** | 19 points for topic heading line; 13 points elsewhere. |

**Headings**

Topic Heading — On the first page of each topic, place topic heading on second line, running it across the width of the page in 18 point bold Modern, mixed case, using 20 point line height. End heading with an underlined, bold tab. Triple space (13 point line height) after the topic heading.

On subsequent pages of the topic, place the topic heading on the inside of the running header.

Running Header — Header begins on the second page of each topic and shows topic on inside and page number on outside. Headings alternate.

Running Footer — Footer begins on the first page of each topic and shows issue date in center and chapter name on outside. Footers alternate.

*Figure C.17*   International Banking System Style Sheet

Side Margin Caption   12 point bold Modern, mixed case.
                      Double space before each heading.

Headings within the   12 point Modern set off into an
body                  in-body side margin caption.

**Page Numbering**       Number each chapter separately.

**Tables**               Position tables flush to the body margin (left margin of
                         25 spaces).

                         Show borders using 2nd width available.

                         Use .06 inch margins within tables.

                         Use 12 point Modern bold for headings; 10 point Modern
                         bold for table text.

                         Capitalize the first word in each heading. If the heading
                         is a phrase, end the heading in a colon. Don't capitalize
                         and punctuate text within the table unless you have to
                         because the text includes sentences.

| For this account type: | Enter: |
|---|---|
| demand deposit, vostro, or time deposit | Q |
| trust | the beneficiary's name |

| If you want to: | Then: |
|---|---|
| accept the given interest rate | Press ENTER. |
| change the interest rate | See your supervisor.<br>▸ If your supervisor allows X, then do Y.<br>▸ If not, do Z. |

*Figure C.17*   *Continued*

**Tabs**                          Tab at 25 for body text.

**Lists**                         Introduce lists with a complete sentence, but end the
                                  sentence with a colon.

                                  Example:  You use the XYZ program to perform these
                                  tasks:

                                  ‣ Entering orders.
                                  ‣ Sorting purchase orders by items ordered.
                                  ‣ Sorting purchase orders by billing address.
                                  ‣ Entering remittance data.

                                  Capitalize the first word in each item and end each item
                                  with a period.

                                  Use 12 point triangular bullet (‣) for first level; dash for
                                  second level (–); and 10 point triangular bullet for third
                                  level (‣).

                                  **Example:**

                                  ‣ List item.

                                      – List item.
                                      – List item.
                                      – List item.

                                          ‣ List item.
                                          ‣ List item.

                                  Double space before and after lists.  Double space first
                                  level items; single space second and third level items.


**Screen Samples**                Use plain box at 2nd available width.  Label screens only
                                  if screen has no online name.  Center label one line below
                                  screen; use 12 point bold Modern.


**Field names**                   Use mixed case 10 point Modern.

                                  Example: Their Ref.    Enter the name . . .


                                              11/5/88                          **IBS Manuals**

*Figure C.17*    *Continued*

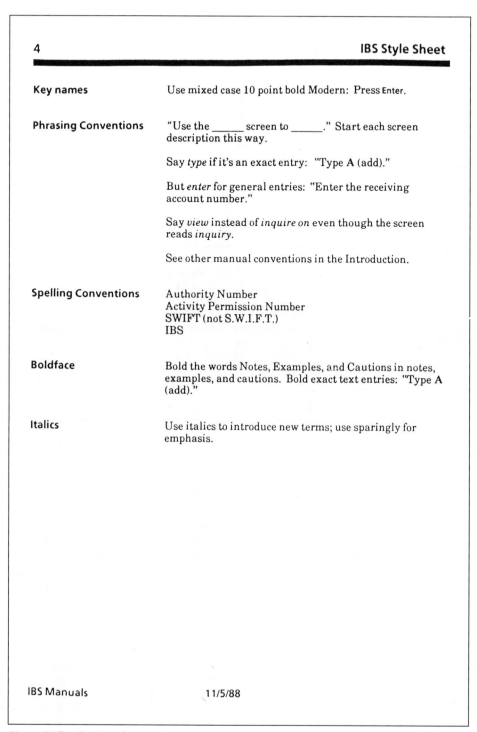

**Key names**                    Use mixed case 10 point bold Modern:  Press Enter.

**Phrasing Conventions**         "Use the _____ screen to _____."  Start each screen
                                 description this way.

                                 Say *type* if it's an exact entry:  "Type **A** (add)."

                                 But *enter* for general entries:  "Enter the receiving
                                 account number."

                                 Say *view* instead of *inquire on* even though the screen
                                 reads *inquiry.*

                                 See other manual conventions in the Introduction.

**Spelling Conventions**         Authority Number
                                 Activity Permission Number
                                 SWIFT (not S.W.I.F.T.)
                                 IBS

**Boldface**                     Bold the words Notes, Examples, and Cautions in notes,
                                 examples, and cautions.  Bold exact text entries:  "Type **A**
                                 (add)."

**Italics**                      Use italics to introduce new terms; use sparingly for
                                 emphasis.

IBS Manuals                              11/5/88

*Figure C.17*   *Continued*

# 15

# Online Documentation

Online documentation is information displayed on a screen to explain how to use the system or device displaying it. The purpose of most online documentation is to provide communication between a computer's software or application program and its users. Status messages, prompts, error messages, help text, and online tutorials are all examples of online documentation.

Different kinds of readers need different types of online documentation. Inexperienced users may need a tutorial, for example, whereas experienced users may need only quick references and help text. Good online documentation not only meets the needs of its readers but accommodates their information-seeking styles as well (Girill, Luk, and Norton 111).

For the technical writer who must devise it, the restrictions of online documentation can be severe. As Annette Bradford puts it:

> Since technical writing represents a special subset of the craft of writing, it has traditionally operated under a tighter set of constraints than other prose forms. Technical writing requires more abundant use of white space, for example, the substitution of visual schemes (like bulleted lists) for verbal ones, and more liberal use of heads and overt pointers . . . . All of the tighter constraints imposed by technical writing

are magnified when the medium of expression moves from the printed page to the display screen. (13)

The primary challenge of creating online documentation—the challenge that separates it from creating printed documentation—is to provide readers with accurate information exactly when and where they need it.

## TYPES OF ONLINE DOCUMENTATION

In its broadest sense, online documentation is anything displayed on the screen—such as menu options, error messages, and prompts—that helps the readers perform a task. Among the most common types of online messages are quick references, requests for assistance, and tutorials.

### Quick References

A quick reference jogs the memory. It allows readers to look up information about a specific topic. For example, if a reader has forgotten the syntax required to enter a particular command, the system displays the appropriate information on request. The needed information is at the fingertips of the reader, who needs only trigger its appearance on the display screen. Quick-reference documentation differs from requests for assistance and tutorials primarily in level of detail and length of the descriptions. Figure 15.1 provides an example of documentation that readers may consult if they forget the number assigned to a function that they want to perform. The displayed information lists these codes and the corresponding function names. In addition, the reader may quickly obtain a description of any of these functions by simply entering its corresponding code number.

Quick-reference information may be displayed graphically as well as verbally; for example, a technical writer might use this type of online documentation to display the layout of function keys on a computer keyboard in graphic form. An example of this type of graphic is shown in Figure 15.2.

### Requests for Assistance

Readers are most likely to request assistance when they get stuck in a program, get lost, or receive a series of unintelligible error messages. This is the most common type of online documentation, and it is context-sensitive; in other words, it provides readers with information based on what the reader is doing when the Help function is accessed. When a reader requests assistance for an error message, such as **OPEN ON ANOTHER TERMINAL**, the system displays information explaining why the error occurred and how to recover from it. For example, the Help message for this error might be this:

> You may not be open on two terminals at the same time. Before opening this terminal, perform a CLOSE on the other terminal. See your Operator's Manual for more information. Press CLEAR to exit HELP.

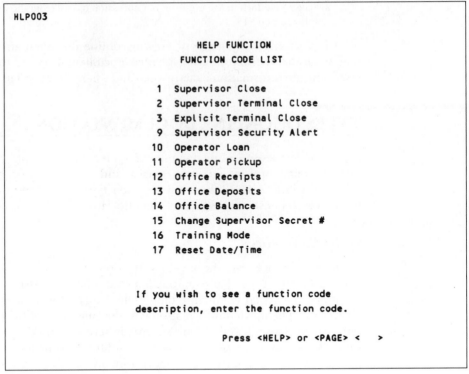

```
HLP003


                        HELP FUNCTION
                      FUNCTION CODE LIST

              1   Supervisor Close
              2   Supervisor Terminal Close
              3   Explicit Terminal Close
              9   Supervisor Security Alert
             10   Operator Loan
             11   Operator Pickup
             12   Office Receipts
             13   Office Deposits
             14   Office Balance
             15   Change Supervisor Secret #
             16   Training Mode
             17   Reset Date/Time



           If you wish to see a function code
           description, enter the function code.


                   Press <HELP> or <PAGE> <    >
```

***Figure 15.1***    A Quick-Reference Display

A Help system should provide readers with the information they need to operate the program efficiently, and it should do so without disturbing or limiting the readers' activities. Leventhal explains requests for assistance as follows:

> Computer programs can use one or more types of Help. The most common type is menu-driven Help, in which a user selects from a list, or menu, of topics. After the user presses the Help function key, a menu is displayed, allowing the user to select a topic. When the user presses the Enter key, the Help information replaces the menu.

> A second type is command-driven Help. Instead of selecting topics from a menu, experienced users save time by entering HELP plus a topic.

> A third type, query-in-depth, allows users to read only the level of detail needed. For example, after viewing the first level of Help information, the user enters MORE to view even more detailed information . . . .

> Still another type is context-sensitive Help. For example, if the user presses the Help function key immediately after an error message, the computer responds with a Help screen explaining the message and an appropriate action. (2)

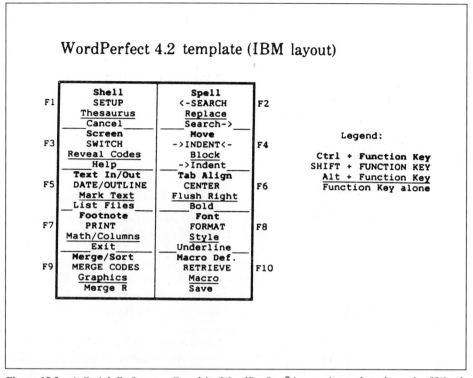

***Figure 15.2***   A Quick-Reference Graphic (WordPerfect® is a registered trademark of Word-Perfect Corporation.)

Help systems may also include sets of screens containing brief summaries of commands, tables for quick reference, and descriptions of system features (Mullins 18), and they may be used to pass the following types of information along to readers:

- The meaning of each option on a menu
- The definition of a highlighted field on a screen
- The steps required to complete a function or task in progress
- The purpose of each function key on a computer keyboard
- The meaning of a status line displayed on the screen

## Tutorials

Online tutorials, also called computer-based training (CBT), can be a set of lessons that users read on the display screen, much like lessons presented in a book, or a set of lessons with online interactive exercises that ask readers to respond to questions displayed on the screen.

Online tutorials are often an ideal way to teach novice users how to use the program. They provide the novice user with some hands-on experience, but they seldom go beyond the basics of a program and therefore are not likely to assist users in exploring its full capabilities (Duffy and Langston 5).

Online tutorials should use a structure that is both obvious and consistent, and they should be designed to give the readers control over their own learning. In other words, they should allow readers to select the lessons they want in the order in which they want to take them.

## Hypertext

Any discussion of online documentation would be incomplete without some mention of *hypertext*. The term *hypertext* refers to a broad, far-reaching application of online documentation—what William Horton, in *Designing and Writing Online Documentation*, calls "a catchword for a grab-bag of various features" (290).

Hypertext is writing designed to provide its reader with many choices, offering numerous informational "paths" that the reader may select. Ken Davis compares hypertext to a well-documented reference work, with detailed foot-notes that provide additional, related information; he uses the example of the *Encyclopaedia Britannica 3*, with its elaborate system of cross-references and varied methods of accessing information (20).

The fact that it is online instead of written documentation means that hypertext can greatly increase the number of possible informational paths and can also allow a reader to select and use various paths much more quickly and easily. According to Davis, a good computer-based hypertext "gives its reader the feel of moving effortlessly through a transparent information environment, like a fish in a sea of knowledge" (20). He provides the following on-the-job example of the use of hypertext:

> Imagine yourself a department manager at an insurance company, going through your morning's electronic mail on your computer system. One memo, from your vice president, concerns a bill introduced in your state legislature that may affect your department. The vice president asks you to study the bill and its implications, and so she has linked her memo to an ever-growing company hypertext on insurance law.
>
> You want to begin by looking at the current bill itself, so you move the cursor on your screen to the name of the bill in the vice president's memo, and you click the button on your mouse. The memo disappears and is replaced by a summary of the bill. As you read the summary, you realize that its third sentence is the potentially important one for you, so you move your cursor to that sentence and click again. The sentence is instantly replaced by the actual text of that section of the bill.
>
> As you read the section, a word troubles you, so you click on the word. A "window" opens on the right side of your screen, with the word's legal definition. You click again, return to the text, and continue reading. An asterisk signals a hidden com-ment, and you click on the asterisk to make the comment visible. It's a note from your corporate attorney, suggesting a precedent in another state's regulations; you click on his note and see the regulation he mentions. And so on. (20)

# DETERMINING WHEN TO USE ONLINE DOCUMENTATION

Documenting a complex system or product in printed form may require several binders of information, and sifting through such a mountain of information may leave readers more confused and frustrated than enlightened. (Many readers, because of such experience in the past, use manuals only as a last resort, when all else fails.) Online documentation, by contrast, provides the reader with specific information exactly when and where it is needed most—on the screen.

Printed documentation and online documentation both have advantages. In determining whether to use online documentation instead of printed documentation or whether to use a combination of both, you need to be fully aware of all the advantages and disadvantages of both.

Printed documentation has the following advantages.

- Readers are familiar with it and comfortable using it.
- It is tangible; readers can touch it, highlight it, and make notes on it.
- It is portable; readers can take it with them.
- It provides retrievability aids—tables of contents, section headings, tabs, cross-references, and subject indexes.
- It is easier to read (Schell 2).
- Its appearance and readability are easy to control with page sizes, colors, graphics, type fonts, and print sizes.

Online documentation has the following advantages.

- It is available when the readers need it; they do not have to leave the screen to go search for a manual.
- It does not force readers to page through material they don't need to read. It can provide the exact information for which readers are looking at the exact moment they need it.
- It can be interactive; the reader learns the system more quickly by using it than by merely reading about it.
- It helps ensure that all readers have the same information; the developer does not have to assume that the reader has purchased the correct manuals and knows where to look for the information.
- It can be kept accurate and up to date more easily because it is distributed with the product and can be updated as the product changes; the production process normally required for printed documentation (typesetting, paste-up, and printing) is not required for online documentation. This benefit may provide greater customer satisfaction and reduce the number of calls to customer support groups.
- It is less expensive to produce, distribute, update, and maintain.
- It takes up no space on a desk or shelf.

According to Mulcahy, "Deciding whether or not documentation should be printed or online involves consideration of the user's background, the goal of the user in performing a specific task, and the constraints of the media" (62).

Factors about users' backgrounds that the technical writer should take into account include their education, learning expectations, level of ability, type of job, experience with computers, and any special or unusual features of their work environment. Such information will help the technical writer determine whether printed or online documentation would be more appropriate. If the readers are new to computers, they may be intimidated by online documentation and feel more comfortable with printed material; in contrast, experts could be annoyed if they had to look up printed information when it would be more accessible as online documentation (Mulcahy 62). The novice, who must learn the program and how to use it, may be best served by a combination of printed and online documentation, whereas the expert is more likely to need interactive problem-solving information of the type ideally suited for online documentation.

In addition to analyzing readers and their tasks, the writer should also analyze the information being presented in deciding whether to use printed or online documentation. Here are some general guidelines (Rockley 59):

- *Quick-reference and tutorial information* is well-suited to online documentation because it allows readers to interact with the computer program and to get help quickly.
- *Installation and service types of information* are better suited to printed documentation because the computer program is usually not operating when this information is needed.
- *Reference material* that includes the theory and philosophy of the design of a product is best presented as printed documentation because online documentation would offer no advantage.

In short, if the readers are experts, if the task is easy, or if the information is simple, online documentation is appropriate. If the opposite is true, online documentation is considerably less likely to be appropriate. For some situations, however, both printed and online documentation might be appropriate. A combination of both could, for example, provide readers with online documentation consisting of summaries, guides, and quick references *and* with printed documentation consisting of more detailed information, graphics, and examples (Waite 1).

## REQUIREMENTS OF ONLINE DOCUMENTATION

Although online documentation requires the same level of audience analysis, research, oganization, formatting, review, and testing as printed documentation, simply placing the printed material online will not satisfy your readers' needs. Online documentation has a number of unique requirements that must be met.

Joseph Dumas asserts, "You will have to be much more clever in organizing, formatting, and choosing your words to create effective online documentation than you would in conveying the same information on paper" (124). The reason is that in designing any online documentation, you must consider the legibility of characters on the screen, the size of the screen, and readers' unfamiliarity with reading on a screen.

- *Screen legibility.* Because of the limited legibility of characters on a screen, readers generally do not read displayed information as quickly as printed information.
- *Screen size.* Most screens are limited to 24 lines of text; at best, screens can display about one-third of an 8½-by-11-inch page. Some of the screen may be used for command lines, status lines, and/or messages, further reducing the amount of space available to display online documentation.
- *User unfamiliarity.* Readers accustomed to printed documentation may avoid using online documentation because it is unfamiliar.

You can overcome these drawbacks to online documentation by designing screen display formats for optimal effect and by choosing the words that go into them carefully.

## Design Considerations

When you design your online documentation, consider how your readers will access and control the information and how the information should be organized.

### Access Methods

The access method is critically important. According to Duffy and Langston, current computer programs use three mechanisms for accessing information. First, the options available to readers may be visible on the screen in the form of a menu, either on a whole screen or in a separate window. Second, a query system may enable readers to request specific help, either by entering a command or by responding to a menu of help options. Third, the program may initiate help, usually when it recognizes that a system error has occurred (11). Whatever access method you choose to use must be simple; readers will avoid a complex one.

Access methods may also provide readers with the ability to move from one part of the documentation to another quickly and easily. For example, you might design documentation that allows your readers to go from one topic to another by simply returning to the menu and selecting another topic. The example shown in Figure 15.3, by contrast, allows the reader to jump directly from one description to another by simply entering a new function code number.

### Organization

Your readers must be able to follow the organization of your online documentation. If they grasp the overall framework of your organization, they will understand and recall the information more readily because they will recognize each component within the structure and hence know where they are at any point. It is somewhat like the reader of a formal report knowing that the report

```
HLP010


                         HELP FUNCTION
                    FUNCTION CODE DESCRIPTION
                      FC 10  OPERATOR LOAN

         Use this function to loan media (cash, etc.) to operators.
         To perform this function:

            o Enter the number of the operator receiving the loan and
              press <Newline>.

            o Enter the dollar amount for each type of media loaned,
              followed by pressing <Newline> after each amount.

         If a field is not required, skip it using <Newline>.
         When all loan amounts are entered for that operator,
         press <ENTER>; then enter another operator number,
         or press <EXIT> to end this function.




                              End of Description
                              Enter New FC or press <EXIT>  <   >
```

***Figure 15.3***    Example of an Easy Access Method

will begin with a summary, which will be followed by an introduction, which will be followed by the body of the report, which will in turn be followed by conclusions and recommendations and finally by appendixes. Because the readers know what to expect, they can read more efficiently (Soderston 72).

Make certain that you give readers the correct level and amount of information. Where one reader may need detailed step-by-step instructions, another may need only an overview. Nothing is more frustrating to readers than to be buried beneath an avalanche of information when they need only a small part of it (Houghton 128).

Here are some pointers to help make your online documentation more usable:

- Include instructions for ending the online documentation, getting more documentation, or moving on to the next topic on the screen so that readers always know what to do next.
- Use white space on the display screen.
- Use emphasis devices carefully; be especially conservative in your use of blinking features and color highlighting because they can be extremely distracting (Bradford 14).

- Format your display so that readers can recognize at a glance what is happening.
- Place messages on the screen with the problem they are addressing (Houghton 131). This will make the messages less disruptive.
- Do not force your readers to make too many selections to reach the information they need.
- Make the form of the interaction between readers and the program consistent in every situation. One way to ensure consistency is to define several possible situations and make certain that in all of them the interaction between readers and the program is the same. If you encounter an exception to the standard (that is, a situation that does not allow for the prescribed interaction), either redesign the interaction or define allowable exceptions to the standard (Rockley 60).

It is also important to establish standards for documentation formats and layouts that are similar to standards for printed documentation (see Chapter 14). For example, the following might be a set of format and layout standards for one type of documentation:

- Highlight titles at the top of each screen.
- Include a general description at the beginning of each document.
- Use one-fourth of the screen for standard callouts (labels) for each type of topic.
- Use three-fourths of the screen for descriptive text.
- Use consistent spacing between paragraphs.
- List related topics at the end of each document.

Other considerations can also make your information more readable:

- Limit each line to no more than 60 characters.
- Start each sentence on a new line rather than presenting information as a solid block of text.
- Place a blank line between paragraphs or blocks of text.
- Avoid hyphenation.
- Use upper- and lowercase letters in text, but use all uppercase letters for screen titles.
- Use several screens; don't pack a screen densely with information just to fit all of the information on one screen.
- If several screens are required to explain a particular topic, let readers know that more information follows by (1) using ellipses to show the continuation of a thought onto another screen, (2) using **Page *x* of *y*** to let the reader know where he or she is in a series of screens, or (3) using a prompt to tell your reader that more information exists (for example, **Press ENTER for more**). At the end of such a series of screens, let readers know when they have reached the end (for example, **Press EXIT to return to the program**).
- Indent paragraphs.
- Make each heading level distinct by (1) using capital letters, (2) using

underlining, or (3) positioning heads differently (for example, left-justified versus centered).

- Make the format of screens consistent; place screen titles, prompts, messages, and other recurring elements in the same place every time. This practice provides readers with a degree of familiarity and assures them that certain types of information can always be found in the same place.
- Use a single-column format.

## Writing Style Guidelines

Remember that text on the screens of most office terminals can be quite difficult to read. This means that your on-screen writing style is especially important. Every line must count in online documentation because brevity is a necessity. In addition, you must always be careful to write at the reader's level of knowledge. Among the things you can do to aid your readers are the following:

- Write online documentation in plain English; avoid jargon.
- Write directly to readers in the second person, addressing them as "you."
- Write in the active voice.
- Do not use the reader's name in interactive messages because that practice has a negative effect on readers.
- Describe procedures in a logical order, usually the order in which readers must execute them.
- Use a conversational tone to make your readers feel more at ease.
- Provide your readers with input examples to help them avoid entry errors.
- Use bullets and numbered lists to make it easy for readers to scan and find important information on the screen.
- Consider using tables and charts instead of straight text.
- Be precise and concise; because of the limited space available on the screen, it is critical that you make maximum use of it.
- Refer to keys, functions, and commands in a consistent way, using the same format each time; this consistency should apply throughout the online documentation and any supporting printed documentation.
- Avoid using abbreviations.
- Avoid using acronyms; if you have to use them, spell them out the first time they appear.

# GETTING YOUR ONLINE DOCUMENTATION INTO THE SYSTEM

Once you have decided to provide online documentation and have made general decisions about the basic design on your online documentation (access method, user control, organization, and so on), you must get the written material into the system. Because the characteristics and capabilities of every system are different, how you accomplish this varies greatly from one system to the next (for example, does your system have windowing capabilities, or will you be replacing a working screen with your online documentation?). We cannot provide step-by-step instructions for every system within this chapter. We can, however, look at some general requirements that would be common to any system.

The first thing you must do is create screens from your text. The layout and design of each screen must be consistent and should be as helpful to your reader as possible. For example, each screen should include some type of navigational aid, such as a title. The format and structure of the text used is often based on the software that will be displaying it. For example, a system without windowing capabilities may require that the screen be divided into a number of two-line records for display purposes. The system may further limit each line to a specific number of alphanumeric characters, usually 60. This allows for a margin of 10 spaces before and after the text.

As the writer, you assign the sequence in which the screens—and the records on each screen—should appear, as well as the navigational aids to be used. You also determine such things as whether the reader will be able to page forward only or to page back and forth (the programmer will implement these decisions in the computer program, but you must make the decisions and then explain to the programmer what you want done).

Before you turn material over to the programmer to integrate into the software, you must convert your text to a computer file that your particular computer system can interpret and process. Then turn your converted file over to the programmer, along with specific instructions about what you would like done.

Figures 15.4 through 15.6 show examples of documentation written for a large computer program. The purpose of this documentation is to explain to retail store supervisors how to perform selected functions on a computer terminal. Most of the supervisors have experience in the retail industry and with computer terminals but are not familiar with this particular retail program.

The Publications Department responsible for documenting this retail application program decided to provide both printed and online documentation. The printed documentation was to serve as both a training tutorial and a reference tool for the reader, and the online documentation was to provide immediate assistance. The same writing style and tone were used in both types of documentation to help the reader feel comfortable with both.

Figure 15.4 illustrates the printed documentation written to support the Operator Loan function. Figure 15.5 illustrates the online text in the form in which the writer submitted it to the programmer (to be integrated into the computer software). Figure 15.6 illustrates the actual screen that is displayed when Help is accessed for the Operator Loan function.

Once the printed documentation has been written and verified, the writer must create the corresponding online documentation. For all the reasons discussed in this chapter so far, the writer reduces three pages of written text to the concise online documentation shown in Figure 15.5. This is not exactly what the reader will see, but what the writer must submit to the programmer (the on-screen version is shown in Figure 15.6). Notice that the title at the top of the

## OPERATOR LOAN (10)

Transfers controlled media to an operator from the office funds.

☞ This function cannot be performed if the Office Balance (FC 14) function is being processed. If an attempt is made to perform this function and processing of the Office Balance function is not yet complete, a message is displayed asking that you try the selected function at a later time.

☞ This function cannot be performed while in training mode.

### FOLLOW THESE STEPS

1. At the TERMINAL OPEN (SUP001) screen, enter the function code and press <ENTER> or follow the menu screens shown in Figure 5-2.

   • The OPERATOR LOAN (SUP024) screen is displayed:

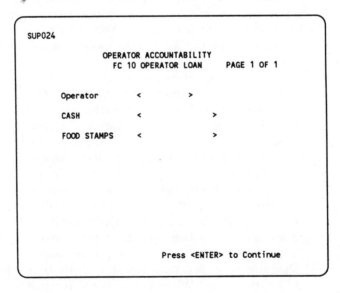

```
SUP024

              OPERATOR ACCOUNTABILITY
              FC 10 OPERATOR LOAN      PAGE 1 OF 1


       Operator      <          >

       CASH          <            >

       FOOD STAMPS   <            >








                     Press <ENTER> to Continue
```

2. Enter the number of the operator and the loan amount. After your entry is made, press <PAGE>.

*Figure 15.4*   Printed Documentation (Courtesy of NCR Corporation)

CHAPTER 5
OPERATOR ACCOUNTABILITY CONTROL
FUNCTIONS

Example:  If you need to loan $500.00 in cash and $50.00 in food
stamps to Operator 4, do the following:

(1) Enter 4 at the Operator prompt and press <New Line>.
(2) Enter 50000 at the Cash prompt and press <New Line>.
(3) Enter 500 at the Food Stamps prompt and press <New
Line>.
(4) Press <New Line> to bypass lines not needed; the <Down
Arrow> to go fields on the same line.
(5) After all entries are made, press <PAGE>.

- The following is an example of what is displayed:

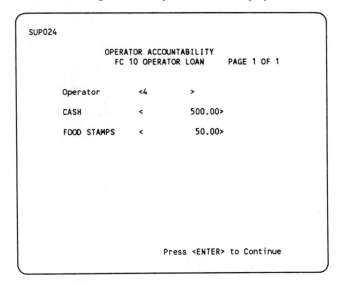

```
 SUP024

                    OPERATOR ACCOUNTABILITY
                    FC 10 OPERATOR LOAN        PAGE 1 OF 1


        Operator        <4          >

        CASH            <       500.00>

        FOOD STAMPS     <        50.00>




                            Press <ENTER> to Continue
```

3. After the entries have been validated using the <PAGE> key,
press <ENTER>.

4. To enter another operator and loan information, repeat Steps 2
and 3.

5. After you have made all loans, press <EXIT> to end this
function.

**Figure 15.4**  *Continued*

- If you accessed this function using Function Codes, the TERMINAL OPEN (SUP001) screen is displayed.
- If you accessed this function using the menu screens, the OPERATOR ACCOUNTABILITY (SUP005) screen is displayed.
- The office and operator totals are adjusted.
- The following is an example of what is printed:

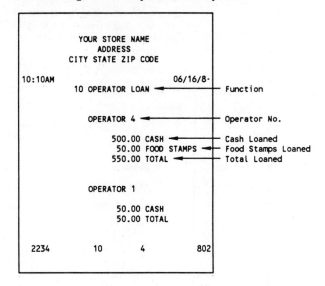

```
              YOUR STORE NAME
                  ADDRESS
            CITY STATE ZIP CODE

10:10AM                       06/16/8-
        10 OPERATOR LOAN  ◄───────────── Function


        OPERATOR 4  ◄─────────────── Operator No.


            500.00 CASH  ◄─────────── Cash Loaned
             50.00 FOOD STAMPS ◄──── Food Stamps Loaned
            550.00 TOTAL  ◄────────── Total Loaned


        OPERATOR 1

             50.00 CASH
             50.00 TOTAL


2234         10        4          802
```

NOTE:   In case of an incorrect entry, this function can be voided. Repeat the function inserting a minus (-) sign in front of the media entries only.

***Figure 15.4***   *Continued*

```
First Record
#············>200100
                        FC 10  OPERATOR LOAN


    200101
    Use this function to loan media (cash, etc.) to operators.
    To perform this function:
    200102


      o Enter the number of the operator receiving a loan and
    200103
        press <Newline).


    200104
      o Enter the dollar amount for each type of media loaned,
        followed by pressing <Newline> after each amount.
    200105


    If a field is not required, skip it using <Newline>.
    200106
    When  all loan amounts are entered for that operator,
    press <ENTER>; then enter another operator number,
    200107
    or press <EXIT> to end this function.
```

***Figure 15.5***    Screen Display

screen provides a navigational aid. Notice also that the screen is divided into one- or two-line records that are numbered sequentially (200100 through 200107).

Figure 15.6 shows the final online documentation screen that describes the Operator Loan function. This screen is displayed when the reader accesses the Help function and requests information for Function Code 10. In this example, the reader can access Help for different function codes (FC) by entering a new function code between the angle brackets ( < > ) at the bottom of the screen.

The Operator Loan (Function Code 10) screen in Figure 15.6 is one in a series of screens that can be displayed when the Help function is accessed by a user of the retail application program. Each screen contains the following information:

- Screen number (this number is raised by one as each additional Help screen is displayed if there are more than one screen)
- Help function code and title
- Description of the function, including the appropriate procedure
- **Press [PAGE] for more Help** if more than one screen is available
- **End of description** or **End of Help** when the end of the Help text is reached
- A message telling the reader what to do next (for example, **Enter new FC or press [EXIT]** or **Press [EXIT] to leave Help**).

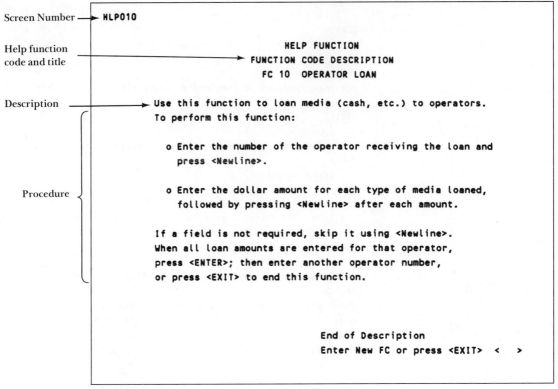

Screen Number

Help function
code and title

Description

Procedure

```
HLP010

                              HELP FUNCTION
                       FUNCTION CODE DESCRIPTION
                         FC 10   OPERATOR LOAN

Use this function to loan media (cash, etc.) to operators.
To perform this function:

    o Enter the number of the operator receiving the loan and
      press <Newline>.

    o Enter the dollar amount for each type of media loaned,
      followed by pressing <Newline> after each amount.

    If a field is not required, skip it using <Newline>.
    When all loan amounts are entered for that operator,
    press <ENTER>; then enter another operator number,
    or press <EXIT> to end this function.




                              End of Description
                              Enter New FC or press <EXIT>  <   >
```

*Figure 15.6*   Help Text

*Checklist 15.1*

# Determining When to Use Online Documentation

Consider the following questions in determining whether printed documentation, online documentation, or a combination of the two will best meet the needs of readers.

☐ What type of product are you describing?
  • Does the product require user interaction, or is it simply computer-driven?
  • How will the reader use the product?

☐ What kind of information do users need in order to use the product correctly?
  • Reference material
  • Procedural instructions

☐ Is it feasible to put this information online, or will users be overwhelmed with the information and revert to printed documentation anyway?

☐ Will placing the information online enhance its usability and usefulness?

☐ Will placing it online aid in the learning of the information?

☐ What is the goal of this information?
  • Quick reference
  • Help in solving problems and responding to errors

☐ Who is the intended audience?
  • Are they computer experts or computer novices?
  • Are they occasional or frequent users? (Will they need long descriptions or simply memory joggers?)
  • How are users going to use the documentation provided—in conjunction with using the product, or as a training tool?

*Checklist 15.2*

## Requirements of Online Documentation

☐ Consider important design decisions for the online documentation
  • How will readers access and control information?
  • How can the material be organized to best serve readers' understanding?
  • What standards should be established?

☐ Establish an appropriate writing style for the online documentation

### WORKS CITED

Bradford, Annette Norris. "Conceptual Differences between the Display Screen and the Printed Page." *Technical Communication* 31.3 (1984): 13–16.

Davis, Ken. "Hypertext for Business Communication: An Introduction and Bibliography." *The Bulletin*, Dec. 1989: 20–21.

Duffy, Thomas M., and M. D. Langston. *On-Line Help: Design Issues for Authoring Systems.* Communications Design Center Report No. 18. Pittsburgh: Carnegie Mellon U, 1985.

Dumas, Joseph S. *Designing User Interfaces for Software.* Englewood Cliffs, NJ: Prentice, 1988.

Girill, T. R., Clement H. Luk, and Sally Norton. "Reading Patterns in Online Documentation: How Transcript Analysis Reflects Text Design, Software Constraints, and User Preferences." *Proceedings of the 34th International Technical Communications Conference* (1987): RET-111–RET-114.

Horton, William. *Designing and Writing Online Documentation.* New York: Wiley, 1990.

Houghton, Raymond C. "Online Help Systems: A Conspectus." *Communications of the ACS* 27 (1984): 126–133.

Leventhal, Charlotte. "Literature Review for On-Line Documentation." Student assignment. Milwaukee, 1987.

Mulcahy, Patricia I. "Making Decisions about Online versus Hardcopy: Considerations for Documentation." *Proceedings of the 34th International Technical Communications Conference* (1987): ATA-62–ATA-63.

Mullins, Mary. "Non-tutorial On-Line Training: Read All about It." *Data Training* Apr. 1988: 18–22, 62.

Rockley, Ann. "Online Documentation: From Proposal to Finished Product. *Proceedings of the 34th International Technical Communications Conference* (1987): ATA-58–ATA-61.

Schell, David A. *Simply Stated* Feb. 1986: 2.

Soderston, Candace. "An Experimental Study of Structure for Online Information." *Proceedings of the 34th International Technical Communications Conference* (1987): ATA-71–ATA-74.

Waite, Bob. "What to Put Online, What to Print, and Why." *Proceedings of the 34th International Technical Communications Conference* (1987): WE-1–WE-2.

## FURTHER READING

Barrett, Edward, ed. *Text, Context, and Hypertext*. Cambridge, MA: MIT Press, 1988.

Bell, Lydia V. "The Design and Implementation of Online Specifications." *Proceedings of the 34th International Technical Communications Conference* (1987): ATA-129–ATA-132.

Brockmann, R. John. *Writing Better Computer User Documentation: From Paper to Online*. New York: Wiley, 1986.

Dansereau, Mary E. "Creating an Online Index." *Proceedings of the 34th International Technical Communications Conference* (1987): ATA-105–ATA-107.

Duffy, Thomas M., et al. *Writing Online Information: Expert Strategies*. Communications Design Center Report No. 37. Pittsburgh: Carnegie Mellon U, 1988.

Fisher, Lou, et al. "Online Information: What Conference Attenders Expect (A Report of a Survey)." *Technical Communication* 34 (1987): 150–155.

Frisch, Kathryn L. "The Convex Info System: An Online Information Index." *Proceedings of the 34th International Technical Communications Conference* (1987): ATA-108–ATA-110.

Garner, Kathleen H. "Checklists for Online Writing." *Proceedings of the 34th International Technical Communications Conference* (1987): ATA-26–ATA-29.

Henderson, Allan, and Annette N. Bradford. "Online Information: A Practical Approach." *Proceedings of the 31st International Technical Communications Conference* (1984): WE-154–WE-157.

Horton, William. "Myths of Online Documentation." *Proceedings of the 34th International Technical Communications Conference* (1987): ATA-43–ATA-46.

———, ed. "The Wired World." *Technical Communication* 36 (1989): 152–153.

McKita, Martha. "Online Documentation and Hypermedia: Designing Learnability into the Product." *IEEE Professional Communication Society Conference Record* (1988): 301–305.

Murray, Philip C. "Hypertext: New Power to the Technical Communicator." *Proceedings of the 36th International Technical Communications Conference* (1989): CC-8–CC-10.

Petrauskas, Bruno F. "Online Documentation: Putting Research into Practice." *Proceedings of the 34th International Technical Communications Conference* (1987): ATA-54–ATA-57.

Ramey, Judith. "Developing a Theoretical Base for On-Line Documentation. Part I: Building the Theory." *Technical Writing Teacher* 13 (1986): 148–159.

Ridgway, Lenore S. "Read My Mind: What Users Want from Online Information." *IEEE Transactions on Professional Communication* (1987): 87–90.

Rubens, Phillip, and Robert Krull. "Application of Research on Document Design to Online Displays." *Technical Communication* 32.4 (1985): 29–34.

See, Edward J. "Documentation: Online or Printed?" *Proceedings of the 34th International Technical Communications Conference* (1987): ATA-51–ATA-53.

Semple, Marline C. "The Electronic Blue Pencil: Editing Online Information." *Proceedings of the 34th International Technical Communications Conference* (1987): ATA-140–ATA-142.

# Appendix

Introduction to Using
the International
Banking System

International Banking System

## Introduction to Using the
# INTERNATIONAL BANKING SYSTEM

## M A N U A L

FIRST
WISCONSIN
NATIONAL
BANK

First Edition
February, 1988

---

FIRST WISCONSIN ● MILWAUKEE          **MEMORANDUM**

To:       International Banking Division Personnel
From:     Information Services - Technical Writing
Date:     February 2, 1988
Subject:  New International Banking System Manual

Here is your new manual, *Introduction to Using the International Banking System*, which covers basic skills all International Banking personnel need. We hope you find it useful as a guide to your new system.

This manual is the first in a series. As additional parts of the International Banking System become functional, you will receive accompanying manuals that explain how to use each new function.

mm

## System Standards

## Using Menus

## Glossary

---

# Table of Contents

## Introduction

## System Overview

## Using a Password

## Logging On and Off

# Introduction

**What's in this section?**

▸ the purpose of this manual and its relation to the other International Banking System manuals

▸ descriptions of the chapters in this manual

▸ an explanation of the manual's page format and other manual conventions

---

# Introduction

**1**

**Purpose of manual**

To provide an overview of the entire International Banking System (IBS) and describe the basic skills needed to use the system. We hope you find this manual useful as a guide to your system.

**Basic services**

As you know, an international banking department provides many different services–often as many services as a domestic bank. The International Banking System assists your work in any of these service areas:

*Letters of credit*

Instruments issued on behalf of a buyer that give the buyer the financial backing of the issuing bank.

*Loans*

Credit extended to businesses and other banks for the financing of inventory purchases and trade.

*Savings*

Deposit instruments such as open and term time deposits, certificates of deposit, and interbank placements.

*Current accounts*

Running accounts that reflect the movement of cash, loan reserves, and settlements between banks.

*Foreign exchange trading*

The buying and selling of currencies in relation to either U.S. dollars or other foreign currencies.

*Payments*

The receipt of payments made via SWIFT, the international bank-to-bank wire service.

*Financial control*

Cash management, individual and global credit risk analysis, and financial control.

# Chapters in this manual

**Chapters in this manual**

Although other manuals will tell you how to use the system to do work in your area of expertise, this manual describes the basic skills in using the system that everyone must have.

Each chapter includes a table of contents and a chapter summary. The chapters are:

*System Overview*
An overview, designed for managers, of the entire system. The overview includes descriptions of the system's components, benefits, and security features, as well as task flows for typical transactions such as payments.

*Using a Password*
Procedures for establishing or changing your personal password.

*Logging On and Off*
Procedures for beginning and ending a session.

*System Standards*
The "rules of the road" for entering amounts, dates, rates, and other information.

*Using Menus*
Procedures for using *menus*, screens from which you can select available options.

*Glossary*
Definitions of banking and data processing terms that appear in these manuals.

*Quick reference card*
The front pocket of this manual contains a card that lists function keys on Tandem terminals and on personal computers. When reading the chapters "Using Menus" or "Logging On and Off," you may find it useful to refer to the Quick Reference Card.

---

# Related manuals

**Related manuals**

In addition to this manual, you will receive one or more of the following manuals, depending on your area of expertise:

*For daily users*

Entering Letters of Credit — Tells how to set up and monitor letters of credit.

Entering Deposits and Loans — Tells how to set up and manage deposit and loan accounts.

Entering Current Accounts — Tells how to set up and manage settlement, interbank placement, and disbursement accounts.

Entering Payments — Tells how to enter and receive payments.

Entering Foreign Exchange Contracts — Tells how to set up and monitor foreign exchange contracts.

Using General Inquiry — Tells how to view and act on previously entered transactions and incoming wires.

*For credit managers*

Interpreting Reports — Describes standard reports in detail and shows how to request customized reports.

*For those who maintain system records*

Entering Customer Records — Tells how to enter customer information, including customer credit lines.

IBS Maintenance Guide — Tells how to perform routine tasks such as changing accounting dates and system rates.

*For operations managers*

## Page format

| | |
|---|---|
| Startup Guide | Describes procedures for setting up an office. |
| Conversion Guide | Describes strategies and procedures for converting existing customer and account information for use in the new system. |

**Page format**

The following information always appears in this format to make it easy for you to locate information:

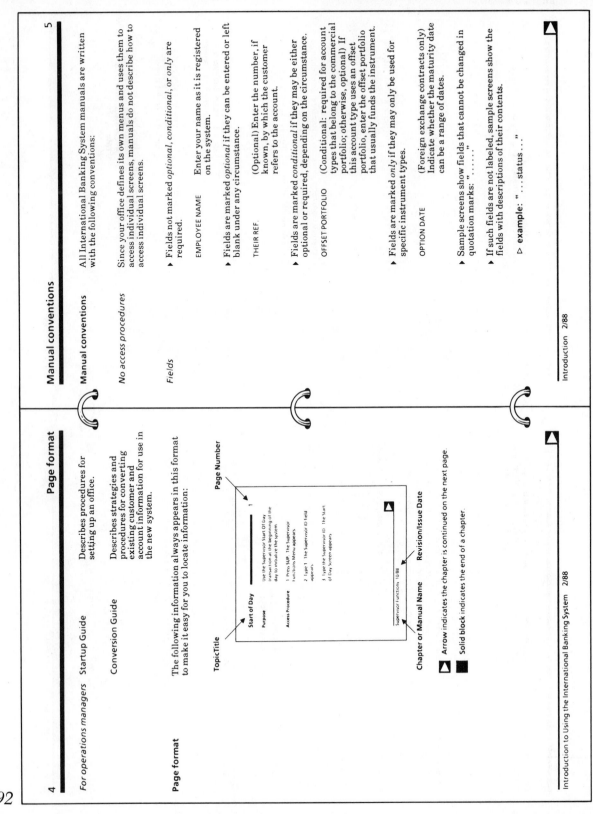

TopicTitle — Page Number

Start of Day

Purpose — Use the Supervisor Start Of Day transaction at the beginning of the day to initialize the system

Access Procedure —
1 Press SUP. The Supervisor Functions Menu appears
2 Type 1. The Supervisor ID field appears
3 Type the Supervisor ID. The Start of Day Screen appears

Supervisor Functions 10/88

Chapter or Manual Name — Revision/Issue Date

▲ **Arrow** indicates the chapter is continued on the next page.

■ **Solid block** indicates the end of a chapter.

---

## Manual conventions

All International Banking System manuals are written with the following conventions:

*Manual conventions*

*No access procedures*

Since your office defines its own menus and uses them to access individual screens, manuals do not describe how to access individual screens.

*Fields*

▶ Fields not marked *optional*, *conditional*, or *only* are required.

EMPLOYEE NAME — Enter your name as it is registered on the system.

▶ Fields are marked *optional* if they can be entered or left blank under any circumstance.

THEIR REF. — (Optional) Enter the number, if known, by which the customer refers to the account.

▶ Fields are marked *conditional* if they may be either optional or required, depending on the circumstance.

OFFSET PORTFOLIO — (Conditional: required for account types that belong to the commercial portfolio; otherwise, optional) If this account type uses an offset portfolio, enter the offset portfolio that usually funds the instrument.

▶ Fields are marked *only* if they may only be used for specific instrument types.

OPTION DATE — (Foreign exchange contracts only) Indicate whether the maturity date can be a range of dates.

▶ Sample screens show fields that cannot be changed in quotation marks: " . . . . . . "

▶ If such fields are not labeled, sample screens show the fields with descriptions of their contents.

▷ **example:** " . . . status . . . "

## System Overview

**What's in this section?**

▶ a brief description of each component in the system

▶ typical work flows for two common transactions

▶ system features and benefits

---

6

*Keys*

▶ Keys are shown in **boldface** as they are defined on the International Banking System.

  ▷ **example:** Press **POST**.

  Refer to your Quick Reference Card for a listing of key names as they appear on your keyboard.

*Your comments*

We welcome your comments on the format, organization, and content of this manual. If you have any suggestions as to how this manual could be made more useful to you, please send us your ideas on the enclosed Reader's Comment Card.

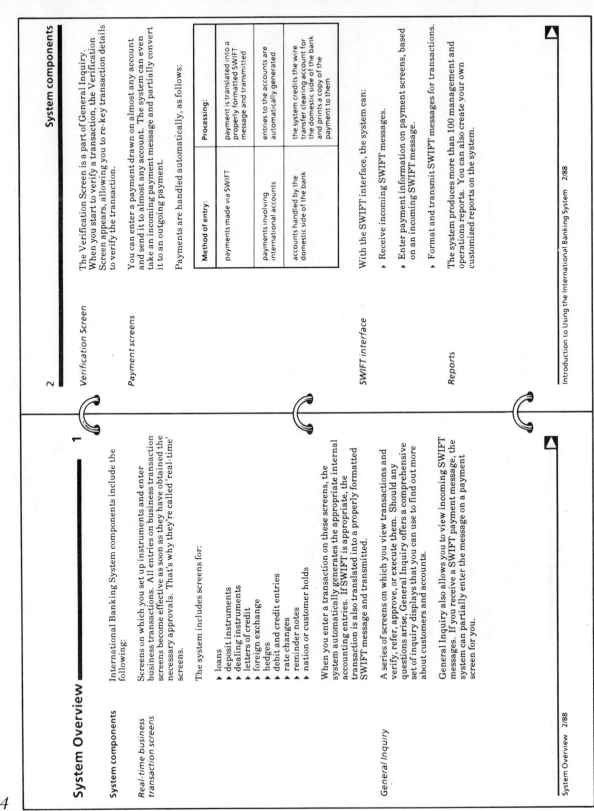

# System Overview

**System components**

International Banking System components include the following:

**Real-time business transaction screens**

Screens on which you set up instruments and enter business transactions. All entries on business transaction screens become effective as soon as they have obtained the necessary approvals. That's why they're called 'real-time' screens.

The system includes screens for:

- loans
- deposit instruments
- dealing instruments
- letters of credit
- foreign exchange
- hedges
- debit and credit entries
- rate changes
- reminder notes
- nation or customer holds

When you enter a transaction on these screens, the system automatically generates the appropriate internal accounting entries. If SWIFT is appropriate, the transaction is also translated into a properly formatted SWIFT message and transmitted.

**General Inquiry**

A series of screens on which you view transactions and verify, refer, approve, or execute them. Should any questions arise, General Inquiry offers a comprehensive set of inquiry displays that you can use to find out more about customers and accounts.

General Inquiry also allows you to view incoming SWIFT messages. If you receive a SWIFT payment message, the system can partially enter the message on a payment screen for you.

---

## System components

**Verification Screen**

The Verification Screen is a part of General Inquiry. When you start to verify a transaction, the Verification Screen appears, allowing you to re-key transaction details to verify the transaction.

**Payment screens**

You can enter a payment drawn on almost any account and send it to almost any account. The system can even take an incoming payment message and partially convert it to an outgoing payment.

Payments are handled automatically, as follows:

| Method of entry: | Processing: |
| --- | --- |
| payments made via SWIFT | payment is translated into a properly formatted SWIFT message and transmitted |
| payments involving international accounts | entries to the accounts are automatically generated |
| accounts handled by the domestic side of the bank | the system credits the wire transfer clearing account for the domestic side of the bank and prints a copy of the payment to them |

**SWIFT interface**

With the SWIFT interface, the system can:

- Receive incoming SWIFT messages.
- Enter payment information on payment screens, based on an incoming SWIFT message.
- Format and transmit SWIFT messages for transactions.

**Reports**

The system produces more than 100 management and operations reports. You can also create your own customized reports on the system.

## Typical work flows

**Customer Records**
Customer records store multiple names and addresses, contacts, wire service IDs, standing settlement instructions, relationships to other customers, and demographic information for each customer.

**Credit Facility Screen**
You can establish revolving credit lines for a customer via the Credit Facility Screen. Credit lines can be shared, if desired, between customers and their parents or affiliates.

**Rate screens**
Rate screens let you maintain three types of system-wide rates which you can change daily: foreign exchange rates, interbank interest rates, and index interest rate symbols. When you change a system-wide rate, the change applies to all transactions using the rate.

**Instruction screens**
Instruction screens define credit management, accounting instructions, authority requirements (who can do what), employee records, and other important options for your office.

**Typical work flows**
The following pages show typical work flows for two common transactions:

▶ Diagram 1: Work Flow for a Payment.

▶ Diagram 2: Work Flow for a New Instrument.

---

### Diagram 1: Work Flow for a Payment

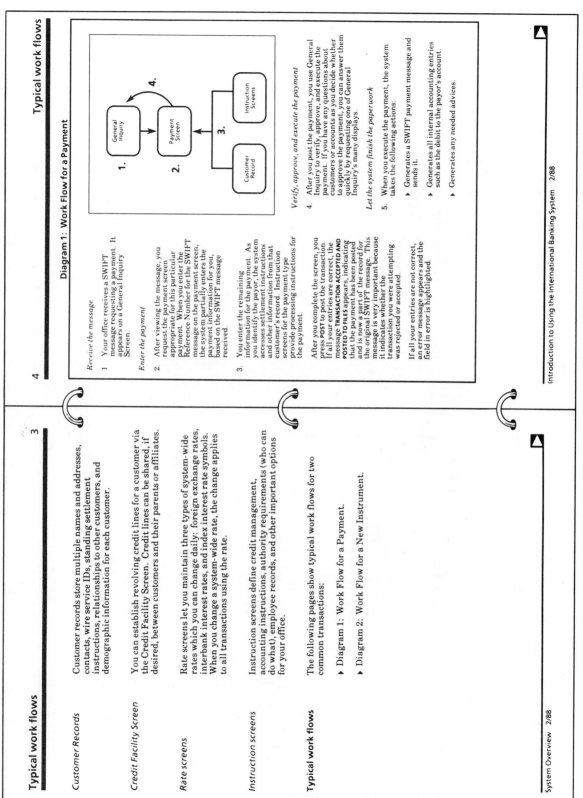

*Receive the message*

1 Your office receives a SWIFT message requesting a payment. It appears on a General Inquiry Screen.

*Enter the payment*

2. After viewing the message, you request the payment screen appropriate for this particular payment. When you enter the Reference Number for the SWIFT message on the payment screen, the system partially enters the payment information for you, based on the SWIFT message received.

3. You enter the remaining information for the payment. As you identify the payor, the system accesses settlement instructions and other information from that customer's record. Instruction screens for the payment type provide processing instructions for the payment.

After you complete the screen, you press **POST** to post the transaction. If all your entries are correct, the message **TRANSACTION ACCEPTED AND POSTED TO FILES** appears, indicating that the payment has been posted and is now a part of the record for the original SWIFT message. This message is very important because it indicates whether the transaction you were attempting was rejected or accepted.

If all your entries are not correct, an error message appears and the field in error is highlighted.

*Verify, approve, and execute the payment*

4. After you post the payment, you use General Inquiry to verify, approve, and execute the payment. If you have any questions about customers or accounts as you decide whether to approve the payment, you can answer them quickly by requesting one of General Inquiry's many displays.

*Let the system finish the paperwork*

5. When you execute the payment, the system takes the following actions:

▶ Generates a SWIFT payment message and sends it.

▶ Generates all internal accounting entries such as the debit to the payor's account.

▶ Generates any needed advices.

## Features and benefits

The International Banking System is:

*Comprehensive*

You can use the system for deposits and loans, letters of credit, foreign exchange, hedges, own acceptances, interbank placements, certificates of deposit, time deposits, and other dealing instruments.

If necessary, you can place a hold on an account, a customer, or even a nation.

The International Banking System fully supports international payments, reimbursement claims, advices-to-receive, and cover payments.

*Integrated*

As you execute each transaction, the system automatically generates any internal accounting entries and any required advices, confirmations, or statements. If the transaction involves a settlement, you can arrange for automatic settlement of the instrument.

If a SWIFT message is appropriate, the transaction is automatically translated into a properly formatted SWIFT message and transmitted for you.

*Fast*

Transactions are processed in "real-time" as soon as they have been executed.

*Tailored to your needs*

Because your office defines even the most basic accounting and processing options, it's easy to adjust the system to meet changing business needs.

*Safe yet flexible*

Risk management is crucial to success in international banking. The International Banking System offers an excellent system of risk management that doesn't limit its flexibility for managers, lenders, or dealers:

▸ Credit management for a full range of credit exposure types, with audit controls for approval of revolving customer credit lines or draws above those lines.

---

### Diagram 2: Work Flow for a New Instrument

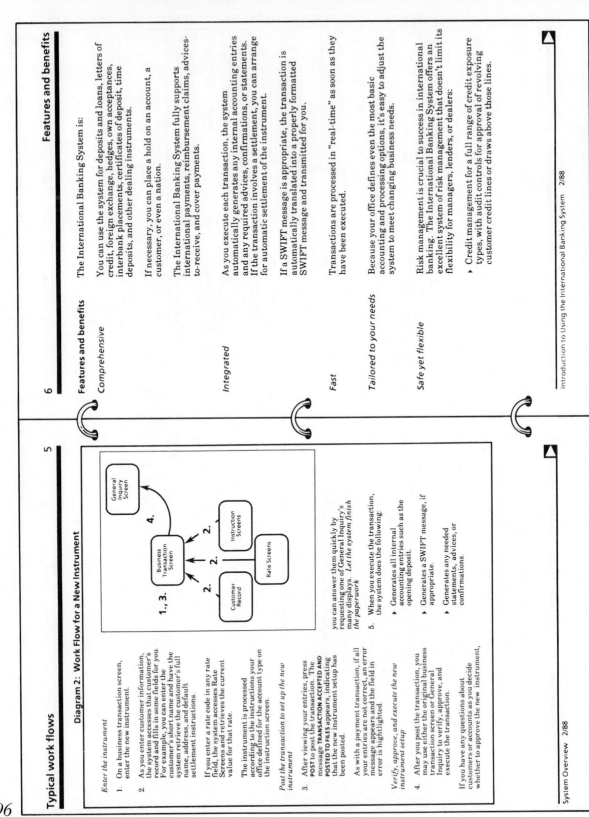

*Enter the instrument*

1. On a business transaction screen, enter the new instrument.

2. As you enter customer information, the system accesses that customer's record and fills in some fields for you. For example, you can enter the customer's short name and have the system retrieve the customer's full name, address, and default settlement instructions.

   If you enter a rate code in any rate field, the system accesses Rate Screens and retrieves the current value for that rate.

   The instrument is processed according to the instructions your office defined for the account type on the instruction screen.

*Post the transaction to set up the new instrument*

3. After viewing your entries, press **POST** to post the transaction. The message **TRANSACTION ACCEPTED AND POSTED TO FILES** appears, indicating that the new instrument setup has been posted.

   As with a payment transaction, if all your entries are not correct, an error message appears and the field in error is highlighted.

*Verify, approve, and execute the new instrument setup*

4. After you post the transaction, you may use either the original business transaction screen or General Inquiry to verify, approve, and execute the transaction.

   If you have any questions about customers or accounts as you decide whether to approve the new instrument,

you can answer them quickly by requesting one of General Inquiry's many displays. *Let the system finish the paperwork*

5. When you execute the transaction, the system does the following:

   ▸ Generates all internal accounting entries such as the opening deposit.

   ▸ Generates a SWIFT message, if appropriate.

   ▸ Generates any needed statements, advices, or confirmations.

## 8

### System security

The International Banking System includes the following security features:

*Passwords*

Each employee must establish a personal password and use it every time he or she logs on. The employee is responsible for all transactions entered while he or she is logged on.

Employees must change their passwords at least every 30 days.

*Consecutive error lockout for passwords*

If an employee makes three consecutive errors while logging on, the employee's password becomes inoperative until a supervisor makes it operative again. The system 'remembers' consecutive errors made on previous days.

*Automatic logoff after inactivity*

To minimize the chance of an unauthorized person using the system, the system automatically logs off terminals that have been inactive for more than a few minutes. (Your office determines the exact amount of time.) A warning appears on the screen before the terminal is logged off. Any incomplete transactions are not saved.

*Authority requirements*

Your office's registration of an employee defines what activities the employee can perform.

▷ **example:** In your office, a letter of credit clerk may be authorized to create letters of credit but not foreign exchange contracts. (Such an employee might still be able to view foreign exchange contracts created by other employees.)

*Authority indexes*

As each employee is registered on the system, he or she is given *authority index numbers* that limit the amount of credit the employee can authorize.

▷ **example:** In your office, a letter of credit clerk may authorize letters of credit for up to $1,000.00, while a credit officer may authorize letters of credit for up to $1,000,000.00.

---

## Features and benefits

### 7

*Safe yet flexible*
(Continued)

▸ The ability to share a revolving credit line between customers and their parents or affiliates, or between the bank's main office, branches, and subsidiaries.

▸ Substantial support for problem loans, including:
- loan restructuring
- reduced-accrual and non-accrual status
- full or partial charge-offs
- recoveries
- refunds to insurers
- continued full accruals as memo postings with no accruals on the balance sheet for reduced-accrual, non-accrual, or charged-off loans
- fixed, floating, or adjustable penalty interest rates
- portfolio segregation
- activity holds and watch lists
- extensive exception reporting

▸ Exposure reporting by nation and area of the world.

▸ Complete management information on profits, yields, volume, backlogs, positions, maturities, and cash flow.

▸ Control over transaction authorizations, especially for large amounts or unusual conditions.

▸ Control of operational risk and security violations.

# Using a Password

**What's in this section?**

▶ definition of a password

▶ description of why a password is important

▶ rules for passwords

▶ how to establish your password

▶ how to change an active password

▶ how to change an expired password

---

# Using a Password

**What is a password?**

A password is a secret word or phrase known only to you that identifies you as an authorized user of the International Banking System.

**Why is it important?**

Your password is important for two reasons: first, because you can't log on without it, and second, because you are responsible for all work done at your terminal while you are logged on. If someone obtains your password or uses your terminal while you are logged on, you are held accountable for the work that person does.

**Password rules**

▶ Never, under any circumstances, let anyone know your password. If anyone asks you for your password, report them to your supervisor or a higher authority.

▶ *Remember your password.* If you forget it, a supervisor will have to delete you from the system, register you again, and then re-assign your authority codes to you before you will be able to use the system.

▶ Use caution when you enter your password. If you make three consecutive errors while entering your password, your password becomes inoperative until a supervisor makes it operative again.

▶ Make your password four to eight characters long. Passwords can include any combination of letters, numbers, blanks, or special characters, but they cannot be completely blank. Do not use an obvious word such as your name, your spouse's name, your pet's name, or your phone number as your password.

▶ Change your password often to keep it secure.

## Establishing your password

### Establishing your password

### Summary of procedure

The first time you use the system, you must establish your personal password before you can do anything else.

Before you can establish your password, your supervisor must register you on the system using some form of your name that everyone on the staff will be able to remember. (Registration consists of assigning you a name and setting other limits to your work on the system.)

After you are registered, you temporarily log on twice with any password, then change that password to your own personal password, using the Password Update Screen that appears during the procedure.

After you successfully establish your password and the message TRANSACTION ACCEPTED AND POSTED TO FILES appears briefly, you are automatically logged off. Your supervisor will then activate your employee record so you can begin using the system.

### Preparation

Make sure your supervisor has registered you on the system and that you have the following information:

▶ Your office's code.

▶ The name by which you are registered on the system.

▶ A personal password that meets the rules on page 1.

---

### Establishing your password

*Step-by-step procedure*

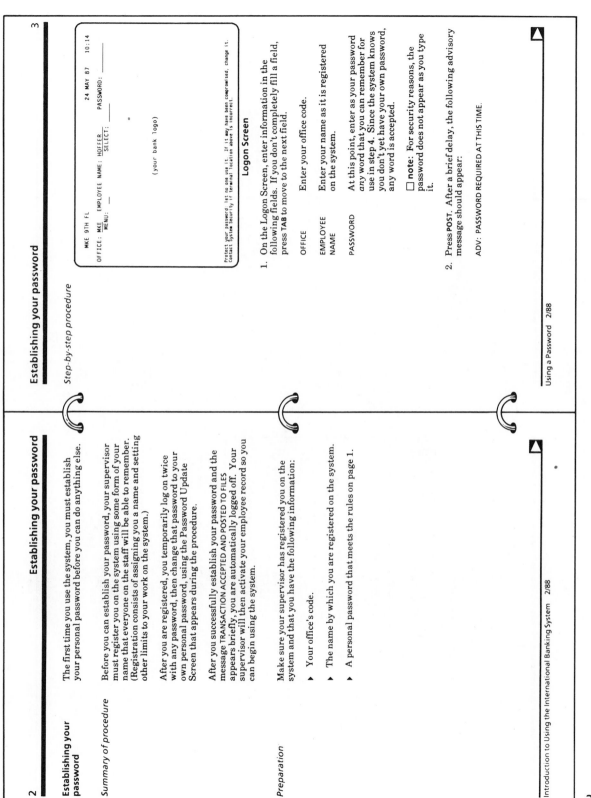

```
MKE 9TH FL                          24 MAY 87   10:14

OFFICE: MKE   EMPLOYEE NAME: HOFFER     PASSWORD: _____
        MENU: _____    SELECT: _____



                    (your bank logo)




Protect your password; let no one use it. If it may have been compromised, change it.
Contact System Security if terminal location above is incorrect.
```

#### Logon Screen

1. On the Logon Screen, enter information in the following fields. If you don't completely fill a field, press **TAB** to move to the next field.

   OFFICE — Enter your office code.

   EMPLOYEE NAME — Enter your name as it is registered on the system.

   PASSWORD — At this point, enter as your password *any* word that you can remember for use in step 4. Since the system knows you don't yet have your own password, any word is accepted.

   ☐ **note:** For security reasons, the password does not appear as you type it.

2. Press **POST**. After a brief delay, the following advisory message should appear:

   ADV: PASSWORD REQUIRED AT THIS TIME.

399

## Changing your password

*Step-by-step procedure*
(Continued)

6. Press POST.

▸ If the message TRANSACTION ACCEPTED AND POSTED TO FILES appears briefly, your new password has been accepted. You are then automatically logged off.

▸ If you entered the password incorrectly, a message appears indicating the nature of the error. Correct the password as indicated by the message that appears. You may also ask your supervisor for assistance.

When the Logon Screen appears, your supervisor can activate your employee record, after which you can log on using your new password.

## Changing your password

You may change your password at any time.

The procedure for changing your password depends on whether the password is active or expired:

▸ If you've changed your password within the last 30 days, your password is active. To change your password, refer to *Changing an Active Password.*

▸ If you haven't changed your password for more than 30 days, your password has expired. To change your password, refer to *Changing an Expired Password.*

---

## Establishing your password

☐ **note:** If you entered any fields incorrectly, the advisory message doesn't appear. Instead, a message appears indicating the nature of the error. Correct the password as indicated by the message. You may also ask your supervisor for assistance.

3. Enter the same password you entered in step 1.

4. Press POST. The Establish Password Screen appears, showing your employee record. The message area of the screen advises that you must change your password.

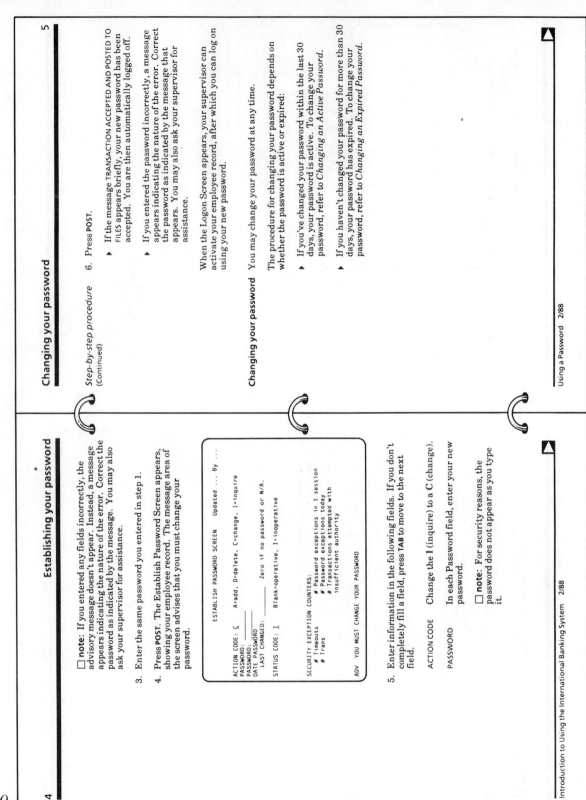

```
      ESTABLISH PASSWORD SCREEN    Updated ... By ...

ACTION CODE: C    A=add, D=delete, C=change, I=inquire
PASSWORD: _____
DATE PASSWORD
LAST CHANGED: _____          Zero if no password or N/A.
STATUS CODE: I    Blank=operative, I=inoperative

SECURITY EXCEPTION COUNTERS:
# Timeouts       # Password exceptions in 1 session
# Trans          # Password exceptions today
                 # Transactions attempted with
                   insufficient authority

ADV  YOU MUST CHANGE YOUR PASSWORD
```

5. Enter information in the following fields. If you don't completely fill a field, press TAB to move to the next field.

ACTION CODE    Change the I (inquire) to a C (change).

PASSWORD    In each Password field, enter your new password.

☐ **note:** For security reasons, the password does not appear as you type it.

## Changing your password

### Changing your password

7

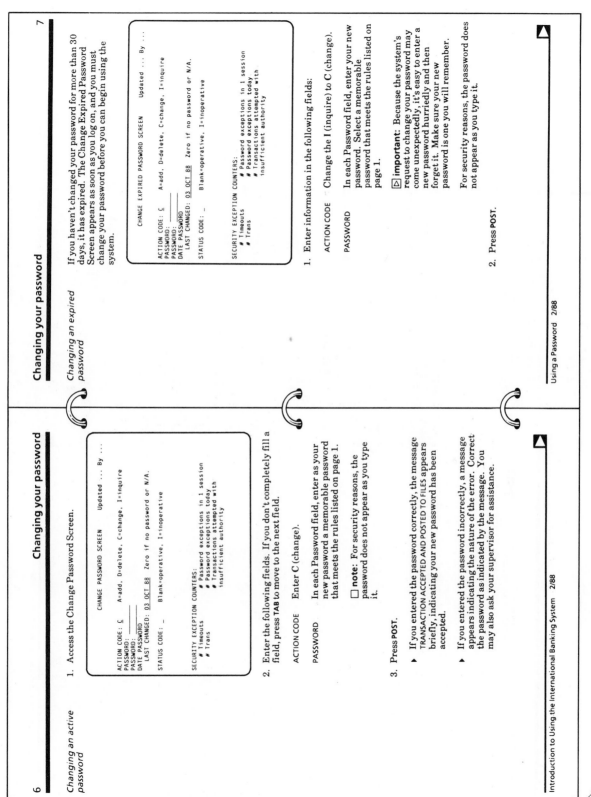

*Changing an active password*

1. Access the Change Password Screen.

```
        CHANGE PASSWORD SCREEN    Updated ... By ...

ACTION CODE: C    A=add, D=delete, C=change, I=inquire
PASSWORD:
PASSWORD:
DATE PASSWORD
   LAST CHANGED: 03 OCT 88  Zero if no password or N/A.

STATUS CODE: _    Blank=operative, I=inoperative

SECURITY EXCEPTION COUNTERS:
  # Timeouts        # Password exceptions in 1 session
  # Trans           # Password exceptions today
                    # Transactions attempted with
                       insufficient authority
```

2. Enter the following fields. If you don't completely fill a field, press **TAB** to move to the next field.

   ACTION CODE   Enter C (change).

   PASSWORD   In each Password field, enter as your new password a memorable password that meets the rules listed on page 1.

   ☐ **note:** For security reasons, the password does not appear as you type it.

3. Press **POST**.

   ◆ If you entered the password correctly, the message TRANSACTION ACCEPTED AND POSTED TO FILES appears briefly, indicating your new password has been accepted.

   ◆ If you entered the password incorrectly, a message appears indicating the nature of the error. Correct the password as indicated by the message. You may also ask your supervisor for assistance.

*Changing an expired password*

If you haven't changed your password for more than 30 days, it has expired. The Change Expired Password Screen appears as soon as you log on, and you must change your password before you can begin using the system.

```
     CHANGE EXPIRED PASSWORD SCREEN    Updated ... By ...

ACTION CODE: C    A=add, D=delete, C=change, I=inquire
PASSWORD:
PASSWORD:
DATE PASSWORD
   LAST CHANGED: 03 OCT 88  Zero if no password or N/A.

STATUS CODE: _    Blank=operative, I=inoperative

SECURITY EXCEPTION COUNTERS:
  # Timeouts        # Password exceptions in 1 session
  # Trans           # Password exceptions today
                    # Transactions attempted with
                       insufficient authority
```

1. Enter information in the following fields:

   ACTION CODE   Change the I (inquire) to C (change).

   PASSWORD   In each Password field, enter your new password. Select a memorable password that meets the rules listed on page 1.

   △ **important:** Because the system's request to change your password may come unexpectedly, it's easy to enter a new password hurriedly and then forget it. Make sure your new password is one you will remember.

   For security reasons, the password does not appear as you type it.

2. Press **POST**.

## Changing your password

8

*Changing an expired password*
(Continued)

▸ If you entered the password correctly, the message TRANSACTION ACCEPTED AND POSTED TO FILES appears briefly, indicating your new password has been accepted.

▸ If you entered the password incorrectly, a message appears indicating the nature of the error. Correct the password as indicated by the message that appears. You may also ask your supervisor for assistance.

---

# Logging On and Off

## What's in this section?

▸ how to access the Logon Screen from either a Tandem terminal or from a personal computer

▸ how to log on

▸ how to log off from either a Tandem terminal or from a personal computer

▸ a definition of an incomplete transaction

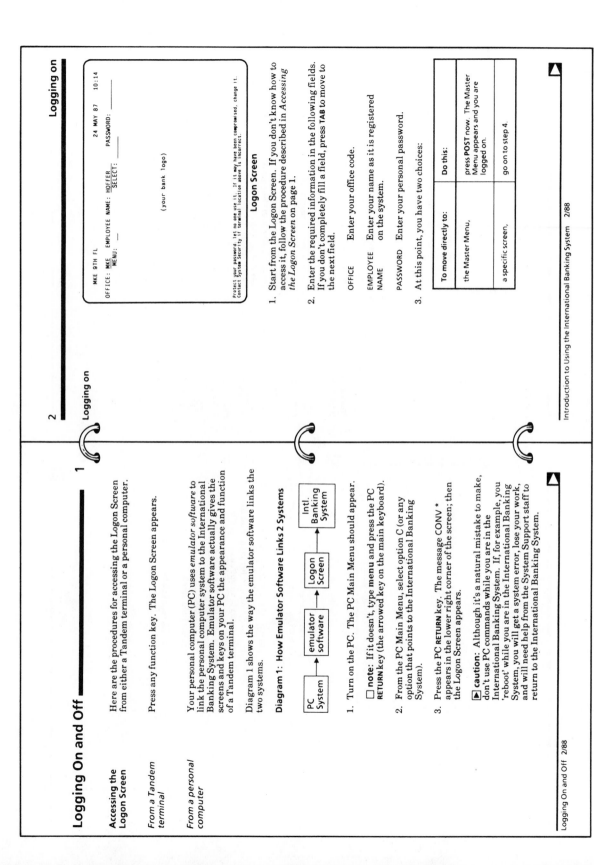

## Logging on

```
MKE 9TH FL                                    24 MAY 87    10:14
OFFICE: MKE  EMPLOYEE NAME: HOFFER       PASSWORD: ___
             MENU: __   SELECT: __




                      (your bank logo)




Protect your password. let no one use it. If it may have been compromised, change it.
Contact System Security if terminal location above is incorrect.
```

**Logon Screen**

1. Start from the Logon Screen. If you don't know how to access it, follow the procedure described in *Accessing the Logon Screen* on page 1.

2. Enter the required information in the following fields. If you don't completely fill a field, press **TAB** to move to the next field.

   OFFICE        Enter your office code.

   EMPLOYEE      Enter your name as it is registered
   NAME          on the system.

   PASSWORD      Enter your personal password.

3. At this point, you have two choices:

| To move directly to: | Do this: |
|---|---|
| the Master Menu, | press **POST** now. The Master Menu appears and you are logged on. |
| a specific screen, | go on to step 4. |

---

# Logging On and Off

**Accessing the Logon Screen**

Here are the procedures for accessing the Logon Screen from either a Tandem terminal or a personal computer.

*From a Tandem terminal*

Press any function key. The Logon Screen appears.

*From a personal computer*

Your personal computer (PC) uses *emulator software* to link the personal computer system to the International Banking System. Emulator software actually gives the screens and keys on your PC the appearance and function of a Tandem terminal.

Diagram 1 shows the way the emulator software links the two systems.

**Diagram 1: How Emulator Software Links 2 Systems**

```
PC   --> emulator --> Logon  --> Intl.
System    software    Screen     Banking
                                 System
```

1. Turn on the PC. The PC Main Menu should appear.

   ☐ **note:** If it doesn't, type **menu** and press the PC **RETURN** key (the arrowed key on the main keyboard).

2. From the PC Main Menu, select option C (or any option that points to the International Banking System).

3. Press the PC **RETURN** key. The message CONV * appears in the lower right corner of the screen; then the Logon Screen appears.

   ▲ **caution:** Although it's a natural mistake to make, don't use PC commands while you are in the International Banking System. If, for example, you 'reboot' while you are in the International Banking System, you will get a system error, lose your work, and will need help from the System Support staff to return to the International Banking System.

*403*

3

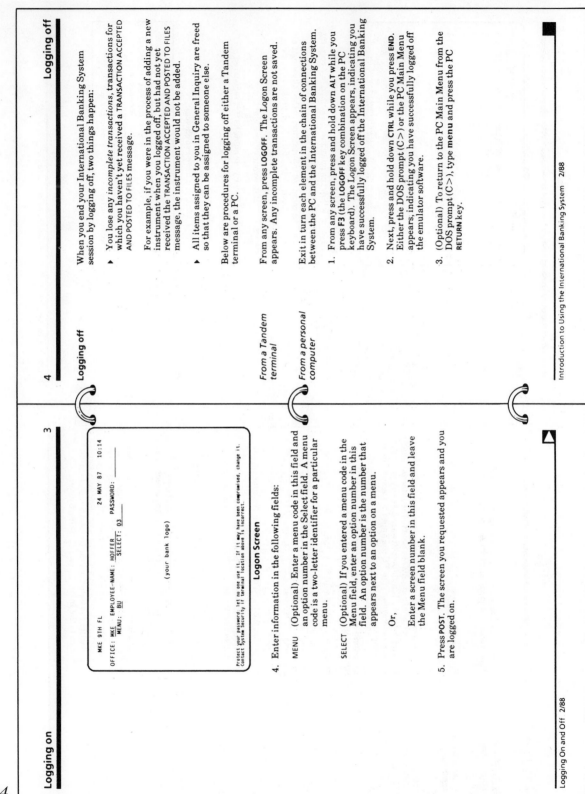

```
MKE 9TH FL                    24 MAY 87    10:14

OFFICE: MKE  EMPLOYEE-NAME: HOFFER   PASSWORD: ___
     MENU: BU      SELECT: 03




              (your bank logo)




Protect your password let no one use it. If it may have been compromised, change it.
Contact System Security if terminal location above is incorrect.
```

**Logon Screen**

4. Enter information in the following fields:

MENU  (Optional) Enter a menu code in this field and an option number in the Select field. A menu code is a two-letter identifier for a particular menu.

SELECT  (Optional) If you entered a menu code in the Menu field, enter an option number in this field. An option number is the number that appears next to an option on a menu.

Or,

Enter a screen number in this field and leave the Menu field blank.

5. Press POST. The screen you requested appears and you are logged on.

---

4

**Logging off**

When you end your International Banking System session by logging off, two things happen:

▸ You lose any *incomplete transactions*, transactions for which you haven't yet received a TRANSACTION ACCEPTED AND POSTED TO FILES message.

For example, if you were in the process of adding a new instrument when you logged off, but had not yet received the TRANSACTION ACCEPTED AND POSTED TO FILES message, the instrument would not be added.

▸ All items assigned to you in General Inquiry are freed so that they can be assigned to someone else.

Below are procedures for logging off either a Tandem terminal or a PC.

*From a Tandem terminal*

From any screen, press LOGOFF. The Logon Screen appears. Any incomplete transactions are not saved.

*From a personal computer*

Exit in turn each element in the chain of connections between the PC and the International Banking System.

1. From any screen, press and hold down ALT while you press F3 (the LOGOFF key combination on the PC keyboard). The Logon Screen appears, indicating you have successfully logged off the International Banking System.

2. Next, press and hold down CTRL while you press END. Either the DOS prompt (C>) or the PC Main Menu appears, indicating you have successfully logged off the emulator software.

3. (Optional) To return to the PC Main Menu from the DOS prompt (C>), type menu and press the PC RETURN key.

404

# System Standards

```
x .office . x   x ..emp. name .. x loc .. x ..screen no. .. x  x..date & time..x
x ...status...x   x .....Screen Name ..x           Updated ...... By ......
```

**Screen headings**  At the top of most screens, you'll see the following fields:

*(top line)*  (Output-only) The top line of every screen shows:

▶ The office code.
▶ The employee's name.
▶ The terminal location.
▶ The screen number (menu number/option number).
▶ The date and time.

**status**  (Output-only) This field shows the current status of a transaction (for example, *pending execution* or *executed*); a message for a transaction; or, for transactions with multiple messages, the number of messages. (Multiple messages appear on the Message Screen, which automatically displays if several messges are generated.)

**Screen Name**  (Output-only) A screen name appears at the top of every screen except the Logon Screen, the General Inquiry Screen, and the Verification Screen.

*Updated ... By ...*  (Output-only) These fields show when and by whom the transaction or instruction was last updated.

---

# System Standards

**What's in this section?**  descriptions of screen conventions that can help you quickly learn how to use the system

# Standard fields

## CloseChain? (Continued)

| Valid entries: | Meaning: |
|---|---|
| (blank) | Normal status. The chain (if there is one) will be closed when the last transaction in the chain has been executed. |
| N | Change the status from normal status to 'remain open' status. The chain will remain open even after the last transaction has been executed. |
| (n) | If the chain is already in 'remain open' status, the system shows a small n in the CloseChain? field. The transaction remains in this status until you enter a Y in the CloseChain? field. |
| Y | Change the status from 'remain open' to normal status. If the last transaction in the chain has already been executed, the chain is closed immediately. Otherwise, the chain will be closed as soon as the last transaction in the chain is executed. |

## Arranged

(Optional) Arranged Date field. This field appears on screens for sales, contracts, or own acceptance transactions. You may either use the field to enter an Arranged Date or leave the field blank (in which case the system enters the current business date as the arranged date).

## Action

Action code field. Action codes tell the system how to process the information you enter on a screen. A list of codes appears on the following page.

Since not all action codes are valid on every screen, each screen description includes a list of valid Action Codes.

△ **examples:** You can't add (A) transactions on the Comprehensive Change Screen.

Several Action codes (CHAIN, HOLD, NEXT, OK, REFER) are only valid in General Inquiry.

---

# Standard fields

```
x .office. . x .. x ..emp. name.. x ..loc .. x .x..screen no .. x  x..date & time..x
x ...status...x    x ....Screen Name ..x        Updated ...... By ......x

Prior 5002631    CloseChain? N Arranged  3 May 87
Action  CEX      Type VOSTRO    Ref 4592298         Their  80094
Cus  Whiley National Bank:Omaha
```

**Standard fields**

On business transaction screens, you'll see the following standard fields:

## Prior

(Optional) If the transaction is linked to an earlier transaction or message, the reference number for the earlier item appears in this field.

Although it's uncommon, you can also use this field to link one transaction to another. To link this transaction to an earlier transaction or message, enter the reference number for the earlier item in this field.

## CloseChain?

(Optional) A *chain* is an association of related messages and transactions. When transactions are chained, no transaction in the chain can be deleted from the file of audit records until all transactions in the chain have been executed.

The CloseChain? field lets you block normal closing of the chain, thus allowing you to maintain audit records for related messages and transactions even after all transactions in the chain have been executed.

△ **example:** You may choose to keep a chain open while you wait for a response from a SWIFT message.

## Standard fields

Action (Continued)

| Action Code: | Meaning: |
|---|---|
| A | Add an instrument, transaction, or instruction record. |
| AEX | Add-and-execute an instrument or transaction. |
| C | Change information on an instruction record or customer record; change static information for an account or instrument. |
| CEX | Change-and-execute a transaction. |
| CHAIN | Link one message or transaction to a previous message or transaction. |
| CLOSE | End a loan participation or sale contract. |
| D | Delete a customer record or an instruction record; release a customer or nation from a Hold status. |
| EX | Execute a transaction. |
| HOLD | Place a transaction, message, customer, or nation into a Hold status. |
| I | View an instrument, transaction, or instruction. |
| NEXT | View the next item in General Inquiry. |
| OK | Approve a transaction. |
| REFER | Refer a transaction or message to another employee or department. |
| REV | Reverse a transaction. |
| VER | Verify a transaction. |
| VEREX | Verify-and-execute a transaction. |

---

## Standard fields

**Type**

A business type code defined by your office. Each business type has a set of instructions that determine accounting, authority, data entry, and many other requirements.

Most business types are account types such as loans or foreign exchanges. However, wire service message types and payment types are also defined as business types.

**Ref**

(Optional) Reference number field; a unique number assigned by the system to each transaction, account, customer, and wire service message.

▶ When you set up a new instrument, the system assigns a reference number to the instrument.

▶ To identify an existing instrument or transaction, you enter its Reference number in this field.

**Cus**

Customer ID field. To identify the primary customer for the instrument or transaction, enter the customer's full name, short name, symbol, or customer number.

After you press **ENTER**, the customer's full name appears in this field.

☐ **note**: On the Outright Instrument Sale Screen, this field is called the Buyer's Name.

**Their**

(Optional) Their Reference Number field. Use this field to enter the number, if known, by which the customer refers to the instrument or transaction.

## Data entry formats

**Data entry formats**

Each of the following items has its own format on the system:

- ▶ dates
- ▶ amounts and currencies
- ▶ rates
- ▶ fields with question marks
- ▶ reference numbers
- ▶ customer IDs

Once you learn an item's format, you can use that format on all system screens.

*Dates*

Enter dates in date, month abbreviation, year format, with or without spaces or leading zeros.

▷ **example:** Use any of the following formats to enter September 9th, 1988:

9SEP88      9 SEP 88
09SEP88     09 SEP 88

Month abbreviations: JAN, FEB, MAR, APR, MAY, JUN, JUL, AUG, SEP, OCT, NOV, DEC.

*Amounts and currencies*

Enter amounts as either whole numbers or as decimals, followed by a Currency code as defined by your office. Amounts without decimals will be read as whole numbers.

▷ **examples:** Enter $5,000.00 as **5000** USD.
Enter £32,588.55 as **32588.55** GBP.

For large, even amounts, you can also use one of these shortcuts:

**M**   thousands
**MM**  millions

▷ **examples:** Enter $5,000.00 as **5M** USD.
Enter £3,000,000.00 as **3MM** GBP.

☐ **note:** You may enter amounts up to nine digits long.

## Data entry formats

Enter whole percents as whole numbers; the system assumes a decimal point at the first position.

▷ **examples:** Enter 11% as **11**

Enter fractions of percents with either a decimal point or a dash and a simple fraction ($\frac{1}{4}$, $\frac{3}{4}$, or $\frac{1}{2}$).

▷ **examples:** Enter 15.75 as either **15.75** or **15-3/4**

For many account types, you can enter an *index interest rate symbol* plus or minus a percentage. Your office defines such symbols for index rates that fluctuate such as the prime rate or the LIBOR rate.

▷ **examples:**   **PRIME**
                   **PRIME + 2**
                   **LIBOR-1.5**

*Rates*

*Fields with question marks*

A question mark indicates a yes/no question. Valid responses:

**Y**     yes
**N**     no
(blank)   bypasses questions that don't apply to the current transaction. Sometimes this response is not valid because an answer is required.

*Account numbers*

The system automatically assigns account numbers to new instruments. So that you can recognize an instrument by its account number, each instrument type is assigned numbers from a distinct range defined by your office.

When you are entering a transaction for an instrument, enter the account number without punctuation.

▷ **example:** 12345678

## System messages

*Reference numbers*

Reference numbers have three parts: the basic reference number, a *transposition check digit* (TCD) that helps verify correct entry of the number, and a *suffix* that lets the system link related reference numbers such as an incoming message and the reply to that message.

The system adds hyphens to longer reference numbers to make them easier to read.

**Parts of a reference number**

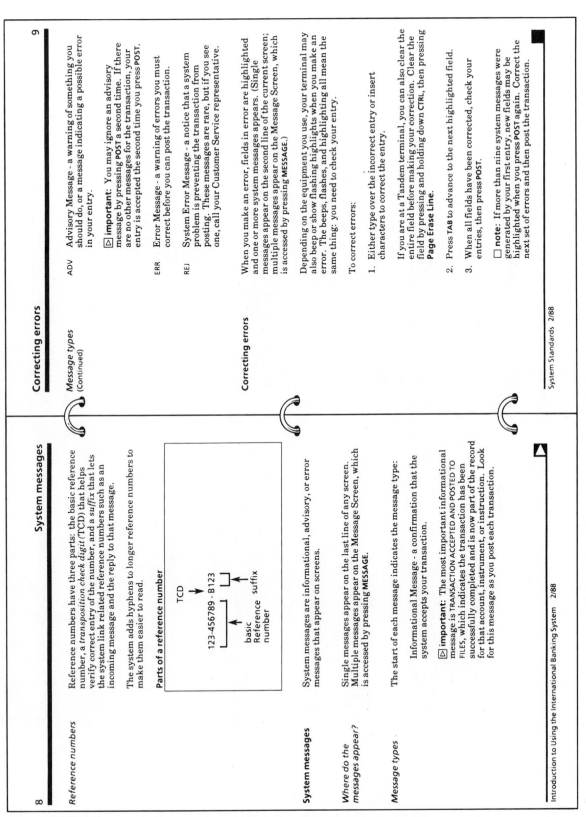

TCD → 123-456789 - B123

basic Reference number          suffix

**System messages**

System messages are informational, advisory, or error messages that appear on screens.

*Where do the messages appear?*

Single messages appear on the last line of any screen. Multiple messages appear on the Message Screen, which is accessed by pressing MESSAGE.

*Message types*

The start of each message indicates the message type:

Informational Message - a confirmation that the system accepts your transaction.

△ **important:** The most important informational message is TRANSACTION ACCEPTED AND POSTED TO FILES, which indicates the transaction has been successfully completed and is now part of the record for that account, instrument, or instruction. Look for this message as you post each transaction.

---

## Correcting errors

*Message types*
(Continued)

ADV   Advisory Message - a warning of something you should do, or a message indicating a possible error in your entry.

△ **important:** You may ignore an advisory message by pressing POST a second time. If there are no other messages for the transaction, your entry is accepted the second time you press POST.

ERR   Error Message - a warning of errors you must correct before you can post the transaction.

REJ   System Error Message - a notice that a system problem is preventing the transaction from posting. These messages are rare, but if you see one, call your Customer Service representative.

When you make an error, fields in error are highlighted and one or more system messages appears. (Single messages appear on the second line of the current screen; multiple messages appear on the Message Screen, which is accessed by pressing MESSAGE.)

**Correcting errors**

Depending on the equipment you use, your terminal may also beep or show flashing highlights when you make an error. The beeps, flashes, and highlighting all mean the same thing: you need to check your entry.

To correct errors:

1. Either type over the incorrect entry or insert characters to correct the entry.

   If you are at a Tandem terminal, you can also clear the entire field before making your correction. Clear the field by pressing and holding down CTRL, then pressing **Page Erase Line.**

2. Press TAB to advance to the next highlighted field.

3. When all fields have been corrected, check your entries, then press POST.

☐ **note:** If more than nine system messages were generated by your first entry, new fields may be highlighted when you press POST again. Correct the next set of errors and then post the transaction.

*409*

# Using Menus

**What's in this section?**

▶ definitions of menus and menu terms

▶ a sample menu structure

▶ examples of the four ways you can use menus

▶ how to return from a screen or submenu to a previous menu

---

# Using Menus

**What are menus?**

Menus are lists of available options from which you select the option you want. Options, which may be either actual screens or other menus, are listed on separate lines with an option number listed beside each line. You select an option by typing its option number in the Select field and pressing POST. After you select an option, a new screen or menu appears.

**Sample menu structure**

Since your office defines its own menus, we can't show you your office's menu structure. However, Diagram 1 on the following page illustrates a typical menu structure, which may include the following items:

▶ A *Logon Screen*, the screen that appears when you first access the system.

▶ A *Master Menu* that lists all other menus.

▶ Up to 60 *submenus* that can list up to 20 options each.

**Menu terms**

▶ The two-letter code to the left of each option in the Master Menu is called a *menu code* or a *menu selection code*.

▶ The numbers listed beside options in the submenus are called *option numbers*.

▶ *Menu numbers* relate to the position of a submenu on the Master Menu. For example, the menu number for the Business Transaction Menu is *1*. The menu number for the Customer Utilities Menu is *2*.

▶ *Screen numbers* consist of a combination of the menu number and option number for a screen. Screen numbers appear on the upper right corner of every screen except the Customer Records screens, which are not numbered.

410

# Four ways to use menus

## Four ways to use menus

There are several ways to access screens via menus:

▶ Go through the Master Menu and submenus.

▶ Enter a screen number on the Logon Screen.

▶ Enter a Menu selection code and option number on the Logon Screen.

▶ On the Master Menu, enter either a menu selection code and an option number, or a screen number.

Follow the examples of each option that appear on the next few pages. All examples assume your office uses the menu structure that appears in Diagram 1.

---

# Sample menu structure

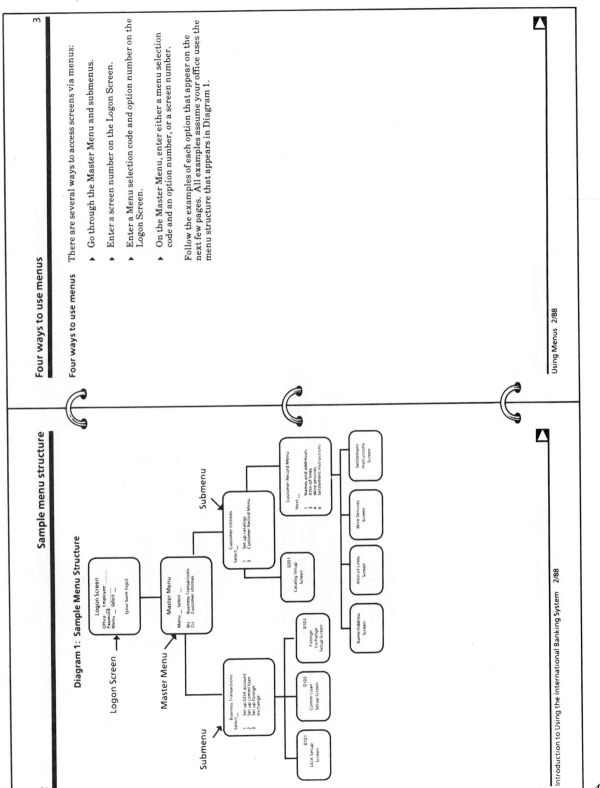

Diagram 1: Sample Menu Structure

Logon Screen →

Master Menu →

Submenu →

Submenu

**Logon Screen**
Office ____ Employee ____
Password ____ Select ____
Menu ____
(your bank logo)

**Master Menu**
Menu ____ Select ____
BU   Business transactions
CU   Customer Utilities

**Business Transactions**
Select ____
1   Set up DDA account
2   Set up comml loan
3   Set up foreign. exchange

**Customer Utilities**
Select ____
1   Set up catalogs
3   Customer Record Menu

**Customer Record Menu**
Next ____
1   Names and addresses
3   Attn-of lines
4   Wire services
6   Settlement instructions

0101 DDA Setup Screen

0102 Comml Loan Setup Screen

0103 Foreign Exchange Setup Screen

0201 Catalog Setup Screen

Name/Address Screen

Attn-of-Lines Screen

Wire Services Screen

Settlement Instructions Screen

*411*

## Four ways to use menus

**Going through menus**

Suppose you want to set up a foreign exchange contract, but you are not familiar with your office's menus.

1. Log on without entering the Menu or Select fields.

```
Logon Screen
Office___ Employee: Smith
Password___
Menu ___ Select ___
        (your bank logo)
```

The Master Menu appears. Of the two submenus listed, *Business Transactions* seems most likely to include a foreign exchange screen.

```
Master Menu
Menu BU Select ___
BU   Business Transactions
CU   Customer Utilities
```

2. In the Menu field, type the menu code for the submenu you want. In this case, the menu code is BU.

3. Press POST. The Business Transactions Menu appears.

```
Business Transactions
Select 3
1   Set up DDA account
2   Set up comml loan
3   Set up foreign
    exchange
```

4. Of the options on the Business Transactions Menu, *Set up foreign exchange* is the most likely option. To select that option, type its option number (3 or 03) in the Select field and press POST. The Foreign Exchange Setup Screen appears.

☐ **note:** If you've accidentally chosen the wrong menu or the wrong screen, press RETURN to return and correct your entry.

---

## Four ways to use menus

**Entering a screen number on the Logon Screen**

Suppose you want to set up a foreign exchange contract, and you have a printed copy of the Foreign Exchange Setup Screen in front of you.

1. Notice the four-digit screen number in the upper right corner of the screen: 0103 (the third option on the first submenu). That is the screen number for the Foreign Exchange Setup Screen.

```
Foreign   0103
Exchange
Setup Screen
```

2. As you log on, enter the screen number, including the leading zero, in the Select field.

```
Logon Screen
Office___ Employee: Smith
Password
Menu ___ Select 0103
        (your bank logo)
```

3. Press POST. The Foreign Exchange Setup Screen appears.

**Entering a menu selection code and option number on the Logon Screen**

Suppose you want to set up a foreign exchange contract.

You know the option to set up a foreign exchange contract is option 3 on the Business Transactions Menu, which has a menu code of BU.

1. As you log on, type BU in the Menu field and 3 or 03 in the Select field.

```
Logon Screen
Office___ Employee: Smith
Password
Menu BU Select 3
        (your bank logo)
```

2. Press POST. The Foreign Exchange Setup Screen appears.

# Glossary

1

**A**

| A (add) | The action code you use to add a new instrument, transaction, customer, or instruction record. |
| --- | --- |
| Account-level credit line | A credit line you establish for a specific instrument (as opposed to revolving credit lines you establish for a customer on the Credit Facility Screen). |
| Account number | A 7- to 10-digit number that identifies a particular instrument. |
| Action code | Your instructions to the system saying how to process the information you enter on a screen. For example, to add a new account or customer, you use the A (add) action code. |
| ADV | The code that identifies an advisory message. |
| | Some advisory messages warn you that you should do something. For example, an advisory message appears if your password has expired. |
| | Other advisory messages warn of a possible error in your entry such as an unusually high exchange rate. The system allows you to ignore this second type of advisory message by pressing **POST** a second time after receiving the advisory message. |
| Advice | A notice sent to customers informing them of debits and credits posted to their accounts. |
| AEX (add-and-execute) | The action code you use to simultaneously add and execute a transaction. |
| Authority indexes | An office-wide scheme that defines what activities each employee can perform. |

---

6

*Using the Master Menu*

Suppose you want to set up a foreign exchange contract, and you know the Foreign Exchange Setup Screen is the third option on the Business Transactions Menu, but you don't know the menu number or menu code for that menu.

```
Logon Screen
Office:____ Employee: Smith
Password:____ Select ___
Menu: ___
           (your bank logo)
```

1. Log on without entering the Menu or Select fields. The Master Menu appears.

```
Master Menu
Menu BU Select   3
BU   Business Transactions
CU   Customer Utilities
```

On the Master Menu you can see that the selection code for the Business Transactions Menu is **BU**. Since you already know the Foreign Exchange Setup Screen's position on that menu, you know it's option number is three.

2. Type **BU** in the Menu field and **3** or **03** in the Select field.

3. Press **POST**. The Foreign Exchange Setup Screen appears.

```
Foreign   0103
Exchange
Setup Screen
```

**Returning to a previous menu**

To return from a screen to a submenu, or to return from a submenu to the Master Menu, press **RETURN**.

**Returning to the previous menu**

413

## Glossary

**Automatic logoff** — A security feature that automatically ends the session at a terminal that has been inactive for more than a few minutes. Also called "timing out."

☐ **note:** Your office determines the exact amount of idle time after which the session automatically ends.

### B, C

**Business Types** — Account types, wire service message types, payment types, and debit/credit classification types defined by your office. For each business type, your office defines a set of instructions that determine accounting, data entry, and many other requirements for the business type.

**C (change)** — The action code you use to change information for an account, transaction, or instruction.

**CEX (change-and-execute)** — The action code you use to simultaneously change and execute a transaction.

**Chain** — An association of related messages and transactions.

**CHAIN** — The action code you use to associate a wire service message with a related transaction.

**Chaining** — The process of relating wire service messages and transactions.

**CLOSE** — The action code you use to close an instrument, account, or customer's Credit Facility.

**Confirmation** — A notice sent to customers informing them in detail about an instrument established according to their instructions.

**Consecutive Error Lockout** — A security feature that prevents you from logging on if you have made more than three consecutive errors while entering your password.

---

## Glossary

**Credit Facility credit line** — A credit line taken from the Credit Facility Screen (as opposed to an account-level credit line you establish for a particular instrument).

**Credit Facility Screen** — A screen on which you establish revolving credit lines for a customer.

**Currency Code** — A three-letter code defined by your office that represents a specific currency. Many offices use the currency codes defined by the International Standards Organization.

**Customer** — An individual, company, or bank that is registered as a customer on the International Banking System. Customer Records are the means by which a customer is registered on the system.

**Customer Record** — A record that stores multiple names and addresses, wire service numbers, standing settlement instructions, and other information about a customer.

**Customer Symbol** — An optional short name by which you can identify a customer. Customer Symbols are defined on the Customer Record.

### D, E

**D (delete)** — The action code you use to delete a customer record or an instruction record; also used to remove a customer or nation from a "report" or a "hold" status.

**DDMMYY format** — The correct format for entering dates: one or two date digits, a three-letter month abbreviation, and a two-digit year designation.

**Employee record** — A record that stores authority requirements, authority indexes, and other employee information.

**ERR** — The code that identifies an error message. Error messages warn you of errors you must correct before you can post the transaction.

## Glossary

**EX (execute)**

The action code you use to execute a transaction.

☐ **note**: If the transaction required officer approval and you have the authority to both approve and execute the transaction, the EX action code will both approve and execute the transaction.

**F, G, H**

**Full name**

The complete customer name. After you enter a customer short name, customer symbol, or customer number in the Cus field and press **POST**, the system replaces your entry with the customer's full name.

**Function keys**

The row of keys you use to perform basic functions on the system such as posting a transaction, printing a screen, clearing fields on a screen, or logging off. Function keys are located at the top of the Tandem keyboard and to the left of the IBM PC keyboard.

**General Inquiry**

A series of screens on which you view transactions and verify, refer, approve, or execute them. Should any questions arise, General Inquiry offers a comprehensive set of inquiry displays that you can use to find out more about customers and accounts.

**HOLD**

The action code you use to place a transaction, message, customer, or nation into a Hold status.

**I, L**

**I (inquire)**

The action code you use to view a screen or customer record.

**Incomplete transaction**

A transaction for which you haven't received a TRANSACTION ACCEPTED AND POSTED TO FILES message. When you log off, you lose any incomplete transactions you were working on.

**Index Interest Rate Symbol**

A symbol, defined by your office, that represents a fluctuating rate such as the prime rate.

---

## Glossary

**Instruction screens**

Screens that define your office's credit management scheme, authority requirements scheme, business types, employee records, valid system codes such as nation codes or cycle codes, and other important office options.

**Internal accounting entries**

Accounting entries the system generates automatically in response to the transactions you enter.

**Item**

A transaction or message.

**Log Off**

To end an International Banking System session. You log off by pressing **LOGOFF**. If you accessed the International Banking System from a personal computer, you must also log off the PC.

**Log On**

To identify yourself as an authorized user by entering your office code, name, and password on the Logon Screen.

**Logon Screen**

The screen you use to log on to the International Banking System.

**M, N**

**Master Menu**

A main menu that lists all available menus.

**Menu**

A list of options from which you select screens or other menus.

**Menu code**

A two-letter identifier for a particular menu.

**Menu number**

A two-digit identifier for a particular menu that corresponds to the order in which the menu appears on the Master Menu.

**Screen number**

A four-digit number that identifies a particular screen. Screen numbers consist of a combination of menu number and option number for a screen.

**Message**

A wire service message or a system message.

## Glossary

| | |
|---|---|
| Message Screen | The screen that displays multiple system messages. |
| NEXT | The action code you use to display the next transaction or message to be worked on. |
| Non-customer | Any party to a transaction that is not a customer on file. |

**O**

| | |
|---|---|
| OK | The action code you use to approve a transaction (either credit approval or exception approval). |
| Option number | The number that appears next to each option in a submenu. |
| Output-only fields | Fields in which the system displays information that you cannot change. |
| Password | A secret word or phrase known only to you that identifies you as an authorized user of the system. |
| Payment screens | Screens on which you enter payments. Payments can be partially entered for you, based on incoming wire service messages. |
| PC | A personal computer. You may be accessing the International Banking System from a personal computer. |

**R, S, T**

| | |
|---|---|
| Rate screens | The screens on which your office maintains these system-wide rates: foreign exchange rates, interbank interest rates, and index interest rate symbols. |
| REFER | The action code you use to refer a transaction or message to another employee or department. The transaction or message is then brought to that person or department's attention as soon as they use General Inquiry. |

## Glossary

| | |
|---|---|
| Reference number | A unique number assigned to each transaction, account, customer, and wire service message on the system. |
| REJ | The code that identifies a system error message. System error messages indicate that a system problem is preventing the current transaction from posting. These messages are rare, but if you see one, please call your Customer Service Representative. |
| REV (reverse) | The action code you use to reverse a transaction. |
| Short name | An abbreviated customer name, defined on the Customer Record, which you can use to identify a customer on transactions screens. |
| Statements | Notices sent to customers informing them of all account activity that took place during a specific period of time. |
| Submenu | A menu that is listed on another menu. |
| TCD | A Transposition Check Digit, a letter assigned to a reference number as a means to check for transposed digits whenever that number is entered on the system. As the system assigns a reference number, it calculates a TCD based on the number itself and on the order in which digits in the number appear. The system appends the TCD to the reference number at that time. Whenever the reference number is entered after that, the system calculates another TCD based on the digits entered and their order. If both TCDs are not identical, the system will detect that digits were transposed as the number was entered. |
| Timing out | The process by which you are automatically logged off your terminal if it has been inactive for more than a few minutes. |

**Transaction**

Any item of business you enter on the system. For example, adding a new account is a transaction; entering a debit or credit is a transaction; changing your password is a transaction.

**TRANSACTION ACCEPTED AND POSTED TO FILES**

The informational message that confirms that the transaction you entered has been completed and is now part of the record for that account, instrument, or instruction.

## V

**VER (verify)**

The action code you use to verify that a transaction entered earlier was entered accurately.

**VEREX (verify-and-execute)**

The action code you use to simultaneously verify and execute a transaction.

**Verification Screen**

A screen, accessed from the General Inquiry Screen, on which you re-key transaction details to verify a transaction.

# Index